Science, Culture and National Identity in Francoist Spain, 1939–1959

Mariciό Janué i Miret • Albert Presas i Puig
Editors

Science, Culture and National Identity in Francoist Spain, 1939–1959

palgrave
macmillan

Editors
Mariciό Janué i Miret
Department of Humanities
Universitat Pompeu Fabra
Barcelona, Spain

Albert Presas i Puig
Department of Humanities
Universitat Pompeu Fabra
Barcelona, Spain

ISBN 978-3-030-58645-4 ISBN 978-3-030-58646-1 (eBook)
https://doi.org/10.1007/978-3-030-58646-1

This Palgrave Macmillan imprint is published by the registered company Springer Nature Switzerland AG.
The registered company address is: Gewerbestrasse 11, 6330 Cham, Switzerland

Contents

Academy of Sciences and Humanities. Memberships: Gesellschaft für Wissenschaftsgeschichte (Society for History of Science and Humanities); History of Science Society (USA); Forum for Human Science in the History of Science Society; Österreichische Gesellschaft für Wissenchaftsgeschichte (Austrian Society for History of Science); Österreichische Gesellschaft für Exilforschung; Sigmund Freud-Privatstiftung, Vienna; Heckmann-Wetzel-Stiftung of the Berlin-Brandenburg Academy of Sciences and Humanities. Editorial Advisory Boards: Isis (Madison), 1990–1993; Revista de Historia de la Psicología (Valencia), 1984–; Psychologie und Geschichte (Heidelberg), 1988–1993; Cuadernos Argentinos de Historia de la Psicología (San Luis), 1994–; *Journal of the History of the Behavioral Sciences* (New York), 1997–; History of Psychology (Washington, D.C.), 1997–; Wiener Zeitschrift für Geschichte der Neuzeit (Vienna, Austria), 1997–; Berichte zur Wissenschaftsgeschichte, 2008–. General Editor (with William R. Woodward), Cambridge Studies in the History of Psychology, published by Cambridge University Press, 1989–2004. He has published fundamental works on topics such as the Sciences in History, Cultures of Knowledge, History of Eugenics in International Comparison, Scientific Change in Times of Political Upheaval, Science and the Public Sphere, Higher Education and the Sciences under Nazism and Afterward, Science and Knowledge Transfer in the Late Habsburg Monarchy; Modern Historiography of the Sciences.

Lorenzo Delgado Gómez-Escalonilla is currently a senior researcher at the National Research Council of Spain (CSIC, Spain). He has a doctorate in Contemporary History from Universidad Complutense de Madrid. He is a research fellow at the Centre d'Histoire des Relations Internationales Contemporaines, Université Paris I-Sorbonne and a researcher participant at the International Visitor Program of the United States. In recent years he has researched Spanish foreign policy in twentieth century, American public diplomacy, and the educational, scientific and military assistance of the United States towards Spain. Among his most recent publications: *Westerly Wind: The Fulbright Program in Spain* (2009); *US Public Diplomacy and Democratization in Spain: Selling Democracy?* (2015); *La apertura internacional de España: Entre el franquismo y la democracia (1953–1986)* (2016); 'The Deployment of U.S. Military Assistance to Franco's Spain: Limited Modernization and Strategic Dependence' (2018); 'El factor exterior en la consolidación y desarrollo de la dictadura'

(2018); *Teaching Modernization. Spanish and Latin American Educational Reform in the Cold War* (2019).

Ramón Tena-Fernández is a teacher and researcher specialized in the field of Social and Legal Sciences. He has worked in various organizations of the Universidad de Extremadura, such as the 'Secretariat of Cultural Activities' and the 'Library, Archives and Information Science'. As a researcher he has worked in the Palace of the Congress of Deputies and in the Ministry of Education, where he has collaborated both in the 'Area of Training and Experimentation' and in the area of 'Institutional Relations' (National Institute for Educational Technologies and Teacher Training, INTEF). He is currently working in the Faculty of Teacher Training, where he focuses on the investigation of Francoist censorship. His results have been published in the following magazines: *Revista Chilena de Literatura*, *Revista de Occidente*; *Confluencia*; *Hispania (AATSP), Taller de letras y Bulletin of Spanish Studies,* all of them classified by CIRC with category A. Likewise, his foreign stays in the CNRS of France and the CIDTFF of Portugal are worth mentioning, since both of them provided him with a scientific background that has allowed him to participate in regional and European research projects, as well as being an online tutor for several official courses of the Ministry of Education.

Marició Janué i Miret is Associate Professor at the Department of Humanities of the UPF. Her current research interests focus on Spanish-German cultural diplomacy in the period 1870–1959. On this context she co-authored and published *España y Alemania: Nuevas investigaciones sobre la historia de las relaciones culturales en el siglo XX* (Ayer 69 (Dossier), 2008) and authored 'Imperialismus durch auswärtige Kulturpolitik: die Deutsch-Spanische Gesellschaft als 'zwischenstaatlicher Verband unter dem Nationalsozialismus' (German Studies Review XXXI, 1, 2008: 109–132); 'Woe Betide Us If They Win!': National Socialist Treatment of the Spanish 'Volunteer' Workers (Contemporary European History, 23, 3, 2014: 329–357); 'Relaciones culturales en el "nuevo orden": La Alemania nazi y la España de Franco' (Hispania LXXV, 251, 2015: 805–822); 'The role of culture in German-Spanish relations during national socialism' (in Fernando Clara; Claudia Ninhos (Ed.). Nazi Germany and Southern Europe, 1933–1945. *Science, Culture and Politics.* UK, Palgrave Macmillan, 2015: 84–104); '*Hispanidad* in the*völkisch* "New Order" in Europe' (in Johannes Dafinger and Dieter Pohl (Ed.). *A New Nationalist Europe under Hitler:*

Concepts of Europe and Transnational Networks in the National Socialist Sphere of Influence, 1933–1945. London and New York, Routledge, 2018: 93–111); and 'Contributing to the cultural "New Order": How German intellectuals attributed a prominent place for the Spanish nation' (in Maria Björkman, Patrik Lundell and Sven Widmalm (Ed.). *Intellectual Collaboration with the Third Reich: Treason or Reason?*, Routledge 2019: 214–229).
More extensive CV information in:

- https://www.upf.edu/web/humanitats/entry/-/-/16442/adscripcion/maria-concepcio-janue
- http://orcid.org/0000-0001-6797-3833
- Research group 'Nexus between science, culture, politics, religion and society'
- List of her publications and open access:

  Scientific Output- UPF
  Academia.edu
  Research Gate

Jorge Luís Marzo is an art historian, doctor in Cultural, Translation and Gender Studies, audiovisual producer, professor at BAU College of Design in Barcelona, and member of GREDITS Research Group. He is member of several editorial boards and scientific evaluation committees. He has developed numerous national and international research projects, in exhibiting, audiovisual and editorial formats, often in relation to image policies. The most recent ones are as follows: *After Post-Truth* (Artnodes, 2019); *La competencia de lo falso. Una historia del fake* (Cátedra, 2018); *La década de 1990. Rewind&Forward. Hacia una competencia pública del arte* (MACBA, 2018); *Espectros* (Virreina, 2017); *FAKE. No es verdad, no es mentira* (IVAM, 2016); *Interface Politics* (BAU, 2016, 2018); *Valor! 10 debates sobre la construcción del valor en la ciudad* (Virreina, 2016, with Joana Masó); *Arte en España (1939–2015). Ideas, prácticas, políticas* (Cátedra, 2015, with Patricia Mayayo); *No es lo más natural. Escritos y trabajos de Octavi Comeron* (BAU, 2014, with Joana Masó and Tere Badia); *No tocar, por favor* (ARTIUM, 2013); *MACBA: la derecha, la izquierda y los ricos* (SUB, 2013); *Videocracia. Ficción y política* (SOY CAMARA, 2013, with Fito Rodríguez).
Web: www.soymenos.net

ORCID: https://orcid.org/0000-0003-1238-8703

Fernando García Naharro is a postdoctoral researcher at the Romanisches Seminar (Institut für Sprache, Literatur und Medien, Europa-Universität Flensburg) (Germany) currently working on the DFG-Project 'Buchmessen als Räume kultureller und ökonomischer Verhandlung'. With a training in Anthropology and Sociology and a PhD in Contemporary History (Universidad Complutense de Madrid), his main research interests include History of the Book and Media, History of Publishing, Science and Technology Studies and Semiotics. He has done several research internships in England (University of Leeds), California (University of California Davis) and Catalonia (Universidad Autónoma de Barcelona). His dissertation, titled *The role of science: Scientific and technical literature under Franco (1939–1966)* (UCM, 2017), was a history of science and technology publishing in Spain, with a special emphasis on the collective production and reception of scientific publications in a dictatorial context.

Antonio Niño is Professor at the Department of Contemporary History of the Complutense University of Madrid since 2005. His first research topic was the history of French Hispanism. Subsequently, he has followed other lines of research such as the Hispano-American movement of the first third of the twentieth century, contemporary Spanish foreign policy and especially the relations of Spain with France, Latin America and the USA, and the cultural dimension of international relations. He has recently directed projects and published works on public diplomacy, propaganda and control of public opinion in the twentieth century. He is a visiting professor to give courses and seminars in 12 American and European universities. He has been coordinator of the master's and the doctorate Program in Contemporary History, taught jointly by seven universities, director of the journal *Cuadernos de Historia Contemporánea*, and member of the scientific board of numerous scientific journals and university publishers.

José M. Pacheco is Professor of Applied Mathematics (retired, honorary) at the Universidad de Las Palmas de Gran Canaria. He has worked in the field of mathematical modelling in Ecology and the Atmospheric Sciences. Since 2005 his field of interest has switched to historical topics, first on Spanish Mathematics in the nineteenth century, and then to the relationship between Mathematics and Spanish history of the twentieth century. He has publications in *Archive for the Exact Sciences*, *Science in Context*, and *Boletim da Sociedade Matemática de Portugal*, as well as contribution to national and international meetings in the field. Since 2014 he is a member of the Real Academia Canaria de Ciencias (Royal Academy of Sciences of the Canary Islands) and is a regular contributor to its journal *Folia Canariensis Academiae Scientiarum*.

Sara Prades Plaza holds a degree in History with final-year prize and European PhD in Contemporary History from Valencia University. She has been granted at CSIC and in the Contemporary History Department at Universitat de València. She has done research, and stays at, from the LSE (UK), under the direction of Paul Preston, in l'EHESS (France), supervised by Jordi Canal, and at the Universidad de Navarra. In her research she has dealt with the construction of the Spanish national identity in the early decades of the Franco regime, focusing on the historical speech of nacionalcatolicismo, so she received the VIII Research Prize for Young Historians of the Asociación de Historia Contemporánea. She has published papers in monographs, congresses and impact journals, among which are *Historia Contemporánea*, *Ayer*, *Historia y Política*, *Bulletin d'Histoire Contemporaine de l'Espagne* and *Historiografías* Currently, she is a professor at the Universitat Jaume I of Castelló.

Albert Presas i Puig holds a PhD from the Technische Universität Berlin. He is Associate professor at UPF, where he lectures on History of Science. He was research fellow at the Max Planck Institute for the History of Science in Berlin. His research interests focus on science, culture and power; science in the European periphery; Science and Francoism; and the history of nuclear energy. He has an extensive publishing curriculum and a long-standing experience in research and project coordination. In the last five years, he has been involved in research projects funded by different Spanish and European institutions. He is actually the coordinator of the Horizon 2020 project 'History of Nuclear Energy and Society' (www.HoNESt2020.eu). Selected publications: (with Espluga Trenc, J., A. Prades, B. Medina, M. Rubio-Varas y J. De la Torre) 'Las dimensiones sociales de la percepción de la energía nuclear. Un análisis del caso español (1960–2015)' (*Revista Internacional de Sociología* 75 (4), 2017) p. 265–286; 'Nuclear energy in Spain and its impact on Physics Research', in M.d.Mar Rubio-Varas and Joseba De la Torre. *The Economic History of Nuclear Energy in Spain* (Palgrave Macmillan, 2017), pp. 97–107; (Ed.) *A Comparative Study of European Nuclear Energy Programs* (Max-Planck-Institut Wissenschaftsgeschichte, Berlin, 201; '*Technoscientific Synergies between Germany and Spain in the Twentieth Century: Continuity amid Radical Change*' (*Technology and Culture*. 51, 1, 2010), pp. 80–98.; (with Baracca, A.; Renn, J.; et al) (Ed.) *Nuclear proliferation: Historical appraisal and present problems* (Max Plank Institut fur Wissenschaftsgeschichte, 2009); 'Science after Versailles and Paris: The scientific relations between Germany and Spain in 20th. Century', in Martin Kohlrausch (Ed.). *Technological Innovation and Transnational Exchange of Knowledge* (*Journal of Modern European History*, 2008, 2), pp. 218–236; (Ed.), *Who is Making Science? Scientists as Makers of Science Policy* (Max Plank Institut fur Wissenschaftsgeschichte, 2008); 'La inmediata posguerra y la relación científica y técnica con Alemania', in Ana Romero y María Jesús Santesmases (Ed.). *Un siglo de política científica en España* (Madrid, Fundación BBVA, 2008) EDITORIAL?, pp. 173–210; '"Germania docet": On a lecture trip to Spain: The scientific relations between Germany and Spain during the Entente boycott (1919–1926)' (*Annals of Science*, 65, 2008: 529–546); 'Deutsche Wissenschaftler und Spezialisten in Spanien im 20. Jahrhundert: Kontinuität und Umbrüche', in Rüdiger von Bruch, (Ed.). *Wissenschaft, Politik und Gesellschaft. Deutschland im internationalen Zusammenhan* (Stuttgart: Franz Steiner, 2007); '"Germania docet": The scientific relations between Germany and Spain during from 1919 until 1926', in Von Bruch, R., (Ed.), *Wissenschaft, Politik und Gesellschaft. Deutschland im internationalen Zusammenhang im späten 19. und im 20. Jahrhundert* (Deutsche Forschungsgemeinchaft, Stuttgart, 2005); 'La correspondencia entre José M. Otero Navascués y Karl Wirtz, un episodio de las relaciones internacionales de la Junta de Energía Nuclear', in Sánchez Ron, J.M., (Ed.), *Cien años de Física Cuántica*, Arbor, Vol 167, No 659–660 (2000), pp. 527–601

Antoni Roca-Rosell currently lectures at the Universitat Politècnica de Catalunya—Barcelona Tech (UPC). He holds a degree in Physics from the University of Barcelona (1975) and a PhD from the Autonomous University of Madrid (1990). In 1996, he co-founded the Centre for the History of Technology of the UPC, which publishes the international journal *Quaderns d'Història de l'Enginyeria* since 1996. In 1991, he was one of the founders of the Catalan Society for the History of Science and Technology (SCHCT), being its president from 1993 to 2009. In 2010, he was elected corresponding member of the International Academy of History of Science. In 2014, he was elected member of the Institut d'Estudis Catalans. His field of research is the history of science and technology in Catalonia and Spain, including the history of engineering and technical education between the XVIII and XX centuries. The background of his research is the comparative perspective and the analysis of the processes of diffusion and appropriation of scientific ideas and technical knowledge to better understand the social impact of science and technology in local and national contexts.

Carolina Rodríguez-López is a tenured professor in the department of modern history and contemporary history at the Complutense University of Madrid (Spain). She has been a visiting scholar at Harvard University and at California State University DH (USA) and a guest researcher at the ZZF (Potsdam, Germany).
Her lines of research include the history of universities and university campuses, academic exile, historiography, tourism, emotions and the history of urban reconstructions after warlike processes.

She is the director of CIAN-Revista de Historia de las Universidades. She is the director of the Complutense Interpretation Center of the University City of Madrid.

Recent publications:

- (2019): *Paisajes de guerra. Huellas, reconstrucción, patrimonio (1939–años 2000)*. Stéphane Michonneau, Carolina Rodríguez-López y Fernando Vela Cossío (eds.). Madrid, Casa de Velázquez-Ediciones Complutense. ISBN: 978-84-909-6228-2.
- (2018): *Hacia el centenario. La Ciudad Universitaria de Madrid a sus 90 años*. Carolina Rodríguez-López y Jara Muñoz Hernández (eds.). Madrid. Ediciones Complutense. ISBN: 978-84-669-3582-1
- (2012): *Reconsidering a Lost Intellectual Project: Exiles' Reflections on Cultural Differences*. Carolina Rodríguez-López y José M. Faraldo (eds.). Cambridge Scholars Publishing. ISBN: 1-4438-3649-4.

Vanessa Tessada Sepúlveda holds a degree in History from the University of Chile and a PhD in History from the University of Valladolid, and is currently as assistant professor at the Autonomous University of Chile. Her lines of research are marked by studies of gender and political and transnational history, pointing to the processes of construction of international spaces of female politicization. Her studies focus on the dictatorial periods of Chile, Spain and Argentina.

Nicolás Sesma is Lecturer (Maître de conférences) in Spanish Civilization at the University of Grenoble Alpes (France) and Researcher at the Casa de Velázquez (Madrid, Spain). His first degree is in History from the University of Zaragoza (Spain), and he holds a PhD in History and Civilization from the European University Institute (Florence, Italy) in 2009 with a thesis about the Institute of Political Studies (Miguel Artola Prize, 2010). He has been researcher at the Residencia de Estudiantes (2001–2003) and the Ortega y Gasset Foundation (2003), visiting researcher at the University of Madison Wisconsin (2006) and postdoctoral researcher at the University of Columbia in New York (2010–2012). In 2004, he was granted with the Prize for Young Researchers from the Spanish Contemporary History Association.

José Soto-Vázquez (PhD in Hispanic Philology) is a professor of Didactics of Language and Literature in the Universidad de Extremadura (Uex). He is the main investigator of the LIJ group and co-directs the magazine *Tejuelo*, included in Esci; Scopus; Erih plus; MLA; DOAJ. He is the author of several books and research chapters in national and international publishers such as Academy of Hispanism, Visor, University of Castilla la Mancha, Dykinson, Schauhoer Verlag or Shaker Verlag. He coordinates the collection 'The Peak of the Stork', which has been translated into six languages and from which more than 20,000 volumes have been published in Germany, France and Spain. He directs the collection 'El Pirata' (Editora Regional de Extremadura), with texts written by Ada Salas or Álvaro Valverde. His work has appeared in journals included in JCR and Scopus: *Paedagogica histórica; Revista Española de Pedagogía; Ocnos; Anuario de Estudios Filológicos; Información, cultura y sociedad; o Investigaciones bibliotecológicas*. He has participated in regional research projects (Reading Habits, 2017), national research projects (Children's Literature, 2012) as well as European (Cultural Routes, 2015).

Gabriela Viadero Carral is a Spanish academic, specialist in nationalism and national identity. She holds a bachelor's degree in Humanities and in Journalism, an MPhil in Contemporary History of Spain and a PhD in

Political Science from Universidad Complutense de Madrid, under the supervision of Professor José Álvarez Junco. She has published several academic and informative articles about the relationship among cinema, nation-building and identity in important journals such as *Revista de Occidente* and is the author of the book *Cinema Serving the Nation* (2016), edited by Marcial Pons Publishing House and awarded with Muñoz Suay 2017 Prize for the best History and Cinema research of the year. In this work, she studied the Spanish imagined community in the feature films produced in Spain between 1939 and 1975 (Francoist regime). In 2017, she was a visiting fellow in the government department of the London School of Economics and Politics, under the supervision of John Hutchinson, expert in nationalism. At present she is dedicated to study the narrative structure of the myths and its relationship with nationalist discourses.

List of Figures

PART I

Theory and Methodology

CHAPTER 1

Introduction: The Usefulness of Science and Culture as 'Nationalization' Tools in the Early Franco Regime

Mariciò Janué i Miret and Albert Presas i Puig

This edited volume examines the role of science and culture as tools for building a national identity during the early Franco regime.[1] By the "early Franco's regime", we refer to the years from 1939, with the military regime firmly established after the end of the Spanish Civil War (1936–1939), until 1959, when an economic 'Stabilization Plan' implied the abandonment of autarkic economic policies and the returning to a more open economic system.[2] At the level of international politics, this first phase of Franco's dictatorship is integrated into two clearly

[1] This volume is part of the framework of the research project titled 'Science, culture, and nation in Spain: from the 1898 "disaster" to the end of Franco's Dictatorship' FFI-HAR 2016-75559 (AEI/FEDER,UE), which has also made its funding possible. The results were discussed in the Nexus–UPF research group (https://www.upf.edu/web/nexus).

[2] Riquer (2010).

M. Janué i Miret (✉) • A. Presas i Puig (✉)
Department of Humanities, Universitat Pompeu Fabra, Barcelona, Spain
e-mail: maria.janue@upf.edu; albert.presas@upf.edu

M. Janué i Miret, A. Presas i Puig (eds.), *Science, Culture and National Identity in Francoist Spain, 1939–1959*,
https://doi.org/10.1007/978-3-030-58646-1_1

differentiated periods: the Second World War (1939–1945), and the consolidation of the Cold War after 1945. However, in relation to Spain, both periods have in common their isolation imposed by Western powers, which was gradually abandoned in the 1950s. This volume, thus, deals with the role that science and culture played in the early Franco regime in the process of definition and implementation of the ideas and policies of *nationalization* in Spain.

While the studies on Franco's regime have emphasized nationalism as one of its most distinctive features, the analysis specifically focused on the nationalization efforts of the regime is limited. The concept of nationalization we are looking at is the complex process of transmitting meanings about the character and relevance of the Spanish nation, namely, ideas, representations, and practices of homogenization linked to the supposedly basic constitutive elements of national identity.[3] This book focuses on the discursive and symbolic dimensions of nationalization through science and culture, rather than on the material efforts, which have already received some attention.[4] In this sense, although we recognize the relevance of material efforts and infrastructures for nationalizing purposes, this volume scrutinizes the role of science and culture in discourses aimed at creating and reaffirming the idea of the Spanish nation. Apart from some very meritorious works on Spanish science and culture,[5] and studies devoted to the development of the idea of the nation,[6] the relationship between the two issues has barely been raised.[7] However, there is enough evidence that science and culture played a role as instruments of nationalization policies and that this is one of the most significant, unexplored aspects of Franco's nationalism.

This edited volume focuses on the aforementioned relationship by exploring four main questions: (1) how the Spanish state used the development in science and culture for the legitimation of the Spanish nation; (2) the ability of science and culture to mobilize state resources in the name of the nation for their own interests; (3) to what extent cultural and

[3] Quiroga (2013).

[4] Pro (2019).

[5] Sánchez-Ron (2008), Romero de Pablos and Santesmases (2008), Gómez and Canales (2009), Gracia and Ruiz Carnicer (2001).

[6] Álvarez Junco (2001), Quiroga and Archillés (2013), Moreno Almendral (2014), Michonneau and Núñez Seixas (2014), Alares López (2017).

[7] An exception, Santana de la Cruz (2009); for engineering Camprubí (2014), for historiography, Prades (2014) and Antolín Hofrichter (2018).

scientific output was determined or affected by the current conceptions of the nation; and (4) what was the influence of the representations and symbols of the nation on the role of scientists and the way they understood their disciplines, as well as their own role.

The contributions in this book are based on the premise that the constitution of the contemporary nation state and the development of science and culture policies are closely interrelated.[8] The contemporary nation state considers knowledge and science as an engine of wealth generation and social development, which, in turn, will determine the relationship among those with political power and society. Already in the twentieth century, the previous trend towards the organization of science in the sense of building scientific policies at the national as well as the international level was confirmed. Thus, the analysis of cultural diplomacy is an ideal means of exploring the links between cultural and scientific fields and their relationship to international political changes and transformations.[9] All this will facilitate the promotion of science in its social and political contexts, as well as its instrumentalization, to reinforce the desired idea of nation state. The question of the significance of science and culture for the national identity has become one of the most important topics on the research agenda of the history and sociology of science.[10]

Spain, despite the impasses, breakouts, and peculiarities of its history, will not fail to accompany this historical cadence.[11] There were initiatives of organization of science, underpinned by different social movements and traditions, all of them participating in the debate on the role of culture and science in shaping the Spanish nation, which aspired to be present in the international policy or intended to be recast under its livelihood. For this reason, it is essential to understand the evolution and interaction of both processes, the role of science and knowledge, and the conceptions about the nation at a fundamental stage of its development and in the light of different political situations.

The development of ideas of nation in the framework of the construction of contemporary nation states has been the subject of study by specialists on nationalism issues for a couple of decades.[12] However, the

[8] Inkster (2009).

[9] Niño (2009), Gienow-Hecht and Donfried (2010). For Spain, Delgado Gómez-Escalonilla (1992) and (2014).

[10] Metzler (2000), Kojevnikov (2004).

[11] Sanz Menéndez (1997).

[12] Smith (1999).

importance and contribution of the science and culture to their significance have hardly been considered.[13] At the same time, although the relationship between science and political power has been the subject of numerous studies, its role in consolidating the nation state has not yet been seriously taken into account. The present volume is well aware of the advantages of understanding the close relationship between science and culture, on the one hand, and the state and the nation, on the other, as a dense network of political, social, cultural, and material relations, in a complex interaction that flows multidirectionally.[14]

Already existing approaches putting science and culture in the focus of historical research on nation-building ideas warn of the risk of hasty acceptance of categories and use of terms such as *science* and *nation* as given, homogeneous, and static. Science and nation do not always appear as univocal concepts at all times and circumstances.[15] It is, therefore, necessary to consider science and nation as a web of references, and to analyse the social relationships and processes in which they are generated and established. Since the end of the nineteenth century, science and technology have not only described the world but have even built it through their applications, technologies, and goals. Hence, it can be considered that the 'construction of nature' and 'nation-building' took place simultaneously.[16] In this line, we understand the generation of science and knowledge linked to the great themes of history in general, and they should therefore be considered by historiography in the analysis of nation-building processes.[17] This requires a cultural perspective on the conditions of production and the power and influence of knowledge, such as the one we propose here for the case of Spain.

In 1904, Max Weber argued that scientific knowledge based on the controllable and reproducible method and on free and universal communication, the principle of rationality of which would extend to the whole world, was opposed to the idea of nation, which he associated with diametrically opposed characteristics.[18] The nation would be governed by emotional identifications and cultural factors, an approach that has been

[13] Exceptions are, among others, Fohrmann (1991), Cañizares-Esguerra (2001), Jessen and Vogel (2002), Krige and Wang (2015).

[14] Höhler (2001).

[15] Pestre (1997).

[16] Haraway (1991).

[17] López Piñero (1992).

[18] Weber (1973).

shared by much of the literature. Just as science, especially modern science, is understood as universal, the nation is conceived as being particular.[19] The investigation of the relationship between these two issues, usually considered separately, responds to the assumption of the innovative research on the idea of nation carried out during the 1990s.[20] Moving away from a positivist interpretation of national historical development, and incorporating sociohistorical studies in the formation of nations in nineteenth-century Europe,[21] those then-new approaches, showed the symbolic character of the nation's construction, based on a society that would rest on a socially constructed idea and tradition. In this way, not only has the same concept of nation been developed but a large number of research topics have also been generated. The nation has come to be seen as a culturally constructed association whose relations will be reaffirmed, staged, stimulated, and symbolized, with its institutionalization in the national state. From this perspective, the analysis of the contribution of science and culture to the construction of the idea of nation and the legitimation of the state, but also how the state favoured the development of science and culture, acquires a special interest. This analysis will benefit from new approaches to cultural history which consider science and nation as cultural phenomena situated in European contemporaneity, related to a specific anti-traditionalism and with utopian moments of national improvement and fullness.[22] Scientific work and cultural development depend on institutional arrangements, social relations, cultural traditions, and legitimization processes that are never completely isolated, or fully autonomous.[23]

Recent historiography confirms the contribution of science to the formation of a 'national culture'.[24] The relationship between nationalist and scientific politics has been considered from the early twentieth century onwards.[25] Work on the popularization of scientific knowledge is relevant, showing its links to nationalist aspirations based on patriotic-nationalist rhetoric, both in the dissemination of knowledge and in the creation of the idea of national progress linked to science.[26] Other studies focus on

[19] Hauge (1996).
[20] Gellner (1983), Anderson (1991), Hobsbawm and Ranger (1992).
[21] Breuilly (1985).
[22] Jordanova (1998).
[23] Golinski (2008).
[24] Raphael (1996), Metzler (2000), Brückweh et al. (2012).
[25] Vom Bruch (1986).
[26] Bensaude-Vincent (1997).

the institutional framework of science, as the relationship between universities and nationalist science.[27] In the realization of the idea, both of the nation and of the universal scientific progress in the person of the scientist and the cultural creator, the State will stylize its researchers as heroes or myths, collaborating in the formation of national identities. Thus, German and French studies focus on the analysis of the involvement of scientists in nationalist propaganda around the First World War.[28] Furthermore, studies dedicated to Nobel laureates have regarded them as a ranking to illustrate scientific capabilities and the potential of different nations.[29] In the Spanish case, an example would be the role of scientists as ambassadors of Francoism during the international isolation.[30] The progressive professionalization of scientific research as a modernizing force in the creation of the nation demands to consider the functionality of the latter to establish and legitimize the agenda of the scientist.[31] However, many of the existing studies do not aim to analyse the complexity of the relations among science, state, and conceptions of the nation, but between nationalism and its concretion. At the same time, only very few studies have been devoted to the relationship between science and nation from the perspective of the new cultural history.[32] Following this perspective, the proposal of this book is to replace the asymmetry of the classical model, according to which the nation state interferes in scientific practice, but not the other way around, with an approach, both analytical and heuristic, to the symmetrical interrelationship between state and science. According to this paradigm, we analyse the relationship between the state and science and culture considering both as organized resource management groups in an interactive pattern similar to a network: science and culture are no longer independent of policymakers. Rather, they are part of general policies, obtaining from them financial support and new goals for scientific agendas, while their results and output provide legitimation to the State political structure. In this vein, one of the contributors to this volume, Mitchell G. Ash, characterizes science as a social subsystem that provides resources to other social subsystems.[33] In return, science appropriates or consumes

[27] Porciani (2000).
[28] Prochasson and Rasmussen (1996), Mommsen (1995), Fell (2000).
[29] Crawford (1992), Friedman (2001).
[30] Presas i Puig (2005), Antolín Hofrichter (2018).
[31] Malet (2009).
[32] Jordanova (1998).
[33] Ash (2002) and his contribution in this book.

resources from other subsystems of society. Ash analyses science and its relationship with the State as set of resources interacting with each other. We must not understand the systems that support science and culture separately from scientific policy, but as an integral part of politics, providing the latter with the necessary resources for its legitimacy. In turn, national scientific and cultural systems reflect the different variants of nation-building.

As various authors have argued, regimes without political freedom have also been aware of the role of culture and science as the backbone of their ideas of 'nation' or 'fatherland'/'motherland'.[34] This is a historically relevant phenomenon in which Franco's Spain was no exception. With respect to the relationship between science and nation under Franco's regime, recent research highlights the existence of the link between 'symbolic order' and 'scientific policy'.[35] Although the parameters, the context, and the actors and their objectives were obviously specific, during the Franco dictatorship, the role of culture and science in the new formation of the state was debated.[36]

At the same time, although the absence of a genuine debate on science and culture during the Franco regime has traditionally been emphasized, recent studies have demonstrated the need to consider cultural and scientific development and its relationship to the idea of nation for a global understanding of Spanish history at this stage as well. Thus, the publications that have appeared in recent years on the Centro Superior de Investigaciones Científicas (CSIC)/Higher Center for Scientific Research, created in 1939—to which several contributions of the volume pay attention—have insisted on the need to consider not only the elements of rupture with the previous stage but also the continuities. [37] In this line, it has been highlighted that some of the main individual actors of the CSIC had been former scholars of the Junta de Ampliación de Estudios e Investigaciones Científicas (JAE)/Board for the Extension of Scientific Studies and Research founded in 1907—although the political-ideological bases of both institutions have been very different. The same can be said

[34] Kojevnikov (2004), Rodríguez López (Ed.) (2016), Saraiva (2016), Janué Miret (2019).

[35] Antolín Hofrichter (2018), Camprubí (2014).

[36] Sanz Menéndez (1997), Herran and Roqué (2012).

[37] Among others Puig-Samper, M.A. (Ed.), (2007). A critic in Nieto-Galan (2008).

in reference to the Instituto de Estudios Políticos (IEP)/Institute for Political Studies, also created in 1939 and also considered in the volume.[38]

This volume assumes the idea of science and culture that integrates the natural and experimental sciences, as well as the social and human sciences and the arts, considering that all of them share the generation of knowledge and its dissemination. Although we do not intend to comprehensively address all possible areas of the relationship between science, culture, and the arts on the one hand, and the idea of nationalization on the other during the first phase of Franco's regime, as a whole, the volume aims to discuss the fundamental aspects for its understanding. The aspects we address can be summarized in the following eight components:

1. The agents in the fields of science and culture, as well as in the State, who played a part as promoters of strategies and subsidizers of science and culture.
2. The institutions that acted as channels for disseminating narratives about the nation, which we can therefore call 'national institutions': academic, scientific, cultural, and artistic, whether official, semipublic, or private.
3. The interrelationship between the following three spheres: (a) agents of scientific and cultural production—scientists, academics, intellectuals, and artists; (b) scientific and cultural institutions, as well as the State itself; and (c) finally, the (self-)identifications of the formers as national representatives in different contexts, positioning themselves in the forefront of the nationalist-scientific patriotism to forge a scientific community for the nation.
4. The attempts to create a 'national science', both in the sense of promoting certain distinctive specialties and in cultivating a specific model of practising science, that is, the cultivation of national scientific landscapes from the state–nation and the science–culture relationships (CSIC, scientific academies, and institutions).
5. The multidimensional relationship between agents, transmitting institutions, state actors and receptors—'consumers' of national narratives—emerging from the different fields of science, culture, and arts, that is, the generation of a scientific interpretation of the nation (e.g. contribution from historical disciplines and social sciences).

[38] Sesma Landrín (2011).

6. The interrelation between the following three levels: (a) political and social transformations, (b) modifications in the narratives about the nation and the concept of the Spanish nation, and (c) changes in scientific and cultural policies.
7. The interchange of material and symbolic resources between political interests and scientific and cultural interests. In this sense, we intend to evaluate the language and symbols borrowed from science and culture to be used as instruments of transmission of national narratives, whether they allude to a supposed pre-existing Spanish national identity or they seek to create such an identity.
8. The links between the efforts to nationalize science and culture on the one hand and to internationalize them on the other, as well as the role of cultural and scientific transfer and exchange, not only as material resources but also as sources of prestige and political power. At this point, it should be considered that the promotion of science and culture was often carried out both at national and international levels by the same institutions. Thus, the fact that the State chose to empower some institutions more than others provides valuable information for the purposes of its project of nationalization. At this level of international relations, we consider the role of scientific and cultural diplomacy, a great driver of identification between science and culture and national images.

We have structured the treatment of the aforementioned issues in the book in five sections. The first one is devoted to the common theoretical and methodological bases that all contributions share, with special stress on the meanings attributed to science, culture, nation, and relations between them. The chapters which make up the second section address significant points in the scientific and cultural policies in Franco's 'New State', including their institutional, ideological, and symbolic elements, as well as a perspective on the evolution of the relationship between Catalan substate nationalism—a relevant phenomenon in twentieth-century Spanish history—and science. Women, and their roles in various areas of the regime's scientific and cultural system, constitute the leading thread of the third section. In relation to this point, it is necessary to keep in mind the mutual relationship between the construction of gender by the nation and the role of gender in shaping national imaginaries, also during Franco's regime.[39] Despite official discourses of the regime's contempt for women's dedication to science and academia—and in general their presence in public sphere—incipient research has highlighted that there were women involved in cultural and scientific activities.[40] Therefore, we are convinced that for a proper exploration of the subject matter of this book, we need to address the still largely ignored role that women played in these areas and their relationship to the prevailing conception of nation and nationalization. Case studies on participation in the nationalization of different scientific disciplines and areas of the arts constitute the fourth section of the book. The volume closes with a fifth section, dedicated to exploring some perspectives on internationalization of culture, science, and technology in Franco's regime, thus offering the tools for a comparison with other Western countries during the Cold War consolidation phase.

In total, there are seventeen contributions that make up the five parts of the book. To the first part, dedicated to the theoretical foundations, belongs this introductory chapter, where we pursue to present the historiographical objectives of the book, its theoretical foundations, the state of the question on the most relevant aspects linked to our topic, and the nature of the contributions it contains. In addition, a contribution by Mitchell G. Ash completes this theoretical part of the book. His chapter considers the interaction processes established between science, the idea of nation, and culture, all of them concepts considered as moving targets. He affirms that the meanings of the terms science, nation (or national identity), and culture have fundamentally changed over time and have varied from one place to another in similar periods of time. In the medium term, national identity and nation itself have become increasingly complicated and ambivalent concepts in an increasingly globalized world. Science and erudition, as well as culture, have also changed radically since the beginning of the nineteenth century. Ash defends the idea that it is impossible to treat any of these three entities as something separate from the others. In his opinion, science is not only results of research or methods but also institutions. For this reason, sciences can also become centres of power. Ash sustains that the relationships between sciences and political power are not as one-sided as is often suggested or assumed. Attending to these interactions, periods of political power predominance, especially in dictatorships, have not always automatically produced pseudoscience. He

[39] Blasco Herranz (2014).
[40] Alcalá and Magallón (2008), Canales (2012), Romero de Pablos (2017).

further argues that from the historical perspective, we must speak of ideological connotations (assignments) of the sciences as discursive resources, or of scientific investigations carried out under different political priorities. In the nineteenth century, at least, and in many European countries in the early twentieth century, natural scientists, humanists, and public intellectuals were educated largely in the same elite high schools and values, and therefore should be considered members of a common culture. Members of all kinds of disciplines produced and propagated *national science*. For Ash, this is an important reason why the supposed fundamental difference between the natural and human sciences may not have been as significant for this particular history as is often assumed. Ash also discusses the term national sciences. He maintains that science is national, international, and transnational, often simultaneously. In this line, he warns that putting the term nation at the centre entails danger, either from a perspective of a single nation with too narrow a focus or from comparisons of a country's scientific policy with that of another, rather than considering the possibility of a transnational circulation of countries, ideas, and people. He concludes that a one-nation approach, which often simply reproduces existing institutional frameworks funded by the state without reflecting on them, cannot do justice to these complexities.

The second part of the book, which focuses on the scientific and cultural policies in Franco's 'New State', includes four contributions respectively devoted to aspects of historiography, symbolical architecture, epistemic communities and science makers, and the role of science and technology in Catalan substate nationalism. Andrés Antolín Hofrichter approaches the history of science under early Francoism, by focusing on its cultural expressions. He examines the narratives, symbols and rituals that surrounded the main institution founded by the new regime in order to promote and politically control scientific research: the Higher Council for Scientific Research (CSIC). Relying on studies on the symbolic dimension of institutions, he understands the High Council as an attempt to institutionally represent the capacity for 'science' of a Catholic nation and, above all, its elite. Antolín argues that the organizational framework of the High Council served to symbolize an allegedly restored 'Christian unity' of science and a specifically Spanish scientific path within the history of modern science. He analyses the organological language and symbology, and the religious-spiritual components used in publications and ceremonies. Following Antolín, they helped shaping a conception of 'science' that referred to a pre-industrial and pre-enlightened era—all this despite the fact that this same institution had been created to incorporate, also, 'technology' and 'natural sciences'. He concludes that, in order to solve this apparent contradiction, the High Council followed a logic of subordination and control: although 'technology' and 'natural sciences' were useful instruments, they also represented a sort of Trojan horse of a foreign 'modernity'. Antolín concludes that this conception soon became unsustainable, as the regime dedicated its imagery to a new technical-industrial developmentalism in the 1950s. The next chapter, authored by Carolina Rodríguez López, studies the plans of Franco's regime to recover, reconstruct, and implement a new political signification at the University City of Madrid. In the process of consolidation after the Spanish Civil War, Franco's regime placed the University as one of its priorities. For those who wanted to establish an *ex novo* regime, it was evident that the university would be one of the pillars on which to settle. However, despite all the effects of victory, the University maintained previous traditions, habits that would be mixed with the Francoist ideas. Franco's regime also inherited spaces that were significant and charged with the memory of war, as the University City of Madrid. Since 1927, especially during the Second Spanish Republic (1931–1939), the campus had been one of the most modern places in Madrid. Its design always tried to adapt the academic novelties to this new space, but after the Spanish Civil War, the campus became one of the scenarios favoured by the Franco regime to extol its ideology and, in short, serve as a political symbol for the New State. Rodriguez explores the symbols, images, and resources that Franco's regime used to leave its mark on the campus (still visible today). The subsequent chapter of this second part of the volume is centred on the policies of nuclear energy during Franco's regime. Albert Presas i Puig shows that the regime reacted quickly to the appearance of nuclear energy in the international policy. Its reaction is exemplified by founding the Junta de Energía Nuclear (JEN)/the Spanish Nuclear Energy Board, which can be compared to the initiatives of most of the countries that were trying to join the development of this new energy source. Presas i Puig considers the performance of the JEN as an epistemic community since it intended to advise the Francoist authorities and guide the development of this energy option until its industrial implementation. In a country without scientific and technological tradition, the generation of this dynamic started from specially trained actors with a special sense of opportunity. This is the role Presas i Puig recognizes in José María Otero Navascués. Following the proposal of Ash, Otero's performance allows Presas i Puig to analyse the relationships established between the industrial development policy of the regime (in the form of economic and institutional support) and Otero's performance offering the regime symbolic resources of modernity and international relationships with the most advanced scientific centres in the world, all at a time when Franco's regime needed to re-establish itself in the new international scenario over the years. The consideration of the JEN as an epistemic community not only helps to understand the actions of some of the actors that defined the Spanish nuclear development policy, but also allows the understanding of the modifications in its function, as a result of the changes in the national and international policy, in the course of the gradual opening of the regime. As part of that scientific political complex, JEN was affected in its original purpose by its desire to be the main actor of the nuclear programme by substantially modifying the political context, and with it, its significance in the policy of economic and technological developments of Franco's regime. At the time of its industrial concretion, nuclear energy acquired a new significance as an element of the new Spanish policy of economic liberalization, proposed by the new cadres of the Spanish administration—the so-called technocrats. They modified the status quo and incorporated new priorities (incorporation into international financing systems approach to Europe), opting once again to reaffirm and shield the interests of Spanish capital in the development of nuclear energy. In this way, both the JEN and Otero were reduced in their function to mere companions and no longer to defining strategies. In the last contribution of this second part of the book, Antoni Roca Rosell studies the role of science and technology in the nationalist debate in substate nationalism in Catalonia after the Spanish Civil War. The development of scientific and technological activity in the period before the Spanish Civil War was dramatically interrupted as a result of the war when an important part of the scientific community identified with the Second Republic and went into exile against the new regime. Some of their members were able to start new scientific careers in countries like Mexico, Venezuela, or Argentina. Those who remained in Spain had to comply with the strict political regulations imposed on university professors and members of scientific institutions. Before the Spanish Civil War, science and technology played a growing role in Catalonia. During the war, Catalan experts made valuable contributions to health care in medical areas which were relevant during the conflagration. Republican propaganda enhanced the importance of scientific culture and technical education. The press and publications of the exiled Catalan community took over this purpose. Some Catalans in exile expressed the desire to pass on Republican experiences to the new generation and to promote science and technology. Some academics went into 'interior exile'; they were expelled from their institutions and obliged to develop a private career, leaving research behind. Franco's regime created a branch of the CSIC in Barcelona, which was also a strongly ideologized institution. The aim of this initiative was to hide the Republican successes. Despite the fact that there were several CSIC centres in Barcelona, scientific activity in Catalonia was significantly impaired. The alliance between science, technology, and the Catalan identity prior to Franco's regime was broken, and it was not replaced by another kind of alliance. In fact, it was probably due to the ideological commitment of science and technology to Franco's regime that these were regarded with scepticism or even hostility by the Catalan nationalist community. In the Catalan exile, plans for a free Catalonia included the wish to promote higher scientific research and universities at the service of the society.

The third part of the book devoted to the role of women contains three contributions. Fernando García Naharro explores the participation of women in a domain of men: the applied sciences in the CSIC. Under dictatorial conditions, there were several constraints imposed on science to abide by specific conventions and laws that shaped the legitimate actors working in the field of science. Under Franco, not all scientists were in the same condition: political repression and legal procedures were in the hands of the dictatorship to manage the reorganization of the brand-new national science. Thus, García Naharro argues that the dictatorship gave the official scientific institutions an important role in the construction of the legitimated scientific knowledge or in demarcating who had a legitimate voice in the field. In so doing, he focuses on the CSIC to analyse the official concept of science that led to the socio-discursive configuration of the scientific field: places and actors associated with representations and classificatory dichotomies that played an active role in shaping the scientific *ethos*. Hence, García Naharro explains that the official discourse of Franco's regime supported the image of the *vir modestus* (a male image of the scholar) within the scientific field with the social and political implications that it should carry with it: beliefs about the nature of the experts and their knowledge claims. The idea of science conceived as a male activity was constructed upon a deep-seated cultural bias transmitted by different channels, using cultural models and metaphors carrying gender meanings that devaluate women and overvalue men. In the second part of this work,

Garcia Naharro answers the question of what kind of women had the legitimacy to be a 'modest witness' under Franco by studying the social profiles of women publishing in scientific journals, occupying research positions or having access to study abroad. It demonstrates that, alongside concerns of expertise and authorship, gender played a relevant role in the construction of those who have a legitimate voice in the field. Therefore, he aims to provide new knowledge about those women of science who worked in fields such as engineering or applied science in Spain during Franco's regime. Women also played a relevant role in the field of the censorship of children and youth literature in Franco's regime. This is the subject that José Soto-Vázquez and Ramon Tena-Fernández research in their contribution to this edited volume. Literary censorship is still one of the least known aspects of Franco's regime, and this work aims to shed some light on the figure of the Reader 22. The authors contrast the proofs of the galley, the opinions of the censors, and the resolutions and erasures that are appreciated in them, as well as interviews with authors and editors, as object of the censorship. Children's and young people's literature maintained the obligation to request the censorship review process throughout Franco's regime, which allows us to understand the zeal of the authorities on the surveillance of texts intended for younger readers. Since Reader 22 was almost always a woman, the analysis is an unprecedented study of the role played by women in censorship in Franco's Spain in the field of children's and youth literature, as well as their academic and personal profiles. The last part of the chapter is devoted to one of the most notable female censors: María Isabel Niño Mas. The last chapter of this third part of the book, authored by Vanessa Tessada, is dedicated to the contribution of the Female Section of Falange—the fascist unique party of the regime—to the Hispanic Community of Nations. The main objective of the chapter is to analyse the cultural influence of the Female Section of fascist party Falange on Latin America, as part of Spain's diplomatic strategies during the first period of Franco's dictatorship. As fascist Spain was banned after the Second World War, it sought strategies to enter the international scene. In this context, the Female Section acted as an agent in international affairs and, together with the Ministry of International Affairs and the 'Institute of Hispanic Culture in Madrid', it actively participated in the construction of the 'Hispanic Community of Nations'. This community was a project of the Ministry to build a spiritual community to improve the role of Spain on the international stage, based on the historical relationship with Latin America and the leadership position that Spain could assume as a natural bridge between Latin America and Europe. The Female Section proposed an archetype of Hispanic women as a contribution to the Community and, based on it, the Female Section put into practice several strategies to approach Latin American women who were pro-Falangist, Catholic, and conservative. The Female strategies included the creation of cultural associations of women in several countries of America; a scholarship system that granted more than a thousand women between 1947 and 1977 facilities to study in Spain; several presentations of the 'Choirs and Dances of Spain' in Latin American cities and the celebration of international congresses, among others. In this chapter, Tessada pays attention to the transnational space that these strategies helped to build during the first phase of Franco's regime in the context of the Hispanic Community of Nations, in which young Latin American women would get acquainted with the ideas of Falangism, antifeminism, Hispanism, and conservative catholic thought.

The fourth part of the volume includes four contributions, respectively, on mathematics, the intellectuals of the so-called Generation of 1948, film-making, and art avant-garde. José Miguel Pacheco Castelao signs the first chapter of this section, which offers an overview on the status of mathematics and mathematicians under Franco's regime between 1940 and 1960. During the first decade, Spain was relatively isolated, first in the aftermath of the Spanish Civil War and the Second World War, then by the isolation decreed on Spain by the United Nations Organization between 1946 and 1950. From 1950 onwards, after Spain and the United States signed agreements leading to the installation of American military bases on Spanish territory (1955–1958), the rigid regime structures were somehow relaxed through 'cooperation programmes' in various fields, including academic interchanges and study scholarships. Under those political conditions, mathematics—or rather, some mathematicians—managed to survive and generate mathematical results despite the regime's scarce interest. Three stages or 'lives' can be observed in the period studied, well correlated with Spanish international policies and the social and economic health of the regime. In addition, an analysis of the two leading Spanish mathematical journals is presented. Sara Prades Plaza analyses in the following chapter the legacy of Charles Maurras' integral nationalism to the intellectuals of the Generation of 1948. Prades considers in detail their national discourse by which they gave an idealized image of the past and tried to rewrite national history to legitimize a politico-cultural project in the present. During the 1940s and early 1950s, the Maurrasian ideology was taken up by the group formed around the Spanish journal *Arbor*. This group may also have been influenced by the integral nationalism of the Acción Española party, which was its ideological antecedent during the Republic. Their discourses vindicate a nation ruled by a monarch, the defence of regional plurality, and the insertion in an eminently Catholic and traditionalist Europe, as they identify with the European political culture of reactionary nationalism. This idea of a nation was the reason for colluding with the less monarchist and more Castilian sectors of the regime because the members of the Generation of 1948 considered themselves the true interpreters of their historical moment. Their work had an academic purpose, but they also exercised a political function, as many of the members of this group held positions of political, cultural, or ideological power. That is why they had no qualms about accepting the idea of Spain's hegemony, which was needed to legitimize Franco's regime after the defeat of fascism during the Second World War. The third contribution of this section, authored by Gabriela Viadero Carral, deals with nationalist stories in the cinema of Franco's regime. Cinema, besides being an artistic expression, is a cultural product, a highly effective tool to communicate ideas. In fact, during the twentieth century, cinema became the most powerful medium for several reasons: its extent (it reaches massive numbers of people), its efficiency (the image is the most expressive means), and its capacity to show as real what is not. Power structures soon realized how powerful cinema was and tried to control and guide it according to their interest. This is the case of Franco's reactionary regime, which controlled cinema through censorship, subsidies, and repression, especially until 1963. Franco's regime was, essentially, a nationalist regime. The Spanish Civil War was interpreted in the following terms: the insurgents defining themselves as the real Spaniards, the ones that defended their nation from enemy invasion. This defence of the Spanish nation became the argument Franco's regime used to legitimate itself, so it is not surprising that many of the films produced under Franco's rule used the Spanish nation idea in the interest of the regime. In the last contribution of this section of the book, Jorge Luís Marzo analyses the role of the Spanish artistic avant-garde of the 1950s. A substantial part of the Spanish abstract and informalist artists of the period undertook a path of collaboration with the dictatorial regime that would last a decade. This anomaly, if we compare it with the attitude of the creative groups with similar styles in other parts of the world, has hardly been analysed by Spanish academia. It was in the late 1990s that the history of art began to address the issue. Among the varied interpretations that were emerging, there was one that was taking shape: the function of tradition as a nationalist narrative, especially the baroque one, which served as a cement for the construction of a platform of common interests. This platform allowed the regime to update its attachment to the essences, as well as project international homologation. It also allowed the artists to provide historical grounds for their works, within the framework of a renewed global competition of artistic brands during the Cold War. The national culture, which gathered around twisted appeals to historical immanence and contemporary update, was promoted as a meeting place between Franco's regime and the avant-garde, creating paradoxes, sometimes insoluble, which contaminated the instruments for their analysis.

The fifth and last part of the book dedicated to the internationalization of science and culture begins with a presentation on French Hispanism and Spanish cultural diplomacy by Antonio Niño Rodríguez. Niño analyses the efforts of the Francoist authorities to repair relations with the French university Hispanism, which had been interrupted by the Spanish Civil War. Since the early twentieth century, Spanish diplomacy had tried to use the University of Sorbonne as a spearhead for an ambitious plan of 'Hispanic intellectual expansion' in Europe. The war interrupted the university exchange, and it could not be resumed in the immediate post-war period because of the quarantine imposed on the dictatorship, as well as the hostile attitudes of the main leaders of the French Hispanism. However, the Spanish authorities were determined to rebuild the Spanish cultural presence in Paris as a sign of political normalization. The following two significant cases are studied: the difficult recovery of the academic exchange with the Institut d'Études Hispaniques of the Sorbonne, achieved in 1958, and the reclamation of control of the Colegio de España in Paris, beginning in 1949. However, the purpose changed and acquired a more defensive slant. The objective of the academic exchanges was no longer to connect the Spanish culture with the great currents of European thought, but instead to present the 'true Spanish tradition' in a medium considered naturally hostile, and to counteract the anti-Spanish sentiments of foreign historiography supposedly influenced by the Black Legend. The collaboration between universities, initially created to foment the prestige of the country on the international stage, was now compromised by the ideological condemnation of the dictatorship. Although the Spanish authorities put effort and resources into their ability to influence the orientation of French Hispanism, they were met with resistance from academic authorities who were reluctant to engage in close academic cooperation for political reasons. The French Hispanism continued to pay attention to the representatives of the culture repudiated by Franco's regime and kept close ties with the republican intellectual exiles. They also accentuated the Latin Americanist orientation of Hispanism, despite formally resuming academic contacts with the Spanish authorities. Following the French case, Marició Janué i Miret analyses the Spanish–German cultural diplomacy between the end of the Second World War and the end of the 1950s. During the Second World War, Germany had reinforced its cultural diplomacy in relation to Franco's Spain. Having as an inescapable prerequisite the ideological agreement, the Nazi authorities sought to attract the most relevant intellectuals and academics. In Spain, its main receptors were the most radicalized sections of the Falange unique party. As for Franco's regime, it only began to reduce its relations with Nazi Berlin, when it had evidence that Germany could lose the war. This chapter asks about the consequences of the end of the war on German–Spanish cultural collaboration. At the beginning of the Cold War, the Spanish–German cultural diplomacy presents the singularity of being conditioned, both for West Germany by its recent National Socialist past and the subrogation of the functions of the German government in the Allied Control Council, and for Spain by the continuity of the dictatorship. As a result, both states had to confront international isolation. The analysis examines the degree of dislocation of mutual cultural diplomacy as a result of these circumstances. Likewise, it examines the elements of change, continuity, and/or adaptation—personal, institutional, and ideological-discursive—once the relations were re-established following the creation of the West German state. In addition, it investigates how the international context and the imperative to redefine its national discourse conditioned the two countries, as well as how the evolution in the place occupied respectively by the two countries in the international balance of powers was reflected in the initiative in mutual cultural relations. Janué concludes that, in perspective, the end of the War did not have a rupture effect in terms of the tradition of German–Spanish cultural cooperation, but rather that of pause. Her analysis aims to contribute to the debate around the limits of denazification and the sociopolitical and cultural continuities between the Europe of fascism and that of the Cold War. The penultimate chapter of the book, authored by Nicolás Sesma-Landrin, deals with the cultural operation and strategies of Franco's regime, with particular emphasis on its desire to become the sole representative of Spanish science and culture in the eyes of the international community, always in competition with the intellectuals of the republican exile. For that purpose, Sesma takes the social sciences as a case study. First, he analyses the history of the creation and consolidation of the social sciences in Spain, a history which reflects accurately the political dynamic of the country at that time, marked by the rivalry between the dominant state religion model, which privileged a catholic approach to social sciences, and the alternative project of liberal-democratic nationalization, promoted by the JAE, which privileged an empirical approach for social sciences. Second, he analyses the situation after the end of the Spanish Civil War, marked by the exile of the main authors of empirical sociology, internationally acknowledged as the main representatives of Spanish culture. Inside Franco's regime, national-catholic social sciences became hegemonic; however, some sectors of the Falange party also tried to appropriate the JAE legacy and its practice of empirical social sciences for their own purposes. This process of appropriation occurred mainly through the Falangist IEP. Finally, he studies the process by which the academic production of Franco's regime succeeded in supplanting the exiled authors as representatives of national culture. One major step for such change was the entry of Franco's Spain into the United Nations Educational, Scientific and Cultural Organization (UNESCO) in 1952. In this sense, Sesma describes the role played by the Francoist permanent delegation to the institution and its main figure, Joan Estelrich, who decided to privilege the participation of the authors coming from the Institute of IEP and of empirical social sciences in the Spanish contributions to the publications and activities of the UNESCO. The last chapter of the book is dedicated to American geopolitics in 1950s' Spain handled by Lorenzo Delgado López-Escalonilla. Delgado analyses the use of science as a tool for American geopolitics in the Spain of the 1950s. After the Second World War, the United States built an R&D system with a large component of public funds and resources, which was consolidated during the Cold War. Nuclear energy and the space race were two of the most outstanding dimensions of the scientific-technological rivalry between the United States and the Soviet Union, which in turn served as a showcase concerning each of the models they represented—capitalism and communism. Simultaneously, the US government looked to science as a tool with which to gain allies and capture the sympathies of relevant groups in other countries. In this sense, the American power sought to strengthen its leadership by promoting training programmes, exchanges of teachers and researchers, and a growing network of international collaboration. Spain joined this process starting in the mid-1950s, following the bilateral agreements that allowed the United States to build military bases in the country. The 'leaders of public life' were the main target groups of these American initiatives, and several programmes for the training of human capital were directed at them. The training of the Spanish army occupied a preferential position. At the same time, especially through the technical assistance programme, professional cadres from various specialties were trained, with priority given to the fields of nuclear technology, aeronautical engineering, and business management. From the Spanish perspective, knowledge transfer was important, along with the establishment of professional contacts, publication of monographs and specialized journals, translation of reference works, and increasing bilateral collaboration in an array of different sectors. In addition to those collaboration programmes, the scientific-technological leadership of the United States was disseminated through its machinery of persuasion in Spain in two key areas, consistent with what was happening globally: atomic energy and, somewhat later, progress in the space race. As Delgado shows, all this prepared the way for an international opening of the country in the following decade.

Thus, the present edited volume attempts to offer a new southern European perspective on the role of science and culture in modern societies as tools of nationalization. It aims to contribute to a better understanding of contemporary Europe, considering scientific and cultural development as engines of economic and social modernization and, as such, sources of political power. Given the particularities of Franco's Spain and the relative scarcity of studies on southern Europe, we hope that the volume will be an important contribution for readers beyond the historians of Spain.

Bibliography

Alares López, Gustavo (2017). *Políticas del pasado en la España franquista (1939–1964): historia, nacionalismo y dictadura*, Madrid, Marcial Pons.

Alcalá, Paloma; Magallón, Carmen (2008). Avances, ruptura y retrocesos: mujeres en las ciencias experimentales en España (1907–2005), in Romero de Pablos, Ana; Santesmases, María Jesús (Eds.). *Cien años de Política Científica en España*, Bilbao, Fundación BBVA: 141–169.

Álvarez Junco, José (2001). *Mater Dolorosa. La idea de España en el siglo XIX*. Madrid: Taurus.

Anderson, Benedict (1991). *Imagined communities: reflections on the origin and spread of nationalism*, London, Verso.

Antolín Hofrichter, Andres. (2018). *Fremde Moderne. Wissenschaftspolitik, Geschichtswissenschaft und Nationale Narrative unter dem Franco-Regime, 1939–1964*, Oldenburg, Gruyter.

Ash, M.G. (2002). „Wissenschaft und Politik als Ressourcen für einander", in: R. vom Bruch, B. Kaderas (Hg.), *Wissenschaften und Wissenschaftspolitik—Bestandaufnahmen zu Formationen, Brüchen und Kontinuitäten im Deutschland des 20. Jahrhunderts.* Stuttgart, Franz Steiner Verlag: 32–51.

Bensaude-Vincent, B. (1997). "In the name of science", in: Krige, J.; Pestre, D. (Eds). *Science in the Twentieth Century*, Amsterdam, Harwood: 319–338.

Blasco Herranz, Inmaculada (2014). „Género y nación durante el franquismo", in Michonneau, Stéphane; Núñez Seixas, Xosé M. *Imaginarios y representaciones de España durante el Franquismo*, Madrid, Casa de Velázquez: 49–71.

Breuilly, John, (1985). *Nationalism and the State*, Manchester, Manchester University Press.

Brückweh, Kerstin; Schumann, Dirk; Wetzell, Richard F.; Ziemann, Benjamin (Hg.) (2012). *Engineering Society. The Role of the Human and Social Sciences in Modern Societies, 1880–1980*, Basingstoke, Palgrave Macmillan.

Camprubí, Lino (2014). *Engineers and the Making of the Franco Regime.* Cambridge, MA., MIT Press.

Canales, Antonio Francisco (2012). "Mujer, franquismo y educación científica", Memorias IX Congreso Iberoamericano de Ciencia, Tecnología y Género, Madrid, OEI, https://scholar.google.es/scholar?cluster=17843197822296898595&hl=es&as_sdt=2005&sciodt=0,5 (accessed 29.6.2020).

Cañizares-Esguerra, Jorge, (2001). *Nature, Empire, and Nation. Explorations of the History of Science in the Iberian World.* Stanford, Stanford University Press.

Crawford, Elisabeth (1992). *Nationalism and internationalism* in Science, 1880–1939. *Four studies of the Nobel population,* Cambridge, Cambridge University Press.

Delgado Gómez-Escalonilla, Lorenzo, (1992). *Imperio de papel. Acción cultural v política exterior durante el primer franquismo*, Madrid. CSIC.

Delgado Gómez-Escalonilla, Lorenzo (2014). "Un siglo de diplomacia cultural española: de la Junta para Ampliación de Estudios al Instituto Cervantes", Documento de Trabajo 12/2014 del Real Instituto Elcano, accesible http://www.realinstitutoelcano.org/wps/wcm/connect/ea1c410045c4aef7a549af5de37d5b31/DT12-2014-Delgado-Siglo-de-diplomacia-cultural-espanola.pdf?MOD=AJPERES&CACHEID=ea1c410045c4aef7a549af5de37d5b31 (accessed 5.2.2021).

Fell, Ulrike (2000). *Disziplin, Profession und Nation: Die Ideologie der Chemie in Frankreich vom Zweiten Kaiserreich bis in die Zwischenkriegszeit*, Leipzig, Leipziger Universitätsverlag.

Fohrmann, Jürgen (1991). *Wissenschaft und Nation: Studien zur Entstehungsgeschichte der deutschen Literaturwissenschaft*. München, Fink.
Friedman, R.M. (2001). *The Politics of Excellence: Behind the Nobel Prizes in Science*, Gordonsville Va, Henry Holt & Co.
Gellner, Ernest (1983). *Nations and Nationalism*, Ithaca, NY, Cornell University Press.
Gienow-Hecht, Jessica C.E.; Donfried, Mark C. (Ed.) (2010). *Searching for a Cultural Diplomacy*, New York, Oxford, Berghahn Books.
Golinski, Jan (2008). *Making Natural Knowledge: Constructivism and the History of Science,* Chicago, University of Chicago Press.
Gómez, Amparo; Canales, Antonio Fco. (Eds.) (2009). *Ciencia y Fascismo*, Barcelona, Laertes.
Gracia, Jordi; Ruiz Carnicer, Miguel Ángel (2001). *La España de Franco (1939–1975). Cultura y vida cotidiana*, Madrid, Síntesis.
Haraway, Donna J. (1991). *Simians, Cyborgs, and Women: The Reinvention of Nature*, New York, Routledge.
Hauge, Hans (1996). "Nationalising Science," in Chartier, Roger; Corsi, Pietro (Eds.). *Sciences et langues en Europe*. European Commission, p. 151–160.
Herran, Néstor; Roqué, Xavier (Eds.) (2012). *La física en la dictadura. Físicos, cultura y poder en España. 1939–1975*, Barcelona, Universitat Autònoma de Barcelona.
Hobsbawm, Eric, Ranger, Terence (Eds.) (1992). *The invention of tradition*, Cambridge, Cambridge University Press.
Höhler, Sabine, (2001). Universalistischer Anspruch und nationale Identitatsbildung im europaischen Vergleich (19. und 20. Jahrhundert). Berichte zur Wissenschaftsgeschichte. Vol. 24, Issue 3, p. 221–223.
Inkster, Ian (2009), „The National Imperative: The State, Science and Technology, and Policy Evolution Circa 1400–2000), in Arvamitis, Rigas (Ed.), *Science and Technology Policy*, vol I, UK, Oxford, EOLSS Publications, 164–181.
Janué Miret, Marició (2019), "Contributing to the cultural "New Order": How German intellectuals attributed a prominent place for the Spanish nation", in Maria Björkman, Patrik Lundell y Sven Widmalm (Ed.), *Intellectual Collaboration with the Third Reich. Treason or Reason?*, London, Routledge, p. 214–229.
Jessen, Ralph; Jakob Vogel (Hg.) (2002), *Wissenschaft und Nation in der europäischen Geschichte*, Frankfurt am Main, Campus Verlag.
Jordanova, Ludmilla (1998). "Science and Nationhood: Cultures of Imagined Communities", in G. Cubitt (Ed.). *Imagining Nations*, Manchester University Press: 192–211.
Kojevnikov, A. (2004). *Stalin's Great Science. The Times and Adventures of Soviet Physicists*, London, Imperial College Press.

Krige, John, Jessica Wang (Ed.) (2015), Nation, Knowledge, and Imagined Futures: Science, Technology, and Nation-Building, Post-1945. *History and Technology*, vol.31, Issue 3, 171–179.

López Piñero, José María, (1992). Introducción, in López Piñero, José María (Ed.), *La ciencia en la España del siglo XIX*, Ayer 7, p. 11–18. Asociacion de Historia Contemporanea and Marcial Pons Ediciones de Historia

Malet, A. (2009). José María Albareda (1902–1966) and the Formation of the Spanish Consejo Superior de Investigaciones Científicas, *Annals of Science*, 66, 3: 307–332.

Metzler, G. (2000). *Internationale Wissenschaft und nationale Kultur. Deutsche Physiker in der internationalen Community, 1900–1960*, Göttingen, Vandenhoeck & Ruprecht.

Michonneau, Stéphane; Núñez Seixas, Xosé M. (2014). *Imaginarios y representaciones de España durante el Franquismo*, Madrid, Casa de Velázquez.

Mommsen, Wolfgang J. (Ed.) (1995). *Kultur und Krieg. Die Rolle der Intellektuellen, Künstler und Schriftsteller im Ersten Weltkrieg*, Oldenbourg, De Gruyter.

Moreno Almendral, R. (2014), "Franquismo y nacionalismo español: una aproximación a sus aspectos fundamentales", *Hispania Nova*, 12, https://e-revistas.uc3m.es/index.php/HISPNOV/article/view/1874/878 (accessed 15.5.2020).

Nieto-Galan, Agustí (2008), "La memoria histórica de la ciencia en España", *Sinpermiso.info*, 7.12.2008 (https://www.sinpermiso.info/textos/la-memoria-histrica-de-la-ciencia-en-espaa (accessed 6.7.2020).

Niño, Antonio (2009), "Uso y abuso de las relaciones culturales en la política internacional", *Ayer*, 75: 25–61.

Pestre, Dominique (1997). "Science, Political power and the State", in Krige, J.; Pestre, D. (Eds.). *Science in the Twentieth Century*, Amsterdam, Harwood, 61–75.

Porciani, Ilaria (2000*). Università e scienza nazionale*, Napoli, Jovene.

Prades, Sara (2014). *España y su historia. La generación de 1948,* Castelló, Publicacions de la Universitat Jaume I.

Presas i Puig, A. (2005). "Science on the periphery. The Spanish reception of nuclear energy: An attempt at modernity?", *Minerva* 43/2 (June): 197–218.

Pro, Juan (2019). *La construcción del Estado en España. Una historia del siglo XIX,* Madrid, Alianza Editorial.

Prochasson, Christophe, Rasmussen, Anne (1996). *Au nom de la patrie: les intellectuels et la première guerre mondiale, 1910–1919*, Paris, La Découverte.

Puig-Samper Mulero, Ángel (Ed.) (2007). *Tiempos de investigación JAE-CSIC, cien años de ciencia en España*, Madrid, CSIC.

Quiroga, Alejandro (2013). "La nacionalización en España. Una propuesta teórica", in Quiroga, Alejandro; Archilés, Ferran (Eds.) (2013). *La nacionalización en España*, Madrid, Marcial Pons (= *Ayer*, 90): 14–38.

Quiroga, Alejandro; Archilés, Ferrán (Eds.) (2013). *La nacionalización en España*, Madrid, Marcial Pons (= *Ayer*, 90).

Raphael, Lutz (1996). „Die Verwissenschaftlichung des Sozialen als methodische und konzeptionelle Herausforderung für eine Sozialgeschichte des 20. Jahrhunderts", *Geschichte und Gesellschaft*, 22:165–193.

Riquer, Borja de (2010). *La dictadura de Franco*, Barcelona, Madrid, Crítica, Marcial Pons.

Rodríguez-López, Carolina (Ed.). (2016). *La Universidad Europea bajo las dictaduras*, Madrid, Marcial Pons (= *Ayer* 101).

Romero de Pablos, Ana (2017). "Mujeres científicas en la dictadura de Franco. Trayectorias investigadoras de Piedad de la Cierva y María Aránzazu Vigón", *Arenal*, 24, 2: 319–348.

Romero de Pablos, Ana; Santesmases, María Jesús (Eds.) (2008). *Cien años de Política Científica en España*, Bilbao, Fundación BBVA.

Sánchez Ron, José Manuel (2008), "La ciencia durante el Franquismo", in Garcia Delgado, José Luís; Fusi, Juan Pablo; Sánchez Ron, José Manuel. *España y Europa*, Barcelona, Crítica; Madrid, Marcial Pons: 491–535.

Santana de la Cruz, Margarita (2009). "Unidad de la patria, unidad de la ciencia: la retórica científica del régimen franquista", in Gómez, Amparo; Canales, Antonio Fco. (Eds.). *Ciencia y Fascismo*, Barcelona, Laertes: 165–184.

Sanz Menéndez, Luís (1997). *Estado, ciencia y tecnología en España (1939–1947)*, Madrid, Alianza.

Saraiva, Tiago (2016), *Fascist pigs: technoscientific organisms and the history of fascism*. Cambridge, MA, MIT Press.

Sesma Landrín, Nicolás (2011). "Importando el Nuevo Orden. El Instituto de Estudios Políticos y la recepción de la cultura fascista y nacionalsocialista en España (1939–1942)", in Gallego, Ferrán, Morente Valero, Francisco (Eds.). *Rebeldes y reaccionarios: intelectuales, fascismo y derecha radical en Europa*, Mataró, El Viejo Topo: 243–280.

Smith, Anthony Douglas (1999). *Ethno-symbolism and nationalism: a cultural approach*. London; New York, Routledge.

Vom Bruch, Rüdiger. 1986. „Krieg und Frieden. Zur Frage der Militarisierung deutscher Hoch-schullehrer und Universitäten im späten Kaiserreich", in Dülffer, Jost, Karl Holl (Hg.). *Bereit zum Krieg. Kriegsmentalität im wilhelminischen Deutschland 1890–1914*. Göttingen, Beiträge zur historischen Friedensforschung.

Weber, Max (1973, 1904). "Die „Objektivität" sozialwissenschaftlicher und sozialpolitischer Erkenntnis," in J. Winckelmann (Ed.). *Gesammelte Aufsätze zur Wissenschaftslehre*, Tübingen, J.C.B. Mohr, p 146–214.

CHAPTER 2

Science, Nation, and Culture: Changing Meanings

Mitchell G. Ash

Introduction

All of the terms in the title of this chapter are what social scientists like to call moving targets. By this I mean that the meanings of the terms science, nation (or national identity), and culture have all changed fundamentally in the nineteenth and twentieth centuries, the period that will be the primary focus of my remarks, and all are contested today as well. Taking the middle term first, national identity and nationhood itself have become increasingly complicated and ambivalent concepts in an increasingly globalized world. But the meanings and content of science and scholarship,

The following is a revised and slightly expanded version of the keynote address to the conference "Cultura y sciencia nacionales en el primer Franquismo," Barcelona, 13 December 2018. Sincere thanks to the volume editors for their constructive suggestions.

M. G. Ash (✉)
Universität Wien, Vienna, Austria
e-mail: mitchell.ash@univie.ac.at

M. Janué i Miret, A. Presas i Puig (eds.), *Science, Culture and National Identity in Francoist Spain, 1939–1959*,
https://doi.org/10.1007/978-3-030-58646-1_2

as well as culture, have also changed radically since the beginning of the nineteenth century. The purpose of this volume is to re-examine the relationships of all of these concepts—and their realizations—during a specific regime at a specific time, so I will offer here only a few ideas and examples of the relationships of these entities to one another, along with brief suggestions about how these considerations might be applied to the case at hand.

The chapter is divided into three parts, corresponding to the three terms in my title. However, as we will soon learn, it is impossible to treat any of these three collective singular entities as something separate from the others.

Science

We begin of course with "science," because that is the main topic of this volume. The history of science literature in English simply takes for granted that the term "science" refers to what are now called the exact natural and life sciences, which implies, first, a certain overlap with the history of medicine, but also and most importantly the exclusion of the social sciences and humanities. It is now also common to speak of "science and technology," thus uniting, or better mashing up, what were once, and for many still are, separate histories of the sciences and of technology. However, informed scholars are well aware that this definition of the term "science" as natural sciences (and mathematics) is of relatively recent date, and that it was not, and is perhaps still not, universally accepted. The first use of the term "scientist" occurs in an unsigned review of a book by Mary Sommerville entitled "On the Connexion of the Sciences," published in *The Quarterly Review* in 1834.[1] The author was the Cambridge philosopher William Whewell, who offered the term as part of his account of the search for a suitable translation for the German term *Naturforscher*; all this was part of the prehistory of The British Association for the Advancement of Science, which Whewell had helped to found. The gradual acceptance of the term "scientist" as the name of the relevant occupation in the English-language literature implied a dual split, one between the English-speaking and the German-speaking worlds on the one hand, and within Anglophone academic culture on the other.

[1] Anon. (Whewell) (1834).

German-speaking authors, in contrast, have continued to use the term *Wissenschaft* to denote the systematic pursuit of knowledge regardless of the object being studied. The object came, and still comes, in the first part of the relevant combined noun, as in *Natur-*, *Geistes-*, or *Kultur-* and later *Sozialwissenschaften*. I note in passing that the roots of the idea of *Wissenschaft* as systematic research lay in classical philology and not in laboratory science. In French, a similarly broad usage was and continues to be maintained, though it was and is not embodied in a single word: "*sciences naturelles*," referring to the study of the external world, was and still is distinguished from "*sciences morales*" or now "*sciences sociales*," pertaining to human affairs. The same might once have been true in Spanish and Italian, at least in the nineteenth century, but in Spanish at least there appears to have been a shift towards the Anglo-Saxon usage, for example in the title of the volume edited by Amparo Gómez and Antonio Canales, *Ciencia y Franquismo* (2009).[2] So we have now located changing meanings at the philological level.

However, semantics alone cannot suffice for our discussion; we must also consider the fundamental shift in the social character of the sciences and of academic life that also began in the early nineteenth century. The first aspect of this shift is the emergence of the research university. This, along with the rise of local, national and international scientific societies, which began already in the late eighteenth century, created institutional settings for scientific research and discussion. The second aspect is the transformation of scientific practices enabled by this institutional transformation, including the introduction of experimental methods into the life sciences, but also the emergence of large-scale editions and other philological projects in the humanities (see later). Such institutional arrangements did in fact lead to the emergence of communities of discourse and practice organized along disciplinary lines. Whether we could or should call these epistemic communities "cultures" is a question to which I will return in the section on "culture."

At the level of institutional frameworks, as well, we need to proceed with care, for two reasons. First, universities have always been multifunctional institutions, and the rise of the research university in Europe in the nineteenth century only altered the mix. Indeed, the certification function of universities remained primary, as it does to this day. What was new, of course, was that the production of a piece of research became a

[2] Gómez and Canales (2009).

requirement of certification at the doctoral level in many, though not all, disciplines or faculties. Second, the rise of extra-university research institutions occurred not following, but in parallel and to some extent in interaction with, that of the research university.

Viewed from this perspective, we must acknowledge that in the nineteenth and twentieth centuries the sciences themselves became institutions, primarily though not entirely subsets of the academic profession. To consider the sciences from a history of ideas perspective alone is therefore insufficient. We must also consider shifting social meanings of science, as well as changing practices of scientific research. That is certainly the case for the study of the relations of the sciences and political power in the twentieth century, which I will now discuss in greater detail.

Once we agree that the sciences cannot be treated only as collections of ideas or research practices, but must also be considered as institutions, precisely because of the rise of the research university, we ought to acknowledge as well that it is inappropriate to consider the sciences as the sphere of truth to be something separate from the realm of political power. Rather, scientific institutions themselves became centres of status and power in the nineteenth and twentieth centuries. At first, the term "chairs" clearly meant one professorship, held by one person. Over time, however, as the research establishment grew, chair-holders acquired research power, and the word "chair" came to mean the research establishment headed by the chair-holder. Research "schools," usually also named after chair-holders, thus became not only intellectual entities or shared communities of thought and practice, but also living embodiments of hierarchically organized research programmes. The process began with the handbook science pioneered by Justus Liebig in chemistry; in this case, students produced discrete laboratory research contributions that were then incorporated into Liebig's handbook, and received doctoral degrees in return.[3] In Spain, the case of Santiago Ramon y Cajal is best known. The work carried out in Cajal's institute was not an example of experimental laboratory science, but rather of anatomical mapping, but it was also conceived from the beginning as a communal enterprise.

Because the funding for these institutions on the Continent generally came from national governments, we might say that with respect to their financial support the sciences in their modern, institutionalized form were "national" from the start. This brings me to the second keyword in my title.

[3] For an account of this exchange relationship, see Brock (2002).

Nation/National Identities/National Science

Scholarship over the past thirty years has undermined any fixed definition of nationhood, and also the equation of the nation with the nation state. Instead, much evidence has been presented in support of Benedict Anderson's now-classical idea that nations are "imagined communities," which are not necessarily limited to states and can indeed be in direct tension or even conflict with existing political arrangements.[4] A significant part of the process of formulating and propagating such "imagined communities" is what Eric Hobsbawm called "the invention of tradition."[5] As the French thinker Ernest Renan put it already in the 1860s, "the possession in common of a rich legacy of memories" is the basis of the national "spirit"—to which he added that the creation of such a legacy also implied "forgetting," even "historical error."[6] Thus, Renan already knew that national communities were legitimated by selective historical memory; linguists today call this mythopeia, the creation of communities of discourse through a common origin story.

Roughly simultaneously with, but in sharp contrast to, Anderson's thesis, Ernest Gellner presented a widely received definition of nationhood as the congruence of ethnicity, language, and state, in order to criticize this ideal; he later examined what he called "the Habsburg dilemma" as an example of what can go wrong when such a congruence does not exist.[7] However, it seems fair to say now that the definition of nationhood that Gellner and Anderson so rightly criticized—one people, one language, one state—is more a normative goal or an ideal type than an empirical description. As is well known, ethnic minorities live even in seemingly ethnically coherent states, and sometimes these minorities regard themselves as nations in their own right. Indeed, we might ask how often the normative ideal just described has ever been achieved outside of small countries like Slovenia. Moreover, seen in the light of modern and recent history, such norms can hardly be considered unambiguously positive. The establishment of ethnically coherent nation states with single predominant languages, if this has been achieved at all, has come often enough at the cost of immense human suffering. Whether such achievements have also come at a cost to scientific quality is a controversial question (more on this later).

[4] Anderson (1991).
[5] Hobsbawm (1990). For relevant examples, see Geary (2002).
[6] Renan (1882/1992).
[7] Gellner (1983, 1998).

Introducing science and scholarship (*Wissenschaft*) into a discussion of the category "nation" appears at first to complicate matters unnecessarily. After all, it is often claimed, rather dogmatically, that science is international, or it is not science. The implicit assumption behind this claim is that sciences are necessarily inter- or transnational because scientific knowledge is, or ought to be, universal. Of course, such claims presuppose the Anglophone definition of science as natural science, which leads us to an apparent paradox: the very claim that "science" is per se international is a creation of the English-speaking world and therefore a "national" invention, or cultural artefact. In any case, ironically, the research network of classical philology in the German mode was then and remains as international as that of mathematics or physics—a point that Anglophone historians of science often fail to notice, due to their restricted view of their discipline. However this may be, the relation of universities, and the sciences practised within or outside them, to nations, nationalism, and national identity is actually a highly complex affair. The complexity increases when we acknowledge that universities have never been the only places in which scientific or humanistic research has been carried out. National geographical or geological surveys, laboratories, and academies of sciences are surely at least equally, perhaps even more importantly, places through which science became instrumental in establishing national identity.[8]

There appears to be a tension here between the local, regional, or national situatedness of scientific institutions and the supposedly universal character of the sciences. In fact, as we know from a great many case studies, the nationalization and the internationalization of knowledge are historical processes that began at roughly the same time and might best be viewed as complementary to one another.[9] It was precisely in this context of simultaneous nationalization and internationalization of knowledge in the nineteenth century that universities and the sciences pursued within them came increasingly to be regarded as loci of national knowledge, in at least two senses. First, explicitly national disciplines came to the fore: modern language studies, "modern" as opposed to general or ancient history, and also social and cultural sciences like ethnology and archaeology. Ralph Jessen and Jakob Vogel cite such disciplines as examples of the

[8] Livingstone (2003), p. 124.
[9] Ash and Surman (2012a).

"scientific construction" of national identities.[10] Of course, speaking or writing in this manner presupposes the broad definition of "sciences" as *Wissenschaften* already mentioned.

Second, as the research imperative came to greater prominence and the educational, professional training, and research functions of universities became more tightly linked, universities came to be increasingly privileged locations for the mobilization of scientific and humanistic knowledge as cultural and political resources for the cultural nation and also for the nation state. Such resource mobilizations could be symbolic, in the shape of prestige or perceived advantage in international competition, but they could also be quite concrete, for example in weapons research or colonialist enterprises. Physiologist and physicist Hermann Helmholtz put the point clearly already in 1862, before German unification: Because "knowledge is power," he asserted in his inaugural address as Vice Rector in Heidelberg, alluding of course to Francis Bacon, nations, meaning nation states, could not fall behind in the competition for the best science, which he defined in this very speech as "the supremacy of intelligence over the world."[11]

There appears to be an ambiguity here: knowledge is produced and cultural resources are mobilized locally, and yet that knowledge is alleged to have universal validity, which can then be appropriated as cultural capital for local political advantage. Such ambiguity might be seen as a problem, or rather as something that is built into the game of power knowledge, whether we like it or not. In any case, the mobilization of science and scholarship as cultural resources for patriotic loyalty and national power politics has been a highly successful strategy for improving the growth and standing of universities. The differences between natural and human sciences in this complex game of power knowledge appear to have been less fundamental than is often supposed. The massive collections of Greek and Latin inscriptions assembled under the direction of August Böckh, Theodor Mommsen, and others under the auspices of the Prussian Academy of Sciences (continuing today under the auspices of the Berlin-Brandenburg Academy of Sciences), and still more the corpora of Medieval manuscripts in the *Monumenta Germaniae Historicae*, an ongoing enterprise of the Bavarian Academy of Sciences, were and still are transnational

[10] Jessen and Vogel (eds.) (2002).

[11] Helmholtz (1862/1995), p. 94. This phrase has also been translated as "the intellectual mastery of nature".

projects, but their success nonetheless redounded as a cultural resource to the prestige of "German" scholarship, in much the same way as the internationally recognized successes of natural and medical science carried out in the German states or in France were, and still are, recorded positively in the cultural accounts of these nations.

Three keywords for the relation of sciences and nation or national identity that I would like to offer here in order to focus the discussion are *co-creation*, the *reproduction*, and the *propagation* of national identity by political and intellectual, in this case scientific and scholarly actors. In each case, a *mobilization of resources* is involved, at multiple levels. Very briefly, I distinguish three types of resources: personnel, called "human resources" today; institutional resources, including financial support, physical facilities, and also the scientific research practices enabled by such support; and intellectual resources (in Pierre Bourdieu's terms, symbolic or cultural capital), including but not limited to the attribution of ideological meanings to scientific ideas, as just described. Such resource mobilizations do not go in only one direction, but are rather *interactive*—that is, universities, sciences, and scholarship can be or present themselves as resources for political goals, and can also be mobilized by political actors.[12]

In the following section I present a few examples of such resource mobilizations from the nineteenth and twentieth centuries, in order to make the changing meanings of "nation" over time and the roles of the sciences clearer.

Sciences and Universities in the "Age of Nationalism"

I must begin with "Germany," or "the German question," for obvious reasons; why I place quotation marks around that name will be clear soon, if it is not already. In 1808, Johann Gottlieb Fichte, then a former professor of philosophy in Erlangen, gave a series of "Addresses to the German Nation" in Berlin.[13] He was responding to Freiherr von Stein's call to mobilize the educated and cultural elites of the nation against French occupation. Fichte located "Germanness" in the supposed continuity of the German language, and cited the Roman historian Tacitus' praise of "Germanic" virtues. For him, speaking soon after the end of the "Holy Roman Empire of the German Nation" in 1807 and long before the

[12] For elaborations of this approach, see Ash (2002, 2010).

[13] Fichte (2009).

establishment of a German nation state in 1871, the "German nation" was thus bound together by culture and ethnicity. Fichte's ethnic vision of German identity, as some now call it, entailed Anti-Semitism, and he was not alone in his opposition to granting citizenship to the Jews. He was soon appointed one of the first professors at the new university in Berlin in 1810, and became rector the next year. However, in that very year Prussian citizenship was granted to the Jews; this was expanded to the Reich in the constitution of 1871. Citizenship in this case meant membership in the German *Staatsnation*; whether it also carried with it "Germanness" in Fichte's sense was another question entirely.

The founding of the University of Berlin and that of the Society of German Naturalists and Physicians (*Gesellschaft deutscher Naturforscher und Ärzte*) in 1822 have been cited in general histories for many years as evidence that Germany was created first as a "cultural nation" (*Kulturnation*), long before the establishment of a unified nation state (*Staatsnation*).[14] This distinction is quite important, and may be relevant beyond German-speaking Europe. However, such conventional claims should be examined with care. The university founded in Berlin in 1810 was named after its nominal patron, the Prussian king Friedrich Wilhelm III; in this respect at least, it continued the long-standing tradition of naming universities after local princes. Friedrich Wilhelm's University, as it was literally called, did not become a national symbol overnight, and the "German" model that was widely emulated later in the century did not emanate from Berlin alone.

What actually triumphed in Germany in 1871 was not liberal nationalism, but a new, Prussia-centred national liberalism, embodied in the often-cited proclamation by world famous physiologist Emil Du Bois-Reymond in a speech as rector of the University of Berlin and president of the Prussian Academy of Sciences in 1870, at the start of the Franco-Prussian war. I cite from the English translation, which Du Bois-Reymond remarkably had published in London the same year: "We, the University of Berlin, quartered opposite the King's palace, are by the deed of our foundation *the intellectual bodyguard of the House of Hohenzollern*."[15] Note that not "German," but rather Prussian patriotism is being evoked here. Nonetheless, this famous quotation can well be interpreted as a classic

[14] For this distinction and its relevance for the Habsburg monarchy, see Feichtinger (2012).
[15] Du Bois-Reymond (1870), p. 31. Emphasis mine.

example of co-creation of national self-consciousness and of science in unified Germany as a national enterprise.

But German national sentiment was not limited exclusively to the territory of the new Reich. In his opening address to the meeting of the Society of German Naturalists and Physicians in Innsbruck in 1869, Hermann Helmholtz proclaimed the following on Austrian soil three years after the battle of Sadowa (in German: Königgrätz) had decided the question of German political nationhood: "In science we need not look to political boundaries; our Fatherland extends so far as the German language, as German industriousness and German courage in the struggle for truth resound."[16] Some of the Austrians in the audience may have seen this a bit differently, but the very fact that the Society held its convention in Innsbruck was surely intended to symbolize and propagate its members' shared allegiance to a language-based epistemic community. The point to be emphasized here is that not only historians and philosophers, but also natural scientists contributed prominently to the formulation of national identity, relying in this case on the idea of the cultural nation (*Kulturnation*).

A contemporary alternative to this ethnic or language-centred nationhood was the concept of the "civic" nation articulated by Ernest Renan, in the work already cited earlier.

This was the concept that was then being enacted in the French Third Republic, mainly though not entirely by secularizing the schools. By this top-down process the Republic made "peasants into Frenchmen," to quote the brilliant book title by Eugen Weber.[17] The political actors in France and elsewhere in Europe clearly understood that nations were made, not found. Less than half of the people on the territory of the Bourbon monarchy at the time of the French Revolution were actually "French." Regionalism long remained the predominant identity there as elsewhere, even as the nation state arose and acquired ever more power. Struggles between nation-building liberals and conservative opponents, especially in the Catholic Church, over control of the schools were endemic throughout this period, in France and elsewhere. Of course, this was also true in nineteenth- and early twentieth-century Spain. Perhaps we might speak here of a parallel or slightly delayed, and ultimately unsuccessful, struggle to reshape the Spanish nation as a secular entity, separate in prin-

[16] Helmholtz (1869/1995), p. 225. Translation revised after consulting the original German.

[17] Weber (1976).

ciple from identification with the king, advanced most vigorously by the Free Teaching Association (*Institución libre de Ensenanza*, ILE) from the 1870s onward. Whether the Board for Advanced Studies (*Junta para la Ampliación de Estudios e Investigaciones Scientíficas*, JAE) in Barcelona, founded in 1907, exemplified a parallel effort in science policy, and whether or how the "nation" trope was evoked in this context are questions worthy of further discussion.[18]

I now turn to the Habsburg monarchy, a multi-ethnic regime in which tensions between the cultural nation (*Kulturnation*) and national state (*Staatsnation*) became particularly acute after the creation of the Dual Monarchy in 1867.[19] But the rise of nationalist movements in the non-Hungarian territories called Cisleithania began long before that, and the Hungarian Academy of Sciences was founded in 1834. How did the Habsburgs confront this issue in the field we would now call science policy?

Immediately following the defeat of the 1848 uprisings, the Habsburgs initiated a series of supranational scientific projects with the clear intention of mobilizing scientific resources to embody the Kaiser's motto *viribus unitis* and thus unify the Empire. The foundation of the Imperial Academy of Sciences in Vienna in 1847 had already pioneered this effort. After 1848 this was followed by the Imperial Geological Institute (founded in 1849) and the Institute for Meteorology and Geodynamics (founded in 1851), both of which soon initiated vast data-gathering networks, including a program to gather temperature measurements and other climatological data from posts throughout the empire and the Geological Survey map of the monarchy.[20] Founded soon thereafter, in 1854, the Institute for Austrian Historical Research became a central node in a network of documentary source collections on which a specifically Austrian "invention of tradition" in the sense defined by Hobsbawm was to be based.[21] In all of these cases the focus was deliberately not placed on the idea of a single-ethnic national state, but on the Habsburg House of Austria, of which the state was seen to be the agent. The term "nation" meant here the dynasty, not a "people." Deborah Coen has recently described such efforts cogently as "imperial and royal science."[22] It should be emphasized

[18] For discussion of the JAE in this context, see Gómez (2015).
[19] Feichtinger (2012).
[20] Klemun (2012), Coen (2013, 2018).
[21] Winkelbauer (2018), pp. 96–110.
[22] Coen (2018), Chap. 3.

that all of these were non- or extra-university projects; links to the universities were established mainly by awarding professorships to the project heads.

Parallel to this the Neo-Absolutist regime carried out university and school reforms that had been proclaimed in 1848 in modified form. The programme, originally formulated by liberal philosopher Franz Exner and others during the 1848 uprisings, was reformulated in a Catholic Neo-Absolutist mode under the ministry of Count Leo von Thun-Hohenstein.[23] In this programme equal standing with the faculties of law, theology, and medicine was accorded to the Philosophical Faculty, and thus also to the natural sciences, as Exner had already proposed. Thun even accepted the appointment of German Protestants like Ernst Brücke, a physiologist, in the natural sciences, but he paid special attention to establishing chairs in what he and others called the "national" disciplines, including modern history and "modern" languages, including Slavic studies.[24] Freedom of research and teaching was not enshrined in law in the monarchy until the liberal-era constitution of 1867, long after Count Thun's resignation; freedom of learning was not legalized even in that document.

One central consequence of this imperial nation-building project was an attempt to enforce German as the primary language of university instruction in all Habsburg domains, including Hungary, even though Hungarian had become an official language of instruction already in 1844. This encountered immediate resistance. The nationalist reactions that emerged in this period and afterwards built upon cultural nationalist programmes that had been propagated by academics and non-academic intellectuals since the early nineteenth century.[25]

A central focus of those programmes was the scientific codification of national vernaculars. As Jan Surman has shown, some of the same scholars who codified the Slavic languages in the early nineteenth century also engaged in efforts to develop scientific terminologies in these same languages.[26] Such codifications were not simple matters of translation from a pre-existing standard, but were undertaken just when, sometimes even before, such standardized scientific terminologies were developed at all.

[23] Aichner and Mazohl (eds.) (2017).

[24] For this policy, see Feichtinger (2010).

[25] On the early debate among academics and intellectuals over the status of the Czech language, see Agnew (2003).

[26] Surman (2012).

The idea of creating or codifying national languages *per se* was not the problem. The controversial issue, rather, was some nationalists' insistence on linguistic "purism" versus "contamination" of the national language by foreign terms, even in the sciences, and on a monolingual rather than a multilingual approach to nationalism in academic instruction. The difficulties that this posed for international scientific communication were already evident at the time. Contemporaries realized that scientific papers published in regional languages were not likely to be noticed at all outside the local community. One way of coping with this problem was to publish the same results in a national and also in a widely read language; new scientific journals published in German, occasionally even in English, were founded in Czech or Polish territories for this purpose.[27] We might call this an effort at knowledge transfer without travel; citation studies show that the success of this strategy was limited.

The most radical result of these language-driven national conflicts was the division of the University in Prague into two institutions, one German and the other "Bohemian" or Czech, in 1882.[28] This is often cited as an example of what national conflicts could achieve—or destroy—depending on the writer's particular viewpoint. However, though plainly spectacular, this case was quite unique. More frequent were conflicts over the use of national languages in teaching and research within existing institutions, which led to serious divisions and often enough to the migration of affected individual scholars and scientists to other places, but not necessarily to institutional break-up.

The Twentieth Century: Radical Ruptures and Continuity in Change

With the outbreak of war in 1914 it became unmistakeably clear that the comfortable belief that it was possible to live in the world of universal science and one's own national culture at the same time with no tension or contradiction was an illusion. A well-known case in point was the manifesto "To the Cultural World!" (*An die Kulturwelt*!), also known as the "Declaration of the 93," co-signed by numerous famous scholars and scientists, many of whom had been liberals in their youth. In this text German

[27] See Surman (2012) and Štrbáňová (2012).

[28] Crawford (1992), pp. 37, 87; Cohen (1996). For an example of the impact, see Vodrážková-Pokorná (2006), esp. pp. 67–73.

scholars and scientists protested against British "cruelty propaganda" alleging deliberate violation of Belgian neutrality and atrocities by German soldiers against Belgian civilians.[29] Subsequent writings of Germans like philosopher Max Scheler on *The Spirit of War and the German War* (1915) or by psychologist Wilhelm Wundt on *The Nations and Their Philosophy* (1917), along with writings by Americans like philosopher George Santillana making similar arguments in the opposite direction, showed that during this conflict patriotism clearly trumped scientific internationalism.

Such texts were not the only route to the mobilization of academics and their knowledge for war. Surely the best-known case of using science to advance warfare (and of warfare to advance science!) in this period is that of Fritz Haber, including both the Haber–Bosch process—chemical synthesis of ammonia from the air, useful for the production of chemical fertilizers and explosives—and a massive research and development programme in poison gas, which he directed at the Kaiser–Wilhelm Institute for Physical Chemistry and Electrochemistry.[30] In the United States, the founding of the National Research Council in 1916 anticipated patriotic mobilization of academic science long before the US entered the war—even though the NRC was and remained a private rather than a government body.

The post-1918 treaty settlements led to a profound change in East Central Europe, as a set of new nation states was created. The extraordinary prestige of science and scholarship in these countries was clearly indicated by the leadership choices of their peoples: philosopher Tomáš Garrigue Masaryk (1850–1937) became the first president of Czechoslovakia; Gabriel Narutowicz (1865–1920), professor of hydroelectric engineering at Zurich Polytechnic, became the first president of the Polish Republic; and historian Mykhailo Hrushevskyi (1866–1934) was elected chair of the short-lived Ukrainian revolutionary parliament Central Rada (*Tsentral'na Rada*), and thus de facto president of the Ukraine. Academics had also been public intellectuals before 1914, but by turning to Masaryk, Narutowicz, and Hrushevskyi a conscious choice was

[29] See, e.g., Ungern-Sternberg (1996). An article published after the war alleged that the great majority of signers later regretted lending their names to the document. Far more scholars and scientists signed another, more broadly worded document with similar intent. Cf. Vom Bruch (2014).

[30] Steinhauser et al. (2011), esp. pp. 26–36.

made of internationally renowned scholars instead of professional politicians.

Still, the old tension between cosmopolitan science with national benefits and what I propose to call monolinguistic nationalism—the vision of one territory, one language, one people—did not disappear. The significant difference from the past was that already existing "national" cultural institutions, including universities, were now backed by states with the power to enforce monolinguistic nationalism and require ethnic minorities to accept it in practice. Whether this monolinguistic nationalism had a catastrophic impact on science and scholarship in East-Central Europe, from which these countries have yet to recover to this day, remains a controversial question. The cases of Holland and Scandinavia suggest that there were alternatives to such policies.

The case of Hungary is a good example of the continuity of tensions between scientific internationalism and monolinguistic nationalism. The political dismissals following the collapse of the "workers' council" government of Béla Kuhn, along with discrimination against Jewish students and academics by the successor regime under "Regent" Admiral Horthy, led to the forced migration of numerous, mainly Jewish and leftist scientists and scholars, among them the chemist Michael Polanyi.[31] However, the reform policies of Count Kuno Klebelsberg, minister of education and religious affairs from 1921 to 1931, appeared to run in a modernizing direction, though Klebelsberg was plainly a nationalist. His commitment to modern science persuaded the Rockefeller Foundation to invest in the University of Szeged and the Institute of Fresh Water Biology in Tihany on Lake Balaton, funding the biomedical science of Albert Szent-György, who was awarded the Nobel Prize for Physiology or Medicine in 1937. But such policies failed to alter the dominant conservative, authoritarian and nationalist views of the Hungarian professoriate. Characteristic of such views was this statement by historian Gyula Kornis in 1936: "Our universities cannot be said to have disregarded the national interest. … The spirit of university autonomy and state power can collaborate with one another as both are motivated solely by the interests of the immortal Hungarian nation."[32] Such sentiments were hardly limited to Hungary.

[31] Frank (2009).

[32] Cited in Péteri (2005), p. 143.

The Nazi era saw the triumph in Germany of a radical transformation of politics and nationhood; "national" became "racial" identity.[33] The so-called Law for the Restoration of the Professional Civil Service, promulgated already on 7 April 1933, made this transformation legal by mandating the dismissal or forced retirement of thousands of civil servants, including hundreds of university teachers, who could not prove that they were of "Arian" descent. Very few German professors protested against this violation of academic traditions. Instead, many spoke out in favour of the new regime, and some presented their own ideas in suitably modified form as candidates for official doctrine. The impact of these dismissals on the sciences, and on the composition of university faculties across disciplines, was immediate, but varied in significant ways. This variation was less a statement about National Socialism than about the presence or absence of structural Anti-Semitism in professorial appointments before 1933.

The impact of this profound political shift on the content of the sciences was, however, ambivalent. Speaking at the 550th anniversary celebration at the University of Heidelberg in 1936, Reich Minister of Education and Science Bernhard Rust denounced the idea of scientific objectivity as a relic of the past "liberalistic" era.[34] Heidelberg was a centre of so-called German" physics propagated by Philipp Lenard and Johannes Stark, but their efforts and those of collaborators to "nationalize" particular disciplines on a "racial" basis, including "German" physics and "Aryan" mathematics, had only limited success. In any case, they came not from the Nazi party, but from physicists trying to decide the debate over relativity and quantum mechanics by political means.[35] Party devotees such as Karl Astel worked to re-establish Jena as a "model" NS-university,[36] but so-called fighting science (*Kämpferische Wissenschaft*) was institutionalized mainly outside the universities, at institutions like Walter Frank's Reich Institute for the History of the New Germany or the "SS-Ahnenerbe."[37]

In military contexts, however, not ideologized versions of Nazi science, but rather technoscience of high quality was required, and it was also

[33] The ground for this had already been prepared during the 1920s, for example by the "Volksdeutsche Forschungsgemeinschaften"; see, e.g., Fahlbusch (1999). For the following, see, among many others, Grüttner (2010).

[34] Rust (1936/1990).

[35] For discussion of these and other cases, see Gordin et al. (2003).

[36] Hossfeld, et al. (eds.) (2003).

[37] See, e.g., Kater (1997).

purchased at high prices.[38] The most spectacular weapons research project of the Nazi era was, of course, the rocketry programme at Peenemünde, funded by the army.[39] Parallel to this, aviation and aerodynamics research was generously funded by the air force. After the war began, the propaganda slogan "science in the service of war" or "the fatherland" (*Wissenschaft im Dienst des Krieges* or *des Vaterlandes*) got reversed in whispers to read: "The war in the service of science."[40] Many basic research programmes were maintained by having them designated "important for the war" (*Kriegswichtig*).

When it came to the actual realization of "national" as "racial" identity—the struggle to conquer "living space" in the East and the "purification" of German "blood"—scholars and scientists across the entire range of academic disciplines took active part. Over 550 humanists took part in a large-scale mobilization of scholarship in the service of the German war effort called the "war effort of the humanities" (*Kriegseinsatz der Geisteswissenschaften*).[41] More significant in practice was the scientific policy advice provided by historians, cultural geographers, and social scientists in support of Nazi occupation policies.[42] Any distinction between "objective" science and science in the service of ideology-driven policy disappeared in the case of eugenics or "race hygiene" in health policy,[43] or the mobilization of agronomy, geography, and land use planning in the service of the massive redistribution of land and people in the German-occupied East projected by Berlin agriculture professor and Nazi multi-functionary Konrad Meyer.[44]

The Case of Spain during the Early Franco Regime

Where does Spain fit into all this? Because the rest of the volume addresses this question, I will limit my remarks here to three issues: the ideological construction and actual impact of the purges; the destruction of the science policy regime of the Republic and the creation of new science policy and funding institutions; funding priorities and impact on research.

[38] For examples, see Maier (ed.) (2002).
[39] Neufeld (1995).
[40] Harwood (2001), p. 141.
[41] Hausmann (1999).
[42] For numerous examples, see Haar and Fahlbusch (eds.) (2008).
[43] Weiss (2010).
[44] Heinemann (2005).

Thanks to the work of Jaume Claret Miranda, E. Otero Carvajal, and others, the dimensions and brutality of the purge of universities after the victory of the Francoists are well known.[45] Criteria for dismissal were clearly political and not "racial" in this case, but we can speak here, too, as in the case of Nazism, of an expanded concept of the political, since religious affiliations or their absence were also considered. The actual impacts varied across disciplines, as they did in Germany, though in rather different ways: at the University of Madrid, for example, from 28.5 per cent in the arts faculty to 50 per cent in the sciences and 60 per cent in the medical faculty were dismissed.[46] Whether this justifies the inference that certain fields or groups of disciplines were actually centres of Republican loyalties before 1936, as they were apparently considered to be by the new powerholders and as has also been suggested in some recent studies, is an interesting question. Perhaps more important, however, is the point that, in contrast to Nazi Germany, the dismissals, once executed, did not all result in permanent exile. The literature contains multiple examples of reinstatements, the reasons for which are not always clearly stated.[47] Coming back to the resource concept articulated earlier, we might speak here of an initial, radical devaluing of "human capital" followed by a partial revaluing of personnel resources over time. Whether this process can be quantified or surveyed in detail beyond the individual cases often cited in the literature, or whether the reinstatements also exhibit variation across disciplines, appear to be open questions.

I turn now to the second resource type mentioned earlier, institutions. Policy towards extra-university institutions, in particular the vicious attacks on the ILE, the destruction of the Board for Advanced Studies (JAE), and the creation of the *Consejo Superior de Investigaciones Científicas* (CSIC), has been a central focus in the literature.[48] Antoni Malet and others have laid out the national Catholic, antimodernist conception of José Maréa Albareda for the CSIC in detail.[49] Perhaps somewhat less attention has been paid to the degree to which this programme was actually carried out. Malet briefly mentions that the CSIC also had bodies for vetting

[45] Claret Miranda (2006); Otero Carvajal (2006); see also Ruíz Carnicer (2005); for a comparison to purges of academics and intellectuals in Nazi Germany and Fascist Italy, see Morente (2006).

[46] Otero Carvajal (2006), pp. 80–82.

[47] See, inter alia, Gómez Rodriguez and Canales Serrano (2009), p. 324.

[48] See, e.g., Gómez Rodríguez and Canales Serrano (2009), Morente Valera (2017).

[49] Malet (2009).

appointments, but it seems unclear whether their actual operations have been studied in detail. Were political allegiances the only operative criteria for appointments or financial support? Did all appointees after the purges subscribe to Albareda's national Catholic programme? Or were practical criteria of various kinds also relevant?

The third resource type mentioned earlier is discursive constructions of sciences, specifically their ideological attributions. The effort to mobilize such discursive resources is already evident in the programmatic statements in the law establishing the CSIC. The announced goal of this entity and of science policy in general was to "restore the unity of sciences" that had allegedly obtained in the sixteenth century, embodied symbolically in the metaphor of the "imperial tree of knowledge" ("*El árbol imperial de la ciencia*").[50] Of course anti-modernism and nostalgia for an imagined golden age were characteristic of national Catholic ideology in general. Some suggest, however, that an ambivalence analogous to that which we have already noted for science under Nazism may have also have obtained in this case. Antoni Roca-Rosell writes of "professionalism and technocracy" in his discussion of Esteve Terradas and science policy in the early years of the regime.[51] Given the investments in soil science, plant genetics, and other such technoscientific endeavours in the service of autarky,[52] as well as the collaboration with German scientists in nuclear research described in detail by Albert Presas i Puig,[53] we might well ask how far the reactionary Catholic nationalist programme articulated by Albareda and others actually reached. In this case, not Albareda, but rather alliances of José María Otero-Navascués, the architect of the Spanish nuclear programme, with Cabrera (whom Presas y Puig calls "the strong man of the regime") and Admiral Carrero Branco were central.[54] The primacy of applied science in the context of autarky policy seems evident, though "autarky" was perhaps less an economic policy than a name for the establishment of political control. Perhaps we might speak of a continuum shifting towards modernization after the abandonment of autarky in the early to mid-1950s.

[50] Ibid. Antolín Hofrichter (2021).
[51] Roca-Rosell (2005).
[52] Camprubí (2010).
[53] Presas i Puig (2005, 2008).
[54] Presas i Puig (2021).

Culture

Finally, I turn now very briefly to the last term in my title, culture. The inflationary use of this term is beginning to provoke counter-reactions, with some justification. Many authors seem, when they write this term, simply to assume that their readers already know what they mean, and therefore do not reflect on the category's precise meaning or content. This is especially true of the nearly ubiquitous references to Snow's "two cultures"—which, by the way, was never intended to be a statement about academic politics. After all, for any concept to claim analytical value it must be possible to state clearly what its limits are. Nonetheless, multiple meanings, or rather uses, of the term "culture" have emerged in the history of science literature, each of which has something going for it.

The most prominent and still useful definition is the one offered by the ethnologist Clifford Geertz, who defined culture as a "network of meanings."[55] In his view meanings could be represented or embodied in two ways: via iconic symbols or via symbolic actions, such as ceremonies or rituals. The first level of meaning-assignment or embodiment via symbols corresponds to the traditional concept of "high" culture, but is equally applicable, of course, to symbolic objects in popular culture (see later). Most important is the inclusion of practices, which opens the door for consideration of science as cultural practice. I will return to this in a moment.

It is important to note that the idea of culture as the assignment of meaning through symbolic objects or actions is not limited to concepts or works of art, but can also extend to buildings. Universities and academies of sciences themselves—their buildings, even sometimes their very existence—can be regarded, and their members have often regarded them, as symbols of nationhood, in this respect as cultural entities. This is certainly the case for the main building of the University of Berlin, located on Unter den Linden, only steps away from the imperial palace (a point to which Du Bois Reymond explicitly referred in his speech of 1871, quoted earlier). More than symbols of prestige and patriotic pride, they have often been *embodiments* of national self-consciousness. The academies of sciences established in Budapest, Cracow, Prague, and elsewhere in the nineteenth century can all be regarded as cultural symbols and active constituents of national identity in (East) Central Europe; they were "national"

[55] Geertz (1973), p. 5.

institutions established long before there were nation states in these places.[56] In the case of the Franco regime, the "university city" campus (*Ciudad Universitaria*) at the Comprehensive University of Madrid, discussed by Carolina Rodríguez López in this volume, appears to be a clear example of an effort to utilize architecture to embody "national science" ideals.[57] In this context we also need to include the institutions employed to create, propagate, and distribute symbolic cultural objects. Benedict Anderson, whose idea of "imagined communities" was cited earlier, also gave weight to the media and spoke of language and media (here: "print capitalism") as fundamental conditions for the creation of "imagined communities." That is, culture meant for Anderson not only ideological words, but also cultural institutions.

This brings us to a second meaning of the term: Culture as the enactment of values, most notably in the realm of what has been called "political culture." I have provided a large number of examples for this usage in the discussion of "nation." This construction of the term "culture" is quite common in the literature on science in the Franco regime, or indeed in modern Spain. At times science itself is denoted as culture in this literature, which is what the actors being described also did.[58] If the intention is to counteract "two cultures" talk and to emphasize the actual interconnections of the sciences and the arts and humanities in Spanish history, as the actors—especially those among the rebels who attacked the sciences—saw them, then this is surely justified. Nonetheless, perhaps it would be useful to historicize this usage, rather than reproducing it, and thus to ask who denoted science as culture in what circumstances, and why. What distinguishes the term "culture" from "ideology" in such contexts is an open question. In my view, it would not be helpful to repeat the endless discussions about what was "Nazi" about science under Nazism by asking what was "Francoist" about science under this regime, but I leave this issue to the experts.

A third, increasingly common meaning of the term "culture" refers specifically to the behavioural dimension of Geertz's conception, just cited. We can divide this into two levels. First, and clearest for our discussion, are ceremonies such as the one enacted in the presentation of the annual report

[56] On the Cracow Academy, see Mazzolini (1995). For literature on the Academies of Sciences in Budapest and Prag, respectively, see Palló (2012); Štrbanová (2012).

[57] Rodríguez López (2021).

[58] See, e.g., Gómez Rodríguez and Canales Serrano (2009).

of the CSIC, which Franco himself always attended, along with the leading cultural officials of the regime, and which was preceded by a mass.[59] In such ritual enactments—here, in the assembly of notables as a physical presence and in the *caudillo*'s speech upon receiving the report as a rhetorical performance—the unity of science policy and "high" politics was deliberately and openly expressed. More common in the history of science literature is a second level, the history of research practices taken for themselves, rather than as theory testing. In this literature references to "cultures (or styles) of scientific (or technical) practice" are common, and have increased recently due in part to the belated reception of Ludwik Fleck's notion of "thought styles" (*Denkstile*) shared by "thought communities" (*Denkkollektive*). According to Hans-Jörg Rheinberger, these practices result in turn in the creation of what he calls "epistemic things" or "preparations," in the laboratory, or more generally "cultural artefacts," which then become the actual objects of scientific research.[60] Perhaps such styles of research practices are what Xavier Roque means when he writes about the clash of "scientific cultures" that led to the failure of Spain's involvement in CERN.[61] More likely in this case is a mixture of meanings, in which culture as an enactment of values (see earlier) at the macro level melds with cultures of research practice at the micro level. The danger in such usages, if they are not reflected upon more precisely, is that they could come close to more traditional discourse referring rather vaguely to "national styles" in science.[62] Whether we need such terms, or whether it is sufficient simply to speak of science in Spain, is an open question.

Fourth and last but not least, I should at least mention the distinction between "high" and "popular" culture, if only to note that this distinction, too, has been radically challenged, and to some extent overcome, in the past decades. Examples of this in the history of science come from the large body of work on the history of popular science versus that of the popularization of science, usefully summarized recently by Augustí Nieto Galan.[63] Following this trend, we might ask: was there a culture of popular science or science popularization in the early Franco regime? If so, how did it look or work, and what sort of "nation" was propagated there?

[59] See Antolín Hofrichter (2021).

[60] Rheinberger (1997); idem. (2006), Chap. 12, pp. 336–349.

[61] Herrán and Roque (2013).

[62] For a more clearly defined usage of the term "national styles" in science, see Harwood (1993).

[63] Nieto-Galan (2016).

Conclusion

Clearly, it is impossible to cover all aspects of this topic in the space available. Regrettably, the perspectives offered here have been entirely too Eurocentric. In closing, let me summarize some of the main points I have tried to make:

(1) All of the key words in my title are moving targets: what science, nation, and culture mean have all changed over time, and the meanings of such terms have also varied from place to place in similar time periods.
(2) Sciences are not only assemblages of ideas and research results or methods, but also institutions. For this reason the sciences, too, can become centres of social status and (research) power. This means that the relations of sciences and political power, though clearly asymmetrical, are not so one-sided as is often suggested or assumed.
(3) Periods of political predominance, especially in dictatorships, have not always automatically produced "pseudoscience." Indeed, the term "pseudoscience" itself is often, though not always, a post hoc attribution with denunciatory rather than analytical intent.[64] Amparo Gómez Rodriguez has written of "science and pseudoscience" in fascist regimes,[65] but it seems wiser from a historian's perspective to speak of ideological mobilizations of sciences as discursive resources (as I tried to do earlier) or of scientific research conducted under different policy priorities. Mark Walker, a leading expert on the sciences under Nazism, puts the current consensus on this issue quite succinctly: "No one political ideology or system is best, or for that matter worst, for supporting science. All of the (totalitarian) regimes investigated here were willing and able to fund and foster scientists and scientific institutions when officials recognized that a particular sort of science might be valuable to them and the policies they wanted to pursue."[66]
(4) Positing essentialist distinctions among types of disciplines in the manner of the "two cultures" distinction is also questionable, espe-

[64] Rupnow, et al. (eds.) (2008).
[65] Gómez (2009).
[66] Walker (2012), p. 375.

cially for the nineteenth and early twentieth centuries. In the nineteenth century, at least, and in many European countries far into the twentieth century, natural scientists, humanists, and public intellectuals were educated largely in the same elite secondary schools, and should thus be regarded as members of a common culture. Members of all types of disciplines produced and propagated "national science" in various guises. This is one important reason why the allegedly fundamental difference between natural and human sciences may not have been as significant for this particular history as is often supposed.

(5) The term "national sciences" is fraught with difficulties. Taking "nation" as a topic brings with it a danger of adopting a too narrowly focused single-nation perspective, or of comparisons that reify the category itself by comparing sciences or science policy in one country with those in another, rather than considering the possibility of transnational circulation of ideas and personnel. As Mark Walker puts it in the article just cited, science is national, international, and transnational, often simultaneously. A single-nation focus, which often simply reproduces existing, state-funded institutional frameworks without reflection on them, cannot do justice to these complexities.

Bibliography

Agnew, Hugh Le Caine (2003): Czechs, Germans, Bohemians? Images of Self and Others in Bohemia to 1848. In: Nancy M. Wingfield (ed.): *Creating the Other: Ethnic Conflict and Nationalism in Habsburg Central Europe*. New York and Oxford: Berghahn Books, pp. 56–77.

Aichner, Christoph; Mazohl, Brigitte (eds.) (2017): *Die Thun-Hohensteinschen Universitätsreformen 1849–1860. Konzeption—Umsetzung—Nachwirkungen*. Vienna: Böhlau-Verlag.

Anderson, Benedict (1991): *Imagined Communities: Reflections on the origin and spread of nationalism*, rev. ed. London: Verso.

Anon. (Whewell, William) (1834): „On the Connexion of the Sciences", by Mrs. Sommerville. *The Quarterly Review*, 51, pp. 54–68.

Andrés Antolín Hofrichter (2021): "The Foreign Modernity": Symbolic order and science policy at the CSIC during early Francoism. In: Maurició Janué i Miret and Albert Presas i Puig (eds.): *Science, Culture and National Identity in Francoist Spain, 1939–1959*. Basingstoke: Palgrave Macmillan.

Ash, Mitchell G. (2002): Wissenschaft und Politik als Ressourcen für einander. In: vom Bruch, Rüdiger; Kaderas, Brigitte (Eds.), *Wissenschaften und Wissenschaftspolitik—Bestandaufnahmen zu Formationen, Brüchen und Kontinuitäten im Deutschland des 20. Jahrhunderts.* Stuttgart: Franz Steiner-Verlag, pp. 32–51.

Ash, Mitchell G. (2010): Wissenschaft und Politik. Eine Beziehungsgeschichte im 20. Jahrhundert. *Archiv für Sozialgeschichte*, 50, pp. 11–46.

Ash, Mitchell G.; Surman, Jan (2012a): The Nationalization of Scientific Knowledge in Nineteenth-Century Central Europe: An Introduction. In: Ash/Surman (eds.) (2012), pp. 1–29.

Ash, Mitchell G. and Surman, Jan (eds.) (2012b): *The Nationalization of Scientific Knowledge in the Habsburg Empire 1848–1918.* Basingstoke: Palgrave Macmillan.

Brock, William H. (2002): *Justus von Liebig: The Chemical Gatekeeper.* Cambridge: Cambridge University Press.

Camprubí, Lino (2010): One grain, one nation. Rice genetics and the corporate state in early Francoist Spain (1939–1952). *Historical Studies in the Natural Sciences*, 40:4, pp. 499–531.

Claret Miranda, Jaume (2006): *El Arroz desmoche. La destrución de la Universidad española por el Franquismo*, 1936–1945. Barcelona: Crítica.

Coen, Deborah R. (2013): *The Earthquake Observers: Disaster Science from Lisbon to Richter.* Chicago: University of Chicago Press.

Coen, Deborah R. (2018): *Climate in Motion: Science, Empire and the Problem of Scale.* Chicago: University of Chicago Press.

Cohen, Gary B. (1996): *Education and middle-class society in Imperial Austria 1848–1918.* West Lafayette, IN: Purdue University Press.

Connelly, John; Grüttner, Michael (eds.) (2005): *Universities under Dictatorship.* University Park, PA: Pennsylvania State University Press.

Crawford, Elisabeth (1992): *Nationalism and Internationalism in Science, 1880–1939. Four Studies of the Nobel Population.* Cambridge: Cambridge University Press.

Du Bois-Reymond, Emil (1870): *On the German War.* London: Bentley.

Fahlbusch, Michael (1999): *Wissenschaft im Dienst der nationalsozialistischen Politik? Die "Volksdeutsche Forschungsgemeinschaften" 1931–1945.* Baden-Baden: Nomos Verlag.

Feichtinger, Johannes (2010). *Wissenschaft als Reflexives Projekt. Von Bolznao über Freud zu Kelsen. Österreichische Wissenschaftsgeschichte 1848–1938.* Bielefeld: Transcript Verlag.

Feichtinger, Johannes (2012): 'Staatsnation', 'Kulturnation', 'Nationalstaat'. The role of National Politics in the Advancement of Science in Austria, 1848–1938. In: Ash/Surman (eds.) (2012), pp. 57–82.

Fichte, Johann Gottlieb (2009): *Addresses to the German Nation*, ed. Gregory Moore. Cambridge: Cambridge University Press. First published 1808.

Frank, Tibor (2009): *Double exile. Migrations of Jewish-Hungarian professionals through Germany to the United States.* Oxford and Frankfurt am Main: Peter Lang.

Geary, Patrick J. (2002): *The Myth of Nations: Medieval Origins of Europe.* Princeton: Princeton University Press.

Geertz, Clifford (1973): *The Interpretation of Cultures.* New York: Basic Books.

Gellner, Ernest (1983): *Nations and Nationalism.* Ithaca, NY: Cornell University Press.

Gellner, Ernest (1998): *Language and Solitude. Wittgenstein, Malinowski, and the Habsburg Dilemma.* Cambridge: Cambridge University Press.

Gómez, Amparo (2015): The 'Social Contract' for Spanish Science before the Civil War. In: Gómez, Amparo; Canales, Antonio Fco.; Balmer, Brian (eds.): *Science Policies and Twentieth-Century Dictatorships: Spain, Italy and Argentina.* Franham/Burlington: Ashgate, pp. 27–58.

Gómez, Amparo (2009): Ciencia y pseudosciencia en los regimens fascistas. In: Gómez, Amparo; Canales, Antonio Fco. (2009), pp. 13–50.

Gómez, Amparo; Canales, Antonio Fco. (2009): *Ciencia y Franquismo. La ciencia de posguerra española.* Barcelona: Laertes.

Gómez Rodriguez, Amparo; Canales Serrano, Antonio F. (2009): The Rebels and the New Spanish Scientific Culture. *Journal of War & Culture Studies*, 2:3, pp. 321–333.

Gordin, Michael; Grunden, Walter; Walker, Mark, Wang, Zinouye (2003). „Ideologically Correct" Science. In: Walker, Mark (ed.), *Science and Ideology: A Comparative History.* London and New York: Routledge, pp. 35–65.

Grüttner, Michael (2005): German Universities under the Swastika. In: Connelly and Grüttner (eds.), pp. 75–111.

Haar, Ingo; Fahlbusch, Michael (eds.) (2008): *Handbuch der völkischen Wissenschaften.* Munich: K.G. Saur.

Harwood, Jonathan (1993): *Styles of Scientific Thought: The German Genetics Community, 1900–1933.* Chicago: University of Chicago Press.

Harwood, Jonathan (2001): German Science and Technology under National Socialism. *Perspectives on Science. Historical, Philosophical, Social*, 5 (1997), pp. 128–151.

Hausmann, Frank-Rutger (1999): Der „Kriegseinsatz" der deutschen Geisteswissenschaften im Zweiten Weltkrieg (1940–1945). In: Winfried Schulze/Otto G. Oexle (eds.), *Deutsche Historiker im Nationalsozialismus.* Frankfurt a. M.: Fischer Taschenbuchverlag, pp. 63–86.

Heinemann, Isabel (2005): Wissenschaft und Homogenisierungsplanungen für Osteuropa. Konrad Meyer, der ‚Generalplan Ost' und die Deutsche Forschungsgemeinschaft. In: Isabel Heinemann; Patrick Wagner (eds.):

Wissenschaft, Planung, Vertreibung. Neuordnungskonzepte und Umsiedlungspolitik im 20. Jahrhundert. Stuttgart: Franz Steiner Verlag, pp. 45–72.

Helmholtz, Hermann von (1862/1995): On the Relation of Natural Science to Science in General (1862). In: idem., *Science and Culture: Popular and Philosophical Essays*, ed. David Cahan. Chicago: University of Chicago Press, pp. 76–95.

Helmholtz, Hermann (1869/1995): On the Aim and Progress of Physical Science (1869). In: idem., *Science and Culture: Popular and Philosophical Essays*, ed. David Cahan. Chicago: University of Chicago Press, pp. 204–225.

Herrán, Nestor; Roque, Xavier (2013): An Autarkic Science: Physics, Culture and Power in Franco's Spain. *Historical Studies in the Natural Sciences*, 43:2, pp. 202–235.

Hobsbawm, Eric (1990): The Invention of Tradition, In.: idem., *Nations and Nationalism since 1780: Programme, myth, reality.* Cambridge: Cambridge University Press.

Hossfeld, Uwe; John, Jürgen; Lemuth, Oliver; Stutz, Rüdiger (eds.) (2003): *„Kämpferische Wissenschaft". Studien zur Universität Jena im Nationalsozialismus.* Cologne: Böhlau-Verlag.

Jessen, Ralph; Vogel, Jakob (eds.) (2002): *Wissenschaft und Nation in der europäischen Geschichte.* Frankfurt am Main: Campus Verlag.

Kater, Michael H. (1997): *Das Ahnenerbe der SS 1935–1945. Ein Beitrag zur Kulturpolitik des Dritten Reiches.* Munich: Oldenbourg.

Klemun, Marianne (2012): 'National Consensus' as Culture and Practice: The Geological Survey in Vienna and the Habsburg Empire (1849–1867). In: Ash and Surman (eds.) (2012).

Livingstone, David N. (2003): *Putting Science in its Place: Geographies of Scientific Knowledge.* Chicago: University of Chicago Press.

Maier, Helmut (ed.) (2002): *Rüstungsforschung im Nationalsozialismus. Organisation, Mobilisierung und Entgrenzung der Technikwissenschaften.* Göttingen: Wallstein Verlag.

Malet, Antoni (2009): José María Albareda (1902–1966) and the formation of the Spanish Consejo Superior de Estudios Scientíficas, in: *Annals of Science,* 66:3, pp. 307–322.

Mazzolini, Renato (1995): Nationale Wissenschaftsakademien im Europa des 19. Jahrhunderts. In: Jordan, Lothar; Kortländer, Bernd (eds.): *Nationale Grenzen und internationaler Austausch. Studien zum Kultur- und Wissenschaftstransfer in Europa.* Tübingen: Mohr, pp. 245–260.

Morente, Francisco (2006): La universidad en los regímenes fascistas: la depuración del professorado en Alemania, Espana y Italia. *Historia Social*, 54, pp. 51–72.

Morente Valera, Francisco (2017): The Falange and Academia: Intellectuals and the Idea of a National-Syndicalist University. In: Gallego, Ferran; Morente, Francisco (eds.): *The Last Survivor: Cultural and Social Projects Underlying Spanish Fascism, 1931–1975.* Brighton, Portland Toronto: Sussex Academic Press, pp. 62–89.

Neufeld, Michael J. (1995): *The Rocket and the Reich. Pennemünde and the Coming of the Ballistic Missle Era.* New York: The Free Press.

Nieto-Galan, Augustí (2016): *Science in the Public Sphere: A History of Lay Knowledge and Expertise.* (City): Routledge.

Otero Carvajal, E. (2006): *La destrucción de la ciencia en Espana: depuración universitaria y franquismo.* Madrid: Editorial Complutense.

Palló, Gábor (2012): Scientific Nationalism: A Historical Approach to Nature in late Nineteenth-Century Hungary. In: Ash & Surman (eds.) (2012), pp. 102–112.

Péteri, György (2005): The Communist Idea of the University. An Essay Inspired by the Hungarian Experience. In: Connelly and Grüttner (eds.), pp. 139–166.

Presas y Puig, Albert (2005): Science on the Periphery. The Spanish Reception of Nuclear Energy: An Attempt at Modernity? *Minerva*, 43:2, pp. 197–218.

Presas y Puig, Albert (2008): Technological Transfer as a Political Weapon: Technological Relations Between Germany and Spain from 1918 to the Early 1950s. *Journal of Modern European History*, 6:2, pp. 218–236.

Presas i Puig, Albert (2021): Epistemic Communities and Science Makers in the Franco Regime: A Study of the Nuclear Energy Board. In Marició Janué i Miret and Albert Presas i Puig (eds.): *Science, Culture and National Identity in Francoist Spain, 1939–1959.* Basingstoke: Palgrave Macmillian..

Renan, Ernest (1882/1992). *What is a Nation?* https://www.humanityinaction.org/files/569-E.Renan-WhatisaNation.pdf. Download 18 June 2019.

Rheinberger, Hans-Jörg (1997): *Toward a History of Epistemic Things: Synthesizing Proteins in the Test Tube.* Stanford: Stanford University Press.

Rheinberger, Hans-Jörg (2006): *Epistemologie des Konkreten. Studien zur Geschichte der modernen Biologie.* Frankfurt am Main: Suhrkamp Verlag.

Roca-Rosell, Antoni (2005): Professionism and Technocracy. Esteve Terradas and Science Policy in the Early Years of the Franco Regime. *Minerva*, 43:2, pp. 147–162.

Rodríguez, López (2021): Carolina: Scenarios of Science and Symbols of the New State: Political Resignification of the University City of Madrid. In Marició Janué i Miret and Albert Presas i Puig (eds.): *Science, Culture and National Identity in Francoist Spain, 1939–1959.* Basingstoke: Palgrave Macmillian..

Ruíz Carnicer, Miguel Ángel (2005): Spanish Universities under Franco. In: Connelly & Grüttner (eds.), pp. 113–138.

Rupnow, Dirk; Lipphardt, Veronika; Thiel, Jens; Wessely, Christina (eds.) (2008): *Pseudowissenschaft. Konzeptionen von Nicht-/Wissenschaftlichkeit in der Wissenschaftsgeschichte*. Frankfurt am Main: Suhrkamp.

Rust, Bernhard (1936/1990): Nationalsozialismus und Wissenschaft (1936). In: Hans-Jochen Gamm (ed.), *Führung und Verführung. Pädagogik des Nationalsozialismus, 3rd ed.* Munich: List-Verlag, pp. 154–158.

Steinhauser, Thomas; James, Jeremiah; Hoffmann, Dieter; Friedrich, Bretislav (2011): *Hundert Jahre an der Schnittstelle von Chemie und Physik. Das Fritz-Haber-Institut der Max-Planck-Gesellschaft zwischen 1911 und 2011*. Berlin: De Gruyter.

Štrbánová, Soňa (2012): Patriotism, Nationalism and Internationalism in Czech science: Chemists in the Czech National Enlightenment. In: Ash & Surman (eds.) (2012), pp. 138–156.

Surman, Jan (2012): Science and its publics. Internationality and National Languages in Central Europe. In: Ash & Surman (eds.) (2012), pp. 30–56.

Ungern-Sternberg, Jürgen and Wolfgang Ungern-Sternberg (1996): *Der Aufruf „An die Kulturwelt!" Das Manifest der 93 und die Anfänge der Kriegspropaganda im Ersten Weltkrieg*. Stuttgart: Franz Steiner Verlag.

Vodrázková-Pokorná, Lenka (2006): *Die Prager Germanistik nach 1882. Mit besonderer Berücksichtigung der bis 1900 an die Universität berufenen Persönlichkeiten*. Frankfurt am Main: Peter Lang.

Vom Bruch, Rüdiger (2014): „Der heilige Sinn des Krieges". Geistige Mobilmachung und professorale Aufrüstung. In: Metzler, Gabriele (ed.), *Die Berliner Universität im Ersten Weltkrieg*. Berlin, pp. 33–60.

Walker, Mark (2012): The ‚National' in international and transnational science. *British Journal for the History of Science*, 45:3, pp. 359–376.

Weber, Eugen (1976): *Peasants into Frenchmen. The Modernization of Rural France 1870–1914*. Stanford: Stanford University Press.

Weiss, Sheila F. (2010): *The Nazi Symbiosis: Human Genetics and Politics in the Third Reich*. Chicago: University of Chicago Press.

Winkelbauer, Thomas (2018): *Das Fach Geschichte an der Universität Wien: Von den Anfängen um 1500 bis etwa 1975*. Göttingen: V & R unipress.

PART II

Scientific and Cultural Policy in the 'New State'

CHAPTER 3

'The Foreign Modernity': Symbolic Order and Science Policy at the CSIC During Early Francoism

Andrés Antolín Hofrichter

The history of science in Spain, like its historiography, has a long tradition. Already in the nineteenth century, 'Spanish science' was one of the preferred topics among patriotic intellectuals in regard to the problems arising from the 'Spanish anomaly' when comparing the country with the emerging technical-industrial powers in Europe. It is not surprising that 'science' became a favourite topic for intellectual disputes, not only in the nineteenth century but also in the twentieth: it was necessary to first define what 'science' was, and, above all, what a 'Spanish science' should comprise.[1] The debates and discourses on science in Spain were based on concepts of identity, whether they were national or regional. In addition to this identity aspect, there was also a temporal component inherited from the Enlightenment that placed civilisations, nations, and even religious

[1] Cf. Juliá (2005): 34*ff.*; Álvarez Junco (2007): 413*ff.*, 443*ff.*

A. Antolín Hofrichter (✉)
Duale Europäische Wirtschaftsakademie, Madrid, Spain

M. Janué i Miret, A. Presas i Puig (eds.), *Science, Culture and National Identity in Francoist Spain, 1939–1959*,
https://doi.org/10.1007/978-3-030-58646-1_3

confessions on an axis formed by two vectors: that of 'progress' and that of historical 'backwardness'. Within this axis, scientific production became one of the greatest manifestations of a modernity that presented itself as a desirable horizon for some and a threat for others.[2]

The twentieth century inherited from the nineteenth century both the identity and temporal dimensions for articulating discourses and founding scientific institutions.[3] The Association for the Advancement of Science (*Asociación para el Progreso de las Ciencias*), the Institute of Catalan Studies (*Institut d'Estudis Catalans*), and the Board for Advanced Studies (*Junta para la Ampliación de Estudios*) are examples of liberal initiatives that emerged in an international historical context in which policies to promote science were based on an inevitable patriotic desire.[4] Studies on the history of science in twentieth-century Spain have often focused on the complex interplay between scientific production and the institutions that should support it.[5] The concern to distinguish between policies that promoted scientific progress and those that prevented it has itself produced a narrative that places, on the one hand, the Board for Advanced Studies as a promoter of a 'Silver Age' following liberal ideas, and, at the other end of the spectrum, the Spanish National Research Council (*Consejo Superior de Investigaciones Científicas*, CSIC), particularly during the founding stage, as the maximum manifestation of a dictatorial science policy.[6] As has been well noted by Amparo Gómez, this vision emerges from a sociology of science that associates good research practices with liberal regimes and bad practices with authoritarian or totalitarian systems, disregarding the experiences of the German *Kaiserreich*, Italian fascism, and even National Socialism.[7]

This article aims to provide a different perspective on the science policy of Francoism. From a strictly historical perspective, 'science', and therefore

[2] As shown by the so-called Polemics of Spanish Science, cf. Álvarez Junco (2007): 383–496; Juliá (2005): 46–57. For a history of science in Spain in the nineteenth and twentieth centuries, the study of Sánchez Ron (1999) remains fundamental.

[3] For the identification of 'science' with a 'Europe' synonymous with 'modernity' in the first third of the twentieth century, see, for example, Juliá (2005): 149–152.

[4] For Spanish institutions, see Romero; Santesmases (eds.) (2008); Gómez; Canales (eds.) (2009); see later for the international institutions.

[5] In addition to those already mentioned, we will only mention recent publications that summarise the general lines of research. See Puig-Samper (2007); Romero; Santesmases (eds.) (2008); Gómez; Canales (eds.) (2009).

[6] See, for example, Otero Carvajal (ed.) (2014).

[7] Cf. Gómez (2009): 13–47.

also 'Spanish science', represented a semantic field open to resignifications through discourses, symbols, and staging—a field that primarily the Catholic elites of the New Regime, political, intellectual, and scientific, sought to coin.[8] In the following sections, we will address the extent to which the great scientific-institutional project of Francoism was an attempt to occupy an extremely powerful semantic space, in which concepts of 'spiritual tradition', 'technology', and 'modernity' overlapped with the national Catholic narrative of the patriotic past. These discursive and symbolic dimensions are particularly important for the history of science during Francoism, since their analysis reveals certain ambiguities, which, added to the institutional and economic problems that afflicted the CSIC, explain the failure of Franco's science policy to institutionalise 'Spanish science' under a new regime.[9] For this, after first providing some methodological notes which are necessary to understand the scope of the perspective selected here, we will analyse the 'construction of a scientific tradition', the hierarchy of knowledge derived from it, and, in a separate section, the specific place occupied by 'technology' within the discourses on science in the CSIC environment.

Perspective and Methodological Notes

From a comparative perspective, the CSIC represents a unique case. Beyond the discourses on Spanish 'exceptionality' and the intellectual and scientific 'deficiencies' of the early years of Francoism, recent research has agreed on the need to place the foundation of this scientific macro-institution in its international context.[10] Both the Board for Advanced Studies (JAE, 1907) and the CSIC were part of a secular movement from which emerged, among others, the *Kaiser-Wilhelm-Gesellschaft* (1911), the *Notgemeinschaft der deutschen Wissenschaft* (1920), the various British councils, the *Consiglio Nazionale delle Ricerche* (1923), and the *Centre National de la Recherche Scientifique* (1939).[11] Although this is true, the interest in the CSIC as an object of historical study does not reside, in the author's opinion, in its features that are comparable to those of the

[8] The studies of Malet (2008): 211–256; ID. (2019) already point in this direction.

[9] Cf. Antolin Hofrichter (2018): 68–116; 321–335.

[10] Cf. Malet (2019).

[11] Cf. Stadler (2004); Vom Bruch (2002): 60–74; Orth (ed.) (2010); Guthleben (2003); Simili; Paoloni (eds.) (2001); Edgerton (2003): 759–776.

international environment. Compared to the institutions mentioned earlier, the CSIC, which was legally founded in November 1939, was particularly unique because of its symbolic load—not to say overload. None of the authoritarian or fascist regimes that emerged in the interwar period made as much effort to institutionalise and represent their scientific body, limiting themselves in most cases to inheriting and reconverting pre-existing institutions. This refers to the fact that one of the basic functions—if not the main one—of the CSIC was not so much to 'coordinate' and 'promote' research in post-war Spain, to use the language of the sources,[12] but rather to represent the ability of a Catholic nation to incorporate, control, and even dominate what was perceived as the most useful conquest of a foreign and threatening modernity. Through highly ritualised discourses, symbols, and plenary sessions, a substantial part of the intellectual, political, and scientific elite ensured that 'science' was incorporated into the new Spain for which they longed.[13] Gathered in plenary sessions, magazines, and institutions, the scientific and intellectual environment of the CSIC imagined itself as a representative of a new Spanish science under the national Catholic ideas.[14]

This said, this article does not consider the Council in its institutional aspect in the classic sense of the term. Rather, we are inspired by approaches developed for the study of institutions from the perspective of the sociology of symbols. These approaches propose to analyse institutional frameworks as symbolic frameworks *in themselves* and to consider symbols and rituals as stabilising elements of the ideas and concepts attached to them.[15] The stabilising function of symbols is particularly important for the analysis of the Council, since it was explicitly created to 'combine the purest spiritual traditions with the demands of modernity'.[16] As we will see, while this 'conjugation' was presented as a very difficult task at the discursive level, it was perfectly possible at the symbolic level, at least until the mid-1950s.

[12] Both terms are already found in the CSIC's founding law. BOE, 28 November 1939: 6669.

[13] Recent research has revealed the importance of symbols in the 'construction of Francoism' and therefore also of its scientific institutions. For this aspect, see Box (2010); Michonneau; Núñez Seixas (eds.) (2014).

[14] In this regard, see also, recently Malet (2019): 124–130.

[15] The results of the research group that primarily studied this aspect in Melville (2001).

[16] BOE, 28 November 1939: 6668.

The Construction of a 'Glorious Scientific Tradition'

The CSIC was officially created in November 1939, thus imposing itself within a series of preliminary projects that had been designed in previous years and during civil conflict.[17] Its initial regulations foresaw the creation of six departments, to which two more would be added years later, and whose main task was to integrate, both administratively and symbolically, the institutes in which research was conducted, across the various 'branches of knowledge'. In this sense, the founding law of the Council involved more than a series of legal and administrative provisions. Like so many other legislative texts of the immediate post-war period, this law represented a declaration of intent, established first of all in its preamble: 'This must also be the noblest ambition of Spain at the present time, which, in the face of past poverty and paralysis, desires to renew its glorious scientific tradition'.[18]

The Council was created with the intention of representing, as this same preamble stated, 'the restoration of the classical and Christian unity of the sciences destroyed in the eighteenth century'. Simultaneously, this new body had to combine 'the purest lessons of the universal and Catholic tradition with the demands of modernity'.[19] However, what was this 'glorious scientific tradition' and how could the CSIC contribute to 'renewing' it? Although this founding law, unlike the subsequent University Law of 1943, did not explicitly mention the restoration of 'ceremonials, emblems, and acts',[20] the design of the Council had an obvious ceremonial and symbolic load, which was evident in its institutional structure, in its publications, in its symbolism, and in the fourteen 'Plenary Sessions' that were held between 1940 and 1958.[21]

To analyse the semantic field surrounding the 'science' presented through the Council, it is convenient to differentiate three dimensions: the temporal, the religious-spiritual, and the organological. Undoubtedly, these three dimensions were interdependent, as was evident in the

[17] We refer here mainly to the Institute of Spain, promoted by Pedro Sainz Rodríguez. Cf. Malet (2008).

[18] BOE, 28 November 1939: 6668. The symbols of the Council were stipulated in BOE, 18 March 1940: 1897–1898.

[19] BOE, 28 November 1939: 6668.

[20] BOE, 31 July 1943: 7408.

[21] For a first approach to the analysis of these Plenary Sessions, see Malet (2008): 211–256.

discourse of the Minister of National Education and President of the Council, José Ibáñez Martín, in the Plenary Session of 1941:

> The imperial tree of Spanish science grew lush in the garden of Catholicity and was not disdained to be the essential fibre and nerve of its trunk, the sacred and divine science, from whose juice all the thick branches were nourished in union. The Spanish theological genius, which flourished to serve the Catholicity of the faith, must also occupy at this supreme moment the first hierarchy of the scientific renaissance. Our current science—in connection with the one that in the past centuries defined us as a nation and as an Empire—wants to be first and foremost Catholic.[22]

The typical exuberant rhetoric of the discourses surrounding the Council in its first decade combined the language of botanical and natural sciences with references to the imperial past and to the Catholic nature of the Spanish nation and science. Even so, it is convenient to analyse these dimensions separately.

Regarding the construction of a scientific tradition in the temporal dimension, the primary function was attributed to the philologist and writer Marcelino Menéndez Pelayo (1856–1912), whose figure was associated with the new institution from the beginning. So much so that one of the largest and costliest editorial projects of the Council consisted of the publication of his complete works, which accounted for a total of 65 volumes from 1940 to 1974 and consumed, to give an example, approximately 60% of the resources available to the entire Jerónimo Zurita Institute of History from 1942 to 1944.[23] The publication of the works of this figure of National Catholicism, now reinterpreted from a Francoist perspective, was in no way an editorial project among others, as shown by the extensive correspondence between the main intellectual founder of the CSIC, José M. Albareda, and his closest collaborator in the first two years, Alfredo Sánchez Bella. The secretary general and the one who would later become director of the Institute of Hispanic Culture paid special attention to this edition and promoted, in the words of Sánchez Bella, a 'propaganda' plan that, through radio and printed media, would associate the figure of

[22] CSIC (1942): 33.

[23] The budget for the National Edition of the complete works of Marcelino Menéndez Pelayo for the years 1942–1944 was 550,000 pesetas and that of the Jerónimo Zurita Institute was 900,000 pesetas for the same period. Calculations made from the budgets published in the Memoirs of the Council for the corresponding years.

'Don Marcelino' and his position in the 'Polemic of Spanish Science' with the foundation of the Council.[24] In accordance with this, Ibáñez Martín maintained in October 1940 that 'the polemic ends today',[25] referring to the inaugural Plenary Session of the Council.

The figure of Menéndez Pelayo even became part of the institutional order of the Council, to name one of the eight Departments, specifically the one that brought together the disciplines of Philology, History, and Art. The remaining Departments were also named after figures considered to be outstanding in Spanish science: Alfonso el Sabio (inorganic sciences), Raimundo Lulio (theology, philosophy, social sciences), Alonso de Herrera (organic sciences), Ramón y Cajal (medicine), Juan de la Cierva (engineering, applied sciences), and, since 1948, Saavedra Fajardo (international studies, later bibliographic) and Quadrado (local studies).[26] The nomenclature of the Departments and Institutes was not only homage but also served to build a national scientific tradition, something that did not go unnoticed in political-scientific circles abroad; thus, in the round table following a lecture by José M. Albareda before the *Arbeitsgemeinschaft für Forschung des Landes Nordrhein-Westfalen* in 1954, the specialist in ecclesiastical history and former member of the *Zentrumspartei* of the Weimar Republic, Georg Schreiber, highlighted this distinctive element:

> Another difference from Germany. It concerns the name of the research institutes [sic, the Departments], their designation, and their flag. In the peninsula, they have decided to preferentially name them after great *scientific personalities*. This is certainly an homage that should be taken for granted. In this support of a great tradition, Spain is far ahead of us. What you, dear speaker, gave us as Departments, is exemplary.[27]

Most of the Institutes were also named after 'great personalities' within their corresponding fields. While this 'homage' may have seemed

[24] General Archive of the University of Navarra, Collection of Albareda, 006/001/075, Letter from Enrique Sánchez Reyes to Alfredo Sánchez Bella of 3 October 1940. In general, see the correspondence between Albareda and Sánchez Bella archived in the General Archive of the University of Navarra, Collection of Albareda, 006/001.

[25] CSIC (1942): 30.

[26] The first six Departments were established in the BOE, 17 February 1940: 1201–1203. The remaining two were institutionalised in the BOE, 24 January 1948: 337–339.

[27] The words of Schreiber are quoted in Albareda (1956): 43 (italics in the original). For more information on this conference, see Presas (1998): 343–357.

Fig. 3.1 Distribution of the dates of birth of the 'scientific personalities' who gave their names to Departments and Institutes of the Council, by centuries; own elaboration

exemplary to Schreiber, the 'scientific tradition' was built along a national Catholic story from which important ambiguities emerged. If we look at the dates of birth of these 'personalities', both for the eight Departments and for the 32 Institutes with their own names (year 1951), we can see the dynamics of the national Catholic story along with their discontinuities (Fig. 3.1):

From the perspective of the national Catholic narrative, this dynamic was perfectly coherent: the thread of personalities starts from the late Middle Ages and peaks in the fifteenth and sixteenth centuries to coincide with the time of imperial and spiritual splendour, only to decline in the following centuries, and rise again in the nineteenth century, honouring figures who, like Balmes and Menéndez Pelayo, but also Antonio de Gregorio Rocasolano and Juan de la Cierva, combined their research activities with their 'Catholic Spanishness'. Regarding the construction of a scientific tradition, on the other hand, this dynamic revealed a fundamental problem: technical-industrial modernity, which the Council promised to 'combine' with this tradition, was, if not excluded, at least relegated to second place, since the number of honoured personalities reached its minimum precisely in the centuries corresponding to the technical-industrial era.

On the other hand, the construction of this tradition involved a religious-spiritual dimension that was evident in the constant references to God, the Catholic religion, and the Church. In this context, on 15 December 1944, the newspaper *ABC* presented the Plenary Session of the Council for that year in which it highlighted the offer of a relic of San Isidoro by the Bishop of Vitoria to the CSIC.[28] The reason for the ecclesiastical hierarchy offering this gift to the main organ of Spanish science was none other than the recent consecration of the *Iglesia del Espíritu Santo*, which was built on the site of the former Auditorium of the Student Residence.[29] In his role as bishop, Carmelo Ballester invoked the patron of the new institution, San Isidoro, who came to represent the communion between country, knowledge, and Catholicism, in accordance with the remaining symbolic order of the CSIC.[30]

Although references to Catholicism were omnipresent in the discourses and even in the architecture itself, it was primarily the Plenary Sessions that presented a liturgical-religious character. They all began with a mass attended by both political and scientific personalities, as well as representatives of leading cultural institutions.[31] Apart from the Department meetings, the Plenary Sessions comprised an opening and a closing session, which were chaired by the Minister of National Education and Francisco Franco as Head of State, respectively. These sessions had a particularly thorough protocol that, inter alia, established the distribution of seats at the presidential table. Therefore, while the personalities that occupied these five or seven seats were rotated according to the representative priorities of each Plenary Session, there was always one high representative of the State (in the centre), one of the Council, and one of the ecclesiastical hierarchy, often the bishop of Madrid-Alcalá. The ritual nature of these Plenary Sessions would climax during the presentation of the various prizes awarded by the Council and, above all, during a ceremony that was called an 'offering' until 1945 and comprised the

[28] N.N.: 'The Minister of National Education presided yesterday at the opening session of the fifth Plenary Session of the Spanish National Research Council', *ABC*, 15 December 1944. The event was recorded and is available at the CSIC (1945): 1–2.

[29] Cf. Guerrero (2007): 287–289.

[30] This 'spiritual Trust' was stipulated in the BOE, 18 March 1940: 1898.

[31] The Memoirs of the Council contain detailed protocols of the programme and the course of each of these Plenary Sessions.

presentation to the Caudillo of works published by the CSIC.[32] The 'service to the country' and to God was manifested, in this logic, through the publications.

At this point, it is particularly important to highlight that the religious symbolism of the Council precisely privileged the spiritual. Not only did theology regain its place as the 'first science', but the whole field of the 'spirit' gained new importance. If the 'glorious scientific tradition' did not come from the natural sciences and technology, at least Spain, in its essence, fulfilled its mission of contributing to the field of the spirit. This line of argument, which was reinforced in the second half of the 1940s thanks to the discourses on Spain as a 'spiritual reserve of the West', would be used in the Council by Catholic intellectuals who, like Rafael Calvo Serer, Florentino Pérez Embid, and José Luís Pinillos, were seeking both internal and external projection.[33]

In this, the religious-spiritual component was intimately associated with a third dimension, the organological one. The predilection of its authorities for a botanical and naturalistic rhetoric when describing the whole of Spanish science and the relationship between its parts was already evident in the previous quotations. The 'tree of science' and its 'branches', the research and its 'fruits', the patriotic spirit as 'juice' flowing through the 'channels' of the 'homeland': far from being a mere rhetorical resource, this language reflected a dichotomous discourse that opposed the 'organic' to the 'mechanical', the 'natural' to the 'artificial', and assigned the first qualities to the Council, and, in general, to the genuinely 'Spanish', and the second to a liberal order prior to 1936/39, and, in general, to a 'foreign' science inherited from the Enlightenment. 'The foreign technology and customs, and the strange and hostile thinking', as Ibáñez Martín explained in 1948 regarding the creation of the Quadrado Department of Local Studies, 'invaded the capital redoubts of our culture'. If 'technology' was foreign, the 'vital pulse' could only come from the 'spirit', so much more present in the most purely patriotic sphere, the local one: 'only the pulsating variety of Spanish provincial life disarticulated the ill-fated Europeanisation of Spain and [...] maintained the privileges of

[32] CSIC (1946): 80.

[33] The desire for external projection is evident, for example, in his correspondence with Alfredo Sánchez Bella, in the General Archive of the University of Navarra, collection of Calvo Serer 001/033/468–1: Letter from Rafael Calvo Serer to Alfredo Sánchez Bella of 4 April 1950. On Calvo Serer and his intellectual circle, cf. Prades (2012).

the Spanish spirit'.[34] In the organological language, the export of this spirit became a 'transfusion',[35] and the Council became a useful instrument to, in the words of Calvo Serer in 1950, put a brake on the 'hypertrophic development of the material sciences'.[36]

It is important to note that the organisational discourse of the authorities described both the entire national scientific community and the relationship between the disciplines. Regarding the scientific community, this discourse expressed the Council's intentions to overcome 'an individualistic, disjointed, miscellaneous research of isolated efforts, a bouquet of scattered flowers',[37] and artificially centralised, above all, in Madrid. Following the motto of José M. Albareda that the 'Council was not a point, it was a map',[38] the CSIC would aspire to 'lay its roots and spread its branches throughout the countryside of the Homeland'.[39] This territorial aspect was attached to an organisational one: words like 'coordination' or 'autonomy', far from being pure euphemisms to describe the use of science to meet the ends of the dictatorship, reflected the line of thought of part of the Catholic intellectuals, even beyond borders. The organological conception of the scientific community implied that the researchers themselves would provide their 'fruits' to the homeland. As Albareda wrote in 1951, in his *Considerations on Scientific Research*, science policy should not respond to 'those great mobilisations of personnel [...], which form large, highly centralised, and rigid companies, mechanical as a machine'. For the author, the Council demonstrated that 'the warmth of the individual has not yet been extinguished from the world and there are still institutions in which the individual exists freely'.[40] It should be noted, however, that this 'individual freedom' did not correspond in any way to a liberal conception. It was rather a conservative concept, according to which the 'spirit' would be 'free', as long as it was not coerced by rationalist

[34] Ibid: 40.

[35] CSIC (1951): 94.

[36] Calvo Serer (1950).

[37] CSIC (1943): 52.

[38] Figure used in the CSIC (1942): 8, as well as in its correspondence, in General Archive of the University of Navarra, Collection of Albareda, 006/001/021-1 and 006/001/036-1, Letters from Albareda to Riviere of 4 June 1940 and 11 July 1940, respectively. Albareda's traditional regionalism rooted in his youth, as highlighted by Malet. Cf. Malet (2009): 307–332.

[39] CSIC (1942): 96.

[40] Albareda (1951): 14.

and mechanistic logics. The researcher was not a 'wheel within the cogwheel' of Spanish science, but rather a part of an organic whole, for which he provided his service 'freely', as highlighted by the head of the Scientific Documentation Service, Juan Roger, in 1953: 'The State, when establishing the Council, indicated its scientific purpose, and it is Spanish science that, governing itself, works autonomously to deliver the result of its research to the nation'.[41]

The organological symbology and discourse also extended, although to a lesser extent, to the relationship between the disciplines. If the 'tree of science' that the Council came to restore had grown in the 'garden of Catholicity', theology would have to be placed back on its 'trunk', so that all the other 'branches' could emerge from it. Perhaps the clearest expression of this order was represented by the Llullian *arbor scientiae* that was established as a symbol of the CSIC from the beginning and was present at all levels, from architectural ornaments to the covers of publications and the letterheads of institutional correspondence.[42]

The organological discourse was also nourished by this symbol, which always opted, as we have seen, for a botanical rhetoric. It even gave its name to the most important publication of the Council, the magazine *Arbor*, which, as announced in the 1944 Plenary Session, had been born to 'exalt the patent harmony of the sciences'.[43]

The representations of this 'tree of science' always placed the Raimundo Lulio Department of theology and philosophy in the trunk. The other 'branches of knowledge' emerged from it and were adapted to the institutional order of the Council, with a Department assigned to each. Although this is evident, it is necessary to highlight one of the main qualities of the symbols in general, and of this one in particular. The symbol offered multiple possibilities of identification, interpretation, and reinterpretation, which would even explain its continuity to this day.[44] However, from the theory of symbols, a second, and no less important, quality emerges: the 'tree of science' allowed the creation of a visual

[41] Roger (1953): 131.

[42] This is all explicitly stipulated in the BOE, 18 March 1940: 1898.

[43] In the words of Ibáñez Martín, CSIC (1945): 45.

[44] On the semantic obscurity and stabilising function of the symbols, see Boyer (2001): 641–643.

coherence between its parts, which did not exist as such. While the image harmonised theology, the humanities, the natural sciences, and technology by incorporating them into an organic whole, the relationship between these disciplines was anything but harmonious at the discursive level, as we will see in the next section.

The 'Corrosive Modernity' and the Dangers of 'Technology'

The construction of this 'scientific tradition' coexisted with the attempt to incorporate the positive sciences and technology, which, in turn, covered much of the semantic field of the term 'modernity'. The founding law of the Council of 1939 proposed not only to restore the 'Christian unity of the sciences' but also to 'finally associate this research activity with the centres of applied science, particularly at this great moment of Spain, in which fostering technology is imposed'.[45] Ten years later, in the Plenary Session that celebrated the 10th Anniversary of the Council, Francisco Franco insisted before a large group of national and foreign guests on the mission for which the highest body of Spanish science was created: 'We understand that there is no greater way of exalting the scientific ideal than to instil a desire for eternity in it, making it the vehicle for a common history that, without neglecting the conquests of technology, places the divine cause of man above all values'.[46] The fact that 'technology' was included in a subordinate sentence was not just a syntactic detail. The 'technology' should be subordinated, literally, to the 'divine cause of man'. This syntax reflected, as we will see, the scepticism of a considerable part of the political and intellectual elite associated with the CSIC.

In national Catholic discourses, the positive sciences and particularly technology were considered a sort of Trojan horse of a foreign modernity: inherent components of a modernity that was at first strange to Spanish history, with useful instruments that were simultaneously dangerous. In his book *España ante el mundo*, published in 1950 under the pseudonym Juan de la Cosa, Luis Carrero Blanco set the starting point for what he

[45] BOE, 28 November 1939: 6668.
[46] CSIC (1951): 94.

described as the secular 'process of isolation' not only in the increase in power of English Protestantism and Freemasonry; the differentiated path that would characterise the history of Spain also began, as highlighted in the same introduction, with technical progress, with 'the discovery of steam':

> From the discovery of steam arises machinism, which creates capitalism and social injustice; from the latter, by logical reasoning, Marxism should sprout, which upon practical implementation, should be transformed into communism, and communism, capitalism, liberalism, and masonry are, with Catholicism, the leading figures of the current human tragedy.[47]

If 'technical progress' had contributed to 'the current human tragedy', Catholicism had preserved true Spain from participating in historical aberrations. From this quote emerges a narrative in which Catholicism, and therefore Spain, would represent a different path from the one that began with the rise of technology and industrialisation. In his *Considerations*, Albareda also followed the logic of a narrative in which history branched into a 'Christian path' and into a path 'corroded' by 'modernity': 'The intimate and anarchic breakdown to which the whole modern process, corrosive to the Christian order, led to the dehumanisation of human values to erect them into independent idols, [...] made Science emerge as one of the myths of pagan restoration'.[48]

This criticism of technology was not a feature of the national Catholic discourses of Francoism but had a long tradition within European conservatism.[49] What distinguished the discourses on science in Spain was not this scepticism but its historical articulation around a national narrative in which the Spanish path would have been on the sidelines of a threatening 'technical civilisation'. For example, although German conservatism would warn about its disruptive effects, it nevertheless assumed that technology was an intrinsic part, for better or worse, of the nation's recent history.[50] This was not the case for the national Catholic line of thought, in which technology appeared as something external and, in essence, alien. This logic structured the discourses of the intellectuals of the Council,

[47] Carrero Blanco (1950): 16.

[48] Albareda (1951): 419.

[49] For the German case, see, for example, Rohrkrämer (1999).

[50] For the place of 'technology' within the representations of the nation in Europe, cf. Vogel (2008): 105–120.

including the contributions of the so-called generation of 1948 to the debate on the 'Problem of Spain'.[51] Perhaps it was Vicente Palacio Atard who best expressed the relationship between national Catholic essentialism, 'technology', and 'modernity'. In his 1949 essay *Derrota, agotamiento, decadencia en la España del siglo XVII*, this historian, who was affiliated to the Jerónimo Zurita Institute of History of the CSIC, concluded that true Spain had been the 'Spain of the defined personality, which has been able to live in the middle of modern Europe without being contaminated by modernity, even though it is crossed by railroads and curdled by laboratories'.[52]

This last conjunction ('even though') indicated a special relationship between 'Spain' and technical-industrial modernity, valid both for the intellectuals of the 'generation of 48' and for the symbolic order of science and its discourses in the CSIC. On the one hand, it expressed a fundamental dilemma between the essences of the nation and positive science and technology. On the other, however, it accepted the possibility, even the need to incorporate these elements into the Spanish historical path. As Pérez Embid would write about the 'Problem of Spain', his generation had 'found a Spain that now faces the technical and economic difficulties that Europe easily solved in the era of machinism and industrialisation'.[53] To overcome these difficulties, it was necessary, in the words of his co-religionist Calvo Serer, to incorporate this 'Europe', that is, the 'prodigious increase in human knowledge in the knowledge of matter and man', and to learn 'from modern technology, that great instrument that today turns against the spirit, and that the latter has to dominate in order to serve man'.[54] For Spain not to be 'contaminated with modernity', it was, therefore, necessary to control and enclose science, and especially technology, to subordinate them to the spirit. It is the ambiguous logic of incorporation, control, and the enclosure of science and technology that makes the symbolic order of the Council understandable: the institution incorporated all these 'branches' and subordinated them to the 'Christian unity of the sciences'.

[51] Of the numerous publications that analyse this debate, the present text only mentions the studies of Saz (2003): 379*ff.*; Juliá (2005): 355*ff.*

[52] Palacio Atard (1949): 179.

[53] Pérez Embid (1949): 151.

[54] Calvo Serer: (1947): 344.

Obviously, not all Spanish Catholic intellectuals shared this scepticism towards 'technology'. The visions of a communion between Catholicism and technology were, however, rare.[55] Positive science and technology as well as Catholicism and spirituality continued to be part of different interpretive categories. Partly as a consequence of this, the discourses on Spanish science also inherited the figure of the 'backwardness' in national science, which was more pronounced in its natural and technical branches. As early as 1945 and quoting the German Hispanicist Karl Vossler, Calvo Serer made it clear that

> [without] denying the participation of the Spanish in naturalistic research, … it is certain that they do not manage to produce figures who are 'hallmarks for Europe', such as Telesio, Bruno, Paracelsus, Bacon, Galileo, Kepler, or Newton. While in Europe the Renaissance, Enlightenment, Romanticism, and Positivism represent a return to Nature with progressively perfected technology and research, 'Spain—Vossler writes—followed these footsteps in a delaying manner, since its special way of being and its own greatness does not lie in this'.[56]

Spiritual greatness and scientific-technical 'backwardness' represented two sides of the same coin, something that the self-proclaimed exponent of that 'greatness' accounted for in Spanish history. Five years later, in his solemn discourse given on the 10th Anniversary of the Council, the art historian Francisco J. Sánchez Cantón would argue within this same logic, but with a darker diagnosis. His essay on *Libros, cuadros y tapices que coleccionó Isabel la Católica* praised the heritage of the 'foundress of Spain' and considered in it the 'germ and [the] prophecy of what was to be the spiritual development of Spain'. However, this heritage, and particularly the collection of books contained in it, also predicted something problematic since 'the shortage of scientific writings was a bad omen that had to occur'.[57] The history of science in Spain was presented as a deficient history, even within a national Catholic discourse that, simultaneously, highlighted the Spanish 'glorious tradition'.

[55] However, some important figures, such as José María Otero Navascués, a promoter of technical research in the CSIC since early Francoism, and later also of nuclear energy, showed a clearly enthusiastic attitude towards the technical branches. Cf. Presas (2008).

[56] Calvo Serer (1945): 32.

[57] CSIC (1951): 67.

The End of the Path: The Construction of 'Scientific Modernity'

If technology as an applied expression of the positive sciences was something alien, the recovery of this delay imposed the exchange with that 'Europe' and that 'West' that were so powerful within the Francoist imagination. The truth is that the Francoist science policy could never establish a purely nationalist discourse on Spanish science.[58] Furthermore, as applied and technical sciences began to take on the meaning of the word 'science' during the 1950s, the CSIC's efforts to represent Spain in the international scientific community also gained strength. The 10th and 25th Anniversary of the Council, celebrated in the spring of 1950 and in the context of the '25 years of Peace', respectively, were a clear manifestation of the growing use of the CSIC as a sort of paradiplomatic bridge to the Western block by the political-scientific sphere of Francoism.[59]

The CSIC was originally an attempt to institutionally represent the capacity for 'science' of a Catholic nation and, above all, its elite. However, for this 'science' to be properly 'Spanish', it had to incorporate all branches of knowledge within a restored 'Christian unity'. This did not happen through the implementation of a new epistemological system, but rather through an order that allowed for the recreation on the institutional, visual, and ritual levels of a 'harmonious' whole that on other levels was difficult, if not impossible, to develop. The organological language and symbology, the religious-spiritual components, and the construction of a 'scientific tradition' throughout the national Catholic narrative of the patriotic past shaped a conception of 'science' that referred to a pre-industrial and pre-enlightened era—all this despite the fact that this same institution had been created to incorporate, also, the 'demands of modernity'. In this, the incorporation of 'technology' and 'natural sciences' followed a logic of subordination, enclosure, and control: although they were useful instruments, they also represented a sort of Trojan horse of a threatening 'modernity' and, in principle, were alien to the Spanish historical path.

[58] The introduction to the first institutional memory already praised the 'Tree of Science that points to Heaven', but 'all this without stopping to look abroad', CSIC (1942): VI.

[59] Cf. Antolin Hofrichter (2018): 94–117; 321–335. See also Delgado Gómez-Escalonilla (2007): 269–277.

If this symbolic load of the Council made it a unique case within the context of the political-scientific institutions of its time, its transition from Spanish 'exceptionality' to international 'normality' was no less remarkable. From 1945, and particularly since the late 1940s, the CSIC has been an institution that represented Spain in the 'Western ecumene', at first privileging the spiritual mission of Spain, and, later, as 'technology' was privileged, moving towards discourses of 'backwardness' and economic utility. The governing institution of science during Francoism went, in just two decades, from representing a historical path linked to national Catholicism to becoming a tentative manifestation of research in natural and technical sciences, and, ultimately, of a technical-industrial modernity to which the regime would gradually dedicate its imagery.[60] This ability to incorporate the 'conquests of modernity' into a national Catholic line of thought was not at all new.[61] However, the main problem was that the positive attitude towards the new discourses on 'science' and 'modernity' made the national Catholic story not only useless but even impossible, at least within the new context of Francoist developmentalism.

References

Albareda, José María (1951): *Consideraciones sobre la investigación científica*, Madrid, CSIC.

Albareda, José María (1956): *Die Entwicklung der Forschung in Spanien. Vortrag vor der Arbeitsgemeinschaft für Forschung des Landes Nordrhein-Westfalen, Sondersitzung am 23. Juni 1954, hrsg. von Leo Brandt*, Köln, Opladen, Westdeutscher Verlag.

Álvarez Junco, José (2007): *Mater Dolorosa. La idea de España en el siglo XIX*, 10th ed., Madrid, Taurus.

Antolin Hofrichter, Andrés (2018): *Fremde Moderne. Wissenschaftspolitik, Geschichtswissenschaft und nationale Narrative unter dem Franco-Regime, 1939–1964*, De Gruyter Oldenbourg, Berlin, Boston.

[60] Particularly striking is the shift of financial and symbolic resources towards applied sciences from the early 1950s to the detriment of the sciences of the spirit. In this regard, see the budgets provided in the Memoirs of the CSIC, as well as the documentation in the General Archive of the Administration, Collection of the Education, Box 31/8811, 'Explanatory memoir on the budget of the Spanish National Research Council for the years 1956/1957', including 16 pp. with budget statistics and 17 pp. with comments and explanations. For an analysis of the relative weight of the sciences of the spirit in the budget and in the Plenary Sessions of the CSIC, see Antolin Hofrichter (2018): 71–80; 149–154; Malet (2019).

[61] Cf. Botti (2008).

BOLETÍN OFICIAL DEL ESTADO: "Ley de 24 de noviembre de 1939 creando el Consejo Superior de Investigaciones Científicas", 28 November 1939: 6668–6671.

BOLETÍN OFICIAL DEL ESTADO: "Orden de 8 de marzo de 1940 disponiendo que el Consejo gozará de la máxima jerarquía en la vida cultural del país", 18 March 1940: 1897–1898.

BOLETÍN OFICIAL DEL ESTADO: "Decreto de 10 de febrero de 1940 regulando el funcionamiento del Consejo Superior de Investigaciones Científicas", 17 February 1940: 1201–1203.

BOLETÍN OFICIAL DEL ESTADO: "Ley de 29 de julio de 1943 sobre ordenación de la Universidad española", 31 July 1943: 7406–7431.

BOLETÍN OFICIAL DEL ESTADO: "Decreto de 9 de enero de 1948 por el que se modifican varios artículos del de 10 de febrero de 1940 y complementarios referentes al Consejo Superior de Investigaciones Científicas", 24 January 1948: 337–339.

Botti, Alfonso (2008): *Cielo y dinero. El nacionalcatolicismo en España, 1875–1975*, 2nd ed., Madrid, Alianza.

Box, Zira (2010): *España, año cero. La construcción simbólica del franquismo*, Madrid, Alianza.

Boyer, Christoph (2001): "Zur spezifischen Symbolizität spättotalitärer Herrschaft", in Melville, Gert (Ed.): *Institutionalität und Symbolisierung. Verstetigungen kultureller Ordnungsmuster in Vergangenheit und Gegenwart*, Köln, Weimar, Wien, Böhlau: 639–658.

Calvo Serer, Rafael (1945): "Valoración europea de la historia española", *Arbor*, 2: 19–47.

Calvo Serer, Rafael (1947): "Una nueva generación española", *Arbor*, 24: 333–348.

Calvo Serer, Rafael: "¿A dónde nos conduce la ciencia?", *ABC*, 5 December 1950.

Carrero Blanco, Luis [pseud. Juan de la Cosa] (1950): *España ante el mundo. Proceso de un aislamiento*, Madrid, Idea.

Consejo Superior de Investigaciones Científicas (Ed.) (1942): *Memoria de la Secretaría General, 1940–1941*, Madrid, CSIC.

Consejo Superior de Investigaciones Científicas (Ed.) (1943): *Memoria de la Secretaría General, 1942*, Madrid, CSIC.

Consejo Superior de Investigaciones Científicas (Ed.) (1945): *Memoria de la Secretaría General, 1944*, Madrid, CSIC.

Consejo Superior de Investigaciones Científicas (Ed.) (1946): *Memoria de la Secretaría General, 1945*, Madrid, CSIC.

Consejo Superior de Investigaciones Científicas (Ed.) (1951): *Memoria, 1949*, Madrid, CSIC.

Delgado Gómez-Escalonilla, Lorenzo (2007): "La dimensión internacional del CSIC", in Puig-Samper, Miguel A. (Ed.): *Tiempos de investigación. JAE-CSIC, cien años de ciencia en España*, Madrid, CSIC: 269–277.

Edgerton, David (2003): "Science in the United Kingdom. A Study in the Nationalization of Science", in Krige, John; Pestre, Dominique (Eds.): *Companion to Science in the Twentieth Century*, New York/London, Routledge: 759–776.

Gómez, Amparo (2009): "Ciencia y pseudociencia en los regímenes fascistas", in Gómez, Amparo; Canales, Antonio F. (Eds.): *Ciencia y fascismo. La ciencia española de posguerra*, Barcelona, Laertes: 13–47.

Guerrero, Salvador (2007): "El conjunto urbano del CSIC en Madrid. Retórica y experimentalismo en la arquitectura española del primer franquismo", in Puig-Samper, Miguel A. (Ed.): *Tiempos de investigación. JAE-CSIC, cien años de ciencia en España*, Madrid, CSIC: 285–291.

Guthleben, Denis (2003): *Histoire du CNRS de 1939 à nos jours*, Paris, Armand Colin.

Juliá, Santos (2005): *Historias de las dos Españas*, 3rd ed., Madrid, Taurus.

Malet, Antoni (2008): "Las primeras décadas del CSIC. Investigación y ciencia para el franquismo", in Romero, Ana; Santesmases, María J. (Eds.): *Cien años de política científica en España*, Bilbao, Fundación BBVA: 211–256.

Malet, Antoni (2009): "José María Albareda (1902–1966) and the formation of the Spanish Consejo Superior de Investigaciones Científicas", *Annals of Science*, 66: 307–332.

Malet, Antoni (2019): "Science and power: Francoist Spain (1939–1975) as a case study", *Centaurus*, 66: 111–132.

Melville, Gert (Ed.) (2001): *Institutionalität und Symbolisierung. Verstetigungen kultureller Ordnungsmuster in Vergangenheit und Gegenwart*, Köln, Weimar, Wien, Böhlau.

Michonneau, Stéphane; Núñez Seixas, Xosé Manoel (Eds.) (2014): *El imaginario nacionalista español en el franquismo*, Madrid, Casa de Velázquez.

Orth, Karin (Ed.) (2010): *Die Deutsche Forschungsgemeinschaft 1920–1970. Forschungsförderung im Spannungsfeld von Wissenschaft und Politik*, Stuttgart, Franz Steiner.

Otero Carvajal, Luis Enrique (Ed.) (2014): *La Universidad nacionalcatólica. La reacción antimoderna*, Madrid, Dykinson-Universidad Carlos III de Madrid.

Palacio Atard, Vicente (1949): *Derrota, agotamiento, decadencia en la España del siglo XVII. Un punto de enfoque para su interpretación*, Madrid, Rialp.

Pérez Embid, Florentino (1949): "Ante la nueva actualidad del ‚Problema de España'", *Arbor*, 14: 149–160:

Prades, Sara (2012): "Las plataformas de acción de la ‚generación de 1948' entre 1944 y 1956", *Historia y política*, 28: 57–82.

Presas, Albert (1998): "Nota histórica. Una conferencia de José María Albareda ante las autoridades académicas alemanas", *Arbor*, 160: 343–357.

Presas, Albert (2008): "La inmediata posguerra y la relación científica y técnica con Alemania", in Romero, Ana; Santesmases, María J. (Eds.): *Cien años de política científica en España*, Bilbao, Fundación BBVA: 173–210.

Rohkrämer, Thomas (1999): *Eine andere Moderne? Zivilisationskritik, Natur und Technik in Deutschland 1880–1933*, Paderborn, Schöningh.

Roger, Juan (1953): *La investigación científica en el mundo*, Madrid, CSIC.

Romero, Ana (2008): "Políticas e instrumentos. De la Junta para la Ampliación de Estudios al Consejo Superior de Investigaciones Científicas", in Romero, Ana; Santesmases, María J. (Eds.): *Cien años de política científica en España*, Bilbao, Fundación BBVA: 107–139.

Sánchez Ron, José M. (1999): *Cincel, martillo y piedra. Historia de la ciencia en España. Siglos XIX y XX*, Madrid, Taurus.

Saz, Ismael (2003): *España contra España. Los nacionalismos franquistas*, Madrid, Marcial Pons.

Simili, Raffaela; Paoloni, Giovanni (Eds.) (2001): *Per una storia del Consiglio Nazionale delle Ricerche*, 2 vol., Roma, Bari, Laterza.

Stadler, Friedrich (Ed.) (2004): *Österreichs Umgang mit dem Nationalsozialismus. Die Folgen für die naturwissenschaftliche und humanistische Lehre*, Wien, New York, Springer.

Vogel, Jürgen (2008): "Mythos Moderne. Die Technik in der nationalen Selbstdarstellung in Europa", in Altenburg, Detlef; Ehrlich, Lothar; John, Jürgen (Eds.): *Im Herzen Europas. Nationale Identitäten und Erinnerungskulturen*, Köln, Weimar, Wien, Böhlau: 105–120.

Vom Bruch, Rüdiger (Ed.) (2002): *Wissenschaften und Wissenschaftspolitik. Bestandsaufnahmen zu Formationen, Brüchen und Kontinuitäten im Deutschland des 20. Jahrhunderts*, Stuttgart, Franz Steiner.

CHAPTER 4

Scenarios of Science and Symbols of the New State: Political Resignification of the University City of Madrid

Carolina Rodríguez-López

When the Spanish War ended in 1939 and the Franco regime began its trajectory, the university became an essential place to promote the principles that would make up the sought-after and desired Francoist nation. In the political programme of the victors of the war, the university, which is the highest level of the educational system, represented an essential component for establishing and disseminating the values with which they wanted to define an entire society. Thus, the university took on political functions that went beyond its role as a generator of knowledge.

This text is based on two publications: RODRÍGUEZ-LÓPEZ, Carolina (2016): 105–130 and RODRÍGUEZ-LÓPEZ, Carolina (2018): 283–314. Thank you to the editors of this manuscript for the comments and suggestions, which have improved the initial version.

C. Rodríguez-López (✉)
Complutense University of Madrid, Madrid, Spain
e-mail: carolinarodriguez@ghis.ucm.es

M. Janué i Miret, A. Presas i Puig (eds.), *Science, Culture and National Identity in Francoist Spain, 1939–1959*,
https://doi.org/10.1007/978-3-030-58646-1_4

Furthermore, in the case of the University of Madrid, the academic institution was inextricably linked to its location in the city. The University City of Madrid, which since 1927 had aimed to contain most of the capital's post-secondary education, was a stable front during the war, and it was always associated with the republican resistance and "They Shall Not Pass", a romantic and internationally known chant that promised the defence of the city from anti-fascism.

As soon as the Francoist army took the city, the campus became part of the discourse with which the Francoist army aimed to cement its victory. In this way, the university, both its contents (movable assets) and its premises (building structures), also became a stage for the development of the allegedly new Francoist science and for the deployment of a whole political symbolism that erased and redefined the legacy of the previous regime—a republic that, although very brief, had brought some of the innovations that Spain had been awaiting for so long. This manuscript takes the University of Madrid and its University City as examples. The campus and the symbols used by the Franco regime to resignify the university will be the points on which this manuscript will concentrate, placing an emphasis on how a space and an institution naturally focused on the creation and dissemination of science ended up being used for the staging of Franco's politics.

Control of the Chancellery and the Foundations of the New Francoist University

When the Spanish War had not yet ended, in the territories occupied by the Francoist army and in which a budding state was emerging, new institutions were created, and the guidelines under which existing ones were to function were presented. This was the case for the University of Madrid. The Spanish University had been immersed in deep debates about its quality, its management and its scientific scope,[1] and those who would go on to win the war developed the first reform projects and named those who would be appointed to lead it when the war ended.[2] The Francoist authorities adhered to the criterion that the university and all its scientific production should be designed to suit them. The advances that the republic was able to undertake in its short duration (changes in study programmes, the incorporation of women into the university, the internationalisation of

[1] Hernández Sandoica, Elena (2008): 45; Hernández Sandoica, Elena (1991): 3–22.

[2] Rodríguez-López, Carolina (2002): 58.

knowledge, etc.) were questioned and, for the most part, eliminated. To a large extent, the university hierarchy, which, at least in Madrid, was closely linked to the republican plan and followed the modernising guidelines that had been promoted since 1874 and 1907 by the ILE and the JAE, respectively, was closely monitored, disabled, and purged so that only the elements that ideologically could collaborate best with the Francoist university plan were kept. The task was greatly facilitated by the fact that the university that the Francoists inherited, purged, and planned still suffered from clear inefficiencies.

The reforms for the university from the fascist side began in 1938, and the meaning of all of them permeated the space, the campus, in which academic life unfolded. Even before the creation of the first Francoist government, on 30 January 1938, when Pedro Sainz Rodríguez[3] was made responsible for the National Education portfolio, some provisions had begun to increase the power of the chancellor. Thus, the chancellors of the universities in an area called "national", which was dominated by the Francoists, had to monitor teaching activities in primary education, looking out for "any manifestation of weakness or orientation opposed to the healthy and patriotic attitude of the Spanish army and people".[4] In addition, the mayors were in charge of informing the chancellors about the teachers who had shown up on 1 September. If mayors needed to appoint a new teacher, it was the chancellors who had to approve the appointment. University chancellors were also in charge of the inspection and administration of primary education, since, as stipulated by the law of 1857, they were the highest authorities in the university districts into which Spain was divided and that included educational establishments of lower levels than the university.

In September 1938, a commission made up of five professors was established to propose the basic terms of the Francoist university reform. Pío Zabala, who a short time later would become the chancellor of the University of Madrid, was the chairman of that commission. The commission worked for several months and, in April 1939, published the result of its work: the Draft Law on University Reform.[5] As a rhetorical objective, it proposed to link the Spanish university with its best traditions, or, as they called it, "the historical revitalisation of the Spanish university" and

[3] Alted Vigil, Alicia (1986): 215–229.

[4] *Boletín Oficial de la Junta de Defensa Nacional de España*, 9, 21 August 1936.

[5] Peset Reig, Mariano (1991): 125–158.

placed the institution in charge of patriotic and religious training. The draft law (or at least the versions of the 1919 guideline published in 1924 and 1928) also promised to recover the university's traditional autonomy, which had established strict hierarchies within the university. The project defined the university as the governing body of educational culture comprising five existing schools (Philosophy and Letters, Sciences, Law, Pharmacy, and Medicine) and those that were created later (Veterinary and Political Sciences and Economics, which emerged a short time later). In these centres, the essential values and purposes of university education would be developed: the Hispanic tradition, the Catholicism "of our imperial thought", a unique and original culture and the professional training of students.

According to the proposed university reform, each university would have a chancellor at the top position who would have a council of ordinary university staff, faculty board members, and special university staff. The chancellor would be appointed by the government from among university professors, and he would serve as a representative of the government both at the university and in the university district. At his side would be the vice chancellor, appointed by the government to replace the chancellor if it should become necessary. The deans, representatives of the chancellor who were also appointed by the government, would be in charge of the schools.

An order issued on 9 March 1940 sought the opinion of the Spanish universities on this draft law, and the University of Madrid, which is the focus of our study, prepared the first report regarding its reception. The person in charge of drafting the project was already running the Chancellery of the University of Madrid by that date; it was Pío Zabala who in 1938 was chairman of the commission that proposed the Draft Law on University Reform. From this position, the first thing that the University of Madrid (and its chancellor) demanded was the fulfilment of university autonomy, which was translated as "a wide margin of pedagogical and economic initiatives, but maintaining discipline in hands of the authorities".[6] For Zabala and for the regime, university autonomy offered a unique opportunity for academic authorities to exercise maximum control and to increase the power of chancellors outside and inside the university. What Zabala demanded for the University of Madrid was what finally determined the general operation of all the Spanish universities, whose chancel-

[6] Letter from Pío Zabala to Sainz Rodríguez cit. in Alted Vigil (1986): 101.

lors accumulated remarkable power. The authoritarian aspect of Francoism was installed in the university, indicating that it was to be an essential component in the formation of the Francoist New State.

Between the publication of the project in April 1939 and the publication of the order of 9 March 1940, there had been important changes in the Ministry of Education. In April, Sainz Rodríguez was dismissed from his post, and shortly after, José Ibáñez Martín took over the Ministry of Education. His agenda prioritised the reform of the university with the aim of "adapting it in its educational and administrative organisation to the principles of the new Spain".[7] The preparatory works for the university reform included the publication of a couple of preliminary projects in 1941 and 1942 and finally resulted in a law that was enacted in late July 1943.

The law does not deviate from some advanced terminal points, and it indicates the pillars on which the regime that emerged from the war was based. The Church was consolidated as a centrepiece of the university. Falange, the only party accepted by the Franco regime, also had a central place, with the compulsory integration of the Spanish University Union-Sindicato Español Universitario (SEU) (for students) and the SEPES (for teachers) into Falange. The Army joined academia to ensure that discipline and sports and military training were provided for students.

The universities would be fully juridical entities, and within them, all power was placed in the hands of the chancellors, something that the chancellor of the University of Madrid had strongly insisted on in his role as the head of the country's central university. The University of Madrid was the only university that included all university fields of study and was the only university in all of Spain (at that time, there were 12) that awarded the doctorate degree. The chancellor, whose title of "president" was replaced with "boss", with its fascist connotations, always had his closest and most dependable authorities under his supervision and all services acted according to his delegation. Thus, the chancellor gained control over the appointment of all university authorities. For example, the chancellor participated in the appointment of the vice chancellor by submitting a list of three candidates to the ministry, and he participated in the appointment of deans, vice deans, and general administrators in the same way. The chancellors accumulated the ability to appoint the chief of the SEU for

[7] Sanz Rodriguez, Pedro (1978): 255.

each university and university district, and all of the initiatives of the Falange in the university district had to be approved by the chancellor.

The chancellor's interventions also transcended the teaching plane, a domain previously claimed by the deans. So, the chancellors would participate in the selection of faculty members, determine the topics that professors could teach in class and appoint associate professors, assistants, and tenured professors. The ability for the person in this role to establish client networks in academia with all these powers is evident. Some of the amendments made to the reform project in its last phase were precisely focused on this accumulation of governing powers, since all the criticism came from people in intermediate positions who saw their authority being threatened. None of these amendments was taken into account, an outcome that is easily understood when we remember that the Madrid chancellor Pío Zabala was in charge of ruling on these criticisms: "All that cantonalist organisation would take the university back to very distant times", he stated".[8]

An organisational chart, functions, and contents were established for the university, marking its hierarchical character and serving as a vessel for the principles that had been designed by the winners of the war. As an educational institution, the university assumed the central function of inculcating these values in the young people who entered into higher studies to become the country's academic and professional elite. The contents (movable assets) of the university were also overturned within its premises (building structures), the physical space that the university occupied in Madrid.

A Stage for the University. The University City of Madrid and Its (Re)Inauguration on 12 October 1943

The University City of Madrid was a project started during the reign of Alfonso XIII. It was used to commemorate his 25 years at the helm of the Spanish crown, and it attracted the most important architects and urban planning experts in Spain at the time. Conceived as an American-style campus on 340 hectares, the University City brought together buildings

[8] AGUCM, General Secretariat, Box 1, "Actas de la Junta de Gobierno de la Universidad de Madrid [Records of the Governing Board of the University of Madrid]", 23 October 1943 and 18 December 1943.

and spaces to encompass a complete academic life: libraries, schools, museums, concert halls, sporting fields, and student residences. It was surrounded by nature and was well connected to the city. Representative architects of the so-called Modern movement (Agustín Aguirre, Manuel Sánchez Arcas, Miguel de los Santos Nicolás, and Pascual Bravo, just to name a few) designed the central buildings of this campus. With its very practical style, clean lines, and little decoration, the space was designed to be comfortable and practical for students and teachers and to have enough space for movement and for hosting events.[9] It had been partially inaugurated in 1933; after 1936, it became a stable war front; and in 1939, the Francoists set their eyes on it as an ideal place to symbolise their new power. The first day of this great production was 12 October 1943.

To announce the full content of the recently approved Spanish University Planning Law and to represent the way in which the university would change, the date of 12 October, Spain's national holiday since 1918, was chosen for the commemoration.[10] This date also served as the inauguration of the partially reconstructed University City of Madrid, in which the ideological contents of the described law could be made visible. According to the chronicles, on that day, the esplanade of the medical complex of the University City of Madrid was overflowing with people who surrounded an 18-metre-high cross dedicated to the memory of the fallen from the Francoist side during the war. The head of state, Francisco Franco, the Minister of National Education, the highest military authorities, the SEU and the Church paid tribute to the students killed in combat for the Francoist side. A wreath of flowers with the legend "The Spanish University to the heroes of University City" summarised the event.[11] The regime identified those it considered to be its heroes at the university. After that opening day, everyone who entered the School of Philosophy and Letters would encounter an inscription in Latin that can be translated as follows: "Stop, you walker. The professors and students of this school whose names you contemplate here sacrificed their lives for the homeland and for faith. Though dead, they still speak, showing the way of virtue and immortality by their admirable example".[12]

[9] Chías, Pilar (1986).

[10] García Sebastiani, Marcela (In press).

[11] *ABC*, 13 October 1943; Rodríguez López, Carolina (2002): 223–281.

[12] Rodríguez-López, Carolina (2016): 107; Pedrezuela, Mario (2008): 623; Bahamonde, Ángel and Cervera, Javier (2000): 249–460.

The University of Madrid, the only university in the capital, would be forever linked to the civil war and its outcome. At the end of the war and for a long time after, the Francoist authorities wanted the academic landscape, conceived in 1927 and turned into a warlike landscape just a decade later, to always be a reminder that there had been a long battle there waiting for Madrid to surrender.[13] In the autumn of 1936 and, especially, in the spring of 1937, the location of University City in the northwest area of Madrid made it the target of Francoist attacks in their attempt to enter the capital, while the republican defenders managed to stop the advances again and again. This situation lasted more than two years and, as a result, the campus was partially destroyed and would always be associated with the war.

Those who later undertook the reconstruction of the campus, which had already been designed and in which most of the work had been carried out with a very pragmatic approach, decided to respect a good part of the original project and immediately went to work to politically resignify the space and leave a physical, institutional, and ideological documentation of the political foundations of the regime. Urgency also prevailed.[14] The sooner the spaces were completed, the sooner the regime would be able to show its capacity to restore normality to the population's daily life. By treating the campus as a stage, the urban conception and the original plans from the late 1920s served to make the new values of the Franco system visible. As Harald Bodenschatz states,

> [w]ith urban design, dictatorships pursued more than a solution to local spatial problems. Through urban design, they communicated their project of a new society, a new state and a new man. They demonstrated where the journey would finally take them and offered their followers spaces for living, working, and relaxing.[15]

The case of the University City of Madrid was not the only university space, nor even the first, in Europe to face formal adaptations imposed by new political imprints. In the first third of the twentieth century, various experiences accumulated in Europe regarding the construction of university campuses which, like that of Madrid, sought to integrate and unite all

[13] Calvo González-Regueral, Fernando (2012); Rodríguez-López, Carolina (2015).

[14] Domenech, Daniel (2018): 218.

[15] Bodenschatz, Harald (2014): 388; Bodenschatz, Harald, Sassi, Piero and Welch Guerra, Max (eds) (2015).

academic activity and daily coexistence within a single venue. The installation of authoritarian regimes in Italy and Portugal (to limit ourselves to the closest ones) made it easier for academic spaces to accumulate a clear political stance, as exemplified by the cases of the Università de la Sapienza in Rome or the Cidade Universitária in Lisbon.[16]

The legislation that had been prepared for the regulation of the university aimed to reclaim the Catholic confession and the spirit of the Falange. With the implementation of these purifying measures from the very beginning[17] and with the systematic granting of powers to the chancellors[18] and the organisation of access to professorships, Francoism was assessing the academic structure with the aim of ensuring that the pre-existing university would cease to exist. Thus, a new university was trying to build itself on the remnants of the previous one and on university students, who now faced exile or retaliation. To try to convince the public that a "new" university was being established, spaces and speeches were filled with symbols representing the new political regime.

Thus, the idea of reclaiming buildings whose construction was well in progress when the war began prevailed. Most of the pre-war projects were taken over, and some of the architects continued in their positions. The campus was then resignified according to the political content of the new regime, so that the reconstruction was presented as a pragmatic exercise as well as a political one. Given the budgets, the circumstances, the accessibility of the materials, their prices, and the urgency with which the new regime wanted to rebuild, it was best to use what existed and resignify it as needed.[19]

To implement this process of political resignification, it was necessary to take note of the production conditions of Franco's urbanism, namely, the institutions, organisations, professional training programmes, and magazines, among other agents, that are activated when designing and implementing a new urban planning space.[20]

[16] Amaral, Samuel (2014): 72–87; Lourenço, Marta; Neto, Maria João (2011): Neto, Maria João; Soares, Clara (2016): 269–289, Pascoal, Ana; Neto, Maria João (2011): 173–194.

[17] Claret Miranda, Jaume (2006): 287–317; Rodríguez López Carolina (2008a): pp. 61–99.

[18] Rodríguez-López, Carolina (2008b):199–210.

[19] Rodríguez-López, Carolina et al. (2015): 92.

[20] Bodenschatz, Harald (2014): 387.

In February 1940, the law reorganising the Construction Board of the University City of Madrid was published. Its preamble again extolled the heroic life of the place during the war years and underlined the importance of the ruins in the reconstruction discourse.[21] The body would be under the patronage of the Chief of State; it would be a juridical entity for all purposes, and it would be made up of a broad range of institutional, political, and academic representatives. Under the presidency of Franco, three vice presidents would serve: the minister of national education, the director general of higher and secondary education and the chancellor of the University of Madrid; these figures would be responsible for a large part of the management of the university. The deans of the five schools of the University of Madrid were integrated into the same body, as were the two new deans that were incorporated in 1944. The participants on the board were the mayor of Madrid, an architect on the Facultative Board of Civil Constructions of the Ministry of National Education; the director general of architecture, a representative of the Madrid Association of Exchange Agents and Stock Exchange and representatives of the army, the Schools of Agricultural Engineers and Architecture, the National *Research Council*, the General Directorate of Devastated Regions, and the SEU.

The board was made up of three commissions, a secretariat, and a technical cabinet: a permanent commission, controlled by the Ministry of National Education, an economic commission entrusted to the chancellor of the University of Madrid, and a third artistic commission. For the development and management of all projects, intervention was available from technicians and professors.

The Construction Board could acquire, possess, and manage all kinds of assets; it had its own budget and could determine the number, placement, dimensions, and purpose of the buildings and sports fields that were to be repaired or constructed. To carry out its work, the board announced competitions for the presentation of projects, which it would also choose. The board appointed the architects who would present their projects, order the execution of the work, supervise the projects, and acquire the furniture and all the fixtures and fittings with which the buildings were to be equipped.

To perform these tasks, the board needed (1) a budget; (2) economic resources, which basically came from the income or interest of its assets; (3) state subsidies; and (4) ongoing donations and legacies. It could also

[21] *Boletín Oficial del Estado*, 17 February 1940.

receive fees from both individual and collective memberships and the income and profits from the foundations it created; additionally, it could sell its publications and enter the amount received from an annual National Lottery drawing. Lottery drawings became the most popular form of propaganda for accelerating the rebuilding of the University City.

At this stage, the architect Modesto López Otero gained special prominence.[22] In 1929, he was entrusted with the project and execution of works for the University City of Madrid. With the arrival of the republic, he was relieved of technical direction and was replaced by Manuel Sánchez Arcas; however, after the Civil War and once the latter had been exiled, López Otero was commissioned again to work on the campus. He then assessed the damages of the war: the supply and evacuation networks had been damaged by the mines, and internal fixtures, furniture, and tiles were barely holding together, but much of the structures remained standing. Hence, "the reconstruction is carried out, therefore, by virtue of a new and meticulous review study of the previous building plans, with new additions of the most modern information, deferring, as far as possible, definitive solutions of details".[23]

The fundamental problems of the reconstruction were of three types: economic, technical, and conceptual. To resolve the first issue, the plans mentioned earlier were activated. To address the technical problems, the architects had to study all the solutions that would help them to solve the infrastructural complications caused by the war and that were made worse by a lack of materials as the European conflict worsened in Spain. From a conceptual point of view, the new Construction Board of the University City decided to incorporate only a few modifications into the previous plans, changes that "should not greatly alter the first vision of the work, in which it is necessary to recognise positive success". The chosen location, set-up, and layout of the buildings were perfectly adapted to the political objectives and their staging, as envisioned by Francoism. Regarding the location, the following aspects stood out: the possibility of collaboration among the different university bodies, the ability to acquire culture while receiving professional and research training, the availability of facilities for comfortable and economical administration, a more intense school coexistence, and, above all, something that the regime was very interested in:

[22] Sánchez De Lerín, Teresa (2000).
[23] *Revista Nacional de Educación* (1941): 6.

the ease of exercising chancellery power and the amplification of the university's presence in Madrid's urban environment.[24]

Four tasks were considered urgent reconstructive tasks: the recovery of the material assets owned by the University City, the custody and use of the buildings, the gathering of all documentation, and the control and care of buildings that were located on the campus but did not belong to the university.[25] These first tasks happened quickly,[26] and vigilance had to be increased due to continuous looting in search of iron pipes and other supplies that could be sold.[27] Offers were also received to employ prisoners, who, according to the system by which sentences could be reduced in exchange for work, were assigned to prison workshops.[28]

These conditions allowed the reconstruction to begin. In the first phase, it involved the Schools of Philosophy, Chemical Sciences, and Pharmacy; the Schools of Architecture and Agronomy; the Thermal Power Plant and part of the Sports Fields; the Ximénez de Cisneros Residential College; and the Pavilion of the Construction Board. The latter, which was where the technical team was located and the location from which the work was directed, was one of the first buildings opened in 1941 (Figs. 4.1 and 4.2).

From that moment, the trickle of openings did not stop. The most important was the one organised on 12 October 1943 due to the number of buildings that had been rehabilitated and the symbolic importance of the opening. On that occasion, Franco gave his speech which, full of references to the war, linked the new campus with the Spanish university tradition. With his words, the campus was forever linked to the memory of the battle; furthermore, his speech helped him to declare himself the winner. The entire university campus was decorated for the occasion. In the place where the *Arco de la Victoria* (triumphal arch) would later be erected, the flags of the Falange and the Movement were placed. On the esplanade of Medicine Square, three grandstands and previously mentioned gigantic

[24] Rodríguez-López, Carolina (2015): 87–93.

[25] AGUCM, D-1770-9.

[26] AGUCM, Personal, 229.

[27] AGUCM, D-1767, 3. "Escrito de construcciones Gamboa y Domingo, SA a Modesto López Otero [Gamboa and Domingo, S.A. construction brief to Modesto López Otero]", 3 February1940.

[28] AGUCM, D-1703, 5, "Escrito de talleres penitenciarios a Junta Constructora de la Ciudad Universitaria [Penitentiary workshop brief to the University City Construction Board]", 7 July 1942.

Fig. 4.1 Opening of the Government Pavilion of the Construction Board of the University City (1941). (Picture: Marqués de Valdecilla Historical Library. Complutense University of Madrid)

Fig. 4.2 Press announcement of the opening. AGUCM. 54-11-30, 1-15 (10) 001 and 54-11-30, 1-15 (12) 001

cross were erected, along with an altar. In front of this altar was a grandstand, accessible by a staircase, where Franco and the government members were seated. It was upholstered in red velvet, and on it, an anagram of "Victor" had been embroidered in gold letters. The opening also featured a military parade, after which all attendees went to the School of Philosophy and Letters for the inauguration of the academic year 1943–44.[29] After a tremendous effort, Madrid had the "clean and bright buildings" that, in Franco's words, served the patriotic development of Spain and sought to end a dire need to "restore everything and create everything" (Fig. 4.3).[30]

[29] s.a. (1943): 103–128; *La Ciudad Universitaria* (1943); s.a. (s.f.).

[30] Amador Carretero, Pilar (1991): 283–302.

Fig. 4.3 Inauguration of University City, October 1943. (Picture: Marqués de Valdecilla Historical Library. Complutense University of Madrid)

Francoist Symbols and the Re-Reading of the Campus

With university law in process, the idealisation projected onto the grounds of the University City made even more sense; the political groups that were permanently settling into the university and the academic powers (as defined by this law) must have a spatial manifestation.[31] The Catholic religion had to be present at the University, and for this purpose, all the finished buildings included premises designed as house chapels. In addition, a prominent place on campus established for the placement of a temple. In the formal layout of the University City, the emphasis on the physical education of the students was notable given the inclusion of sports fields in

[31] Pérez De Urbel, Fray Justo (1943): 55–62 and Agucm, 135/10–13, "Acta de la reunión de la comisión permanente de la Junta de la Ciudad Universitaria [Minutes of the meeting of the permanent commission of the University City Board]", 17 June 1942.

the design.[32] The Falangist Movement was also very present on the campus. The José Antonio Residential College was designed, followed later by the *Casa del SEU* (House of the SEU), in honour of the founder of the fascist party in Spain. Just like "in towns, the church and the town hall were rebuilt", and next to them, the party house was erected,[33] such efforts ensured that the ideological forces of Francoism would be present at the university. In the design of the José Antonio Residential College, the characteristics of Falangist architecture in Madrid can be appreciated. The combination of red brick and granite was reminiscent of El Escorial, and it also made the SEU coat of arms clearly visible from many points on the campus. The project was signed by José Luis Arrese Magro and José Manuel Bringas Vega in 1948, and the work ended in 1953 (Figs. 4.4 and 4.5).

Fig. 4.4 José Antonio Residential College, today the Chancellery of the Complutense University of Madrid. (Picture by José Luis González)

[32] Moscardó, Javier (1943): 63.
[33] Chías, Pilar (1986): 194.

Fig. 4.5 Building plan by Modesto López Otero. *S.f.* Historical Heritage of the Complutense University of Madrid

The drawing of symmetrical lines along which to distribute the schools throughout the campus led to the true expression of the acting powers at the university. These lines were to be crowned by the buildings of the Paraninfo (Assembly Room) and the Chancellery; all routes led to these buildings, which underlined the strong power of the chancellor over university routines. The Chancellery was to house the office of the chancellor, and its surroundings were meant to accommodate any academic ceremony organised by the university. To enhance the ceremony, the building could be accessed through an esplanade with various levels that was flanked by columns, friezes, and reliefs. Modesto López Otero had already designed it in 1928, and now he only had to equip it with the symbols of the regime and imagine large crowds gathering around him, in accordance with fascist taste. The Assembly Room and the Chancellery were never built, but they were still present in the discourse and in all the ideal and idealised projects.

Once the ideological signs included in the law were transferred to the architectural plan, an attempt was made to incorporate other elements that further elevated the symbolic reading into the university space. For this effort, the artistic commission of the Construction Board of the University City was deployed. The commission declared that one of its first intentions was to make the University City one of the "most beautiful parks in Madrid with its gardens and monuments and its most artistic corners". The construction of three dedicated monuments to the Cardenal Cisneros (founder of the first University in Madrid, in Alcalá de Henares, in 1499), Alfonso XIII (the king who promoted the creation of the campus in 1927), and Franco (for obvious political reasons at the time) was authorised. A monument to the combatant youth, monuments to Primo de Rivera and the embodiment of wisdom, Minerva, were also planned. In 1948, when all these plans had still not been carried out, the minister also commissioned the erection of a bust of Minister Ibáñez Martín.[34] In addition, it was proposed that, at least for a time, the ruins of the University City should be maintained as a permanent memorial of the war. For example, Count Ciano went there for a walk when he was on an official visit to Spain. The remains of José Antonio Primo de Rivera also passed through these remains in 1939, when they were moved from Alicante to the Monastery of El Escorial.[35]

The section of road that linked the university campus facilitated the creation of a didactic tour, a kind of political journey that evoked mythical past moments and realities that would serve ideologically to reconstruct the national collective imagination.[36] The tour began on the site of the model prison where the headquarters of the ministry in charge of the Air Force was built. For the design of the Ministry of Air, the best model of sixteenth-century Herrerian art was used; at that time, the Escorial Monastery symbolised the splendorous stage of the Spanish Empire in America. Luis Gutiérrez Soto, the architect of this building, made numerous trips to Germany and North Africa to seek inspiration from buildings that had been erected by Albert Speer and Troost.[37] Although the building was not completed until 1958, we know from Soto's own words that

[34] AGUCM. 54/11.30, 1.3, 1.4 y 54/11–32, 1, 5, 7. Agucm, 135/10–4, "Acta de la Comisión Artística de la Junta Constructora de la Ciudad Universitaria [Reports of the Artistic Commission of the University City Construction Board]", 3 December 1942, 16 February 1943, 13 April 1943, 7 June 1943, 3 November 1943, 9 December 1943, 15 February 1944 and 22 January 1948; MARQUÉS DE LOZOYA (1953): 31.

[35] Ros, Samuel (1940).

[36] Domenech (2018): 223; Box, Zira (2010).

[37] Baldellou, Miguel Ángel (1973): 63.

the German architect Paul Bonatz, who had been to Spain on several occasions, also had considerable influence on his final project (Fig. 4.6).[38]

The campus would be entered and exited through its most symbolic element, the triumphal arch, which can still be admired today.[39] The arch summarises how the reconstruction was understood by the regime and its desire for the public to always remember that a battle had been fought on the campus. Consecutive projects overlapped until, in 1956, the arch took the form we know today, although it never was inaugurated.[40] In 1942, the Board of the University City commissioned Modesto López Otero to carry out the project; for this objective, he made pencil drawings of the main elements of the future gate. The project was paralysed due to budgetary issues and urban redevelopment in the area. It was resumed in the

Fig. 4.6 Ministry of Air. (Photo: José Faraldo)

[38] Medina Warmburg, Joaquín (2004): 34.
[39] s.a. (1942): 95–96.
[40] Bonet Correa, Antonio (1981): 28.

years of World War II, but the war forced the Francoist authorities to review the ideological approaches to the monument and to soften the fascist references. At that time, the inscriptions were changed to commemorate the construction of the University City, although references to Franco remain current: "founded by the generosity of the king/restored by the leader of the Spanish/the headquarters of Madrilenian studies/flourished in the presence of God." (Figs. 4.7 and 4.8).

The itinerary continued towards Hispanic America through a visit to the *Casa de América* (House of America) and its museum. The SEU had an aesthetic presence in the building that would house both it and in the statue of José Antonio, as determined by Falange. Next, Catholic influences were powerfully represented in the form of a temple and allegorical references to the intimate, confessional nature of Francoism. In the background, the principle of authority, so marked at the legislative level, gave an unquestionable preponderance to the chancellor and to his corresponding building, the Chancellery.

This tour was evident in a layout that was presented at the opening ceremony of the University City in 1943. The layout, with an approximate

Fig. 4.7 Presentation of the layout of the *Arco de la Victoria. S.f.* Marqués de Valdecilla Historical Library. Complutense University of Madrid

Fig. 4.8 Present appearance of the *Arco de la Victoria*. (Photo by José Luis González)

dimension of 5×5 metres, was an ideal projection that included both the buildings that had already been designed and were partly rebuilt in 1943 and those referred to here, the ones with which the Franco regime proclaimed its ideological position and showed how it wanted to be perceived and presented using the university space (Fig. 4.9).

Conclusions

To expose and showcase the principles on which the new organisation and conception of the Spanish University was based, the Franco regime relied on the design and structure of the previous university and, above all, its well-marked political spaces that served as the setting in which the regime's new institutional proposal was presented. The disciplines, spaces,

Fig. 4.9 Layout of the University City of Madrid according to reconstruction plans. 1943. Historical Heritage of the Complutense University of Madrid. (Photo by Leyre Mauleón)

buildings, and functions proved to be adequate for signifying the new political era and the defining symbols of the Francoist nation.[41]

The fact that the University City of Madrid was also where the battle of Madrid was fought gave even greater symbolic strength to the dual process of reappropriating and signifying spaces. In the University City, the heritage from the past was deconstructed, and the most advanced aspects of the previous scientific cultural policy was retained; for example, the regime dismantled the legacy of the Board for Advanced Studies and the Student Residence by imposing the Ximénez de Cisneros Residential College on the campus. It was also possible to reuse the rationalist and practical design of the campus to make it the stage for the political representation of the new powers of the state: the Church, the Falange, and the army, among others. Thus, in both its immediate use and its post-war design, the University City was impregnated with the visual and allegorical elements of the dictatorship. It became a preferred space for masses, civil

[41] Rodríguez-López, Carolina (2018).

ceremonies, and military parades; the buildings that housed SEU, the Church, and centres for the promotion of so-called Hispanic culture were placed on campus and, more importantly, accessed through an arch commemorating the Francoist victory in Madrid.

The symbolic capital that a space conceived for scientific practice acquired within the political discourse was clearly determined by the Spanish Civil War, specific events on the campus during the war, and the subsequent Franco victory there. To this day, many of the symbols that were conceived then and laden with such political burdens and meanings are still part of the academic and daily landscape of the Madrid campus.

Acronym

AGUCM	Archivo General de la Universidad Complutense de Madrid (General Archive of the Complutense University of Madrid)
JAE	Junta para Ampliación de Estudios (Board for Advanced Studies)
ILE	Institución Libre de Enseñanza (The Free Institution of Education)
SEPES	Servicio Español de Profesorado de Enseñanza Superior (Spanish Service of Higher Education Professors)
SEU	Sindicato Español Universitario (Spanish University Union)

References

Alted Vigil, Alicia (1986): "Notas para la configuración y el análisis de la política cultural del franquismo en sus comienzos: la labor del Ministerio de Educación Nacional durante la guerra", in FONTANA, Josep (ed.): *España bajo el franquismo*, Barcelona, Crítica: 215–229.

Amador Carretero, Pilar (1991): "La Universidad española de los cuarenta. Discursos de inauguración y apertura de curso de la "Ciudad Universitaria de Madrid", in CARRERAS ARES, Juan José y RUIZ CARNICER, Miguel Ángel (eds.): *La universidad española bajo el régimen de Franco (1939–1975)*, Zaragoza, Institución Fernando el Católico: 283–302.

Amaral, Samuel (2014): "Augusto y Mussolini: la presencia de la antigua Roma en la Roma fascista", Actas de y comunicaciones del Instituto de Historia Antigua y Medieval /10: 72–87.

Bahamonde, Ángel y Cervera, Javier (2000): *Así terminó la guerra de España*, Madrid, Marcial Pons.

Baldellou, Miguel Ángel (1973): *Luis Gutiérrez Soto*, Madrid, Dirección General de Bellas Artes.

Bodenschatz, Harald (2014): "Urban design for Mussolini, Stalin, Salazar, Hitler and Franco (1922–1945)", Planning Perspectives, 29:3: 381–392.

Bodenschatz, Harald, Sassi, Piero, y Welch Guerra, Max (eds) (2015): *Urbanism and Dictatorship. A European Perspective*, Berlin, Bauverlag.

Correa, Antonio (1981): *Arte del franquismo*, Madrid, Cátedra.

Box, Zira (2010): *España, año cero: la construcción simbólica del* franquismo, Madrid, Alianza.

Calvo González-Regueral, Fernando (2012): *La guerra civil en la Ciudad Universitaria*, Madrid, Ediciones La Librería.

Claret Miranda, Jaume (2006): *El atroz desmoche. La destrucción de la Universidad española por el franquismo, 1936–1945*, Barcelona, Crítica.

Chías, Pilar (1986): *La Ciudad Universitaria de Madrid. Génesis y realización*, Madrid, Editorial Universidad Complutense.

Domenech, Daniel (2018): "The National Revolution Architecture: Rooted Modernism in the Spanish New State (1939–1959", *Fascism. Journal of Comparative Fascist Studies*, 7: 213–240.

García Sebastiani, Marcela (In press): *12 de octubre: 100 años de hispanoamericanismo e identidades transnacionales*, Madrid, Ediciones Complutense.

Hernández Sandoica, Elena (1991): "Cambios y resistencias al cambio en la Universidad española (1875–1931)", in GARCÍA DELGADO, José Luis and TUÑÓN DE LARA, Manuel (dirs.): *España entre dos siglos (1875–1931): continuidad y cambio*, Madrid, Siglo XX: 3–22.

Hernández Sandoica, Elena (2008): "La Universidad de Madrid en el primer tercio del siglo XX", in LÓPEZ-RÍOS MORENO, Santiago and GONZÁLEZ CÁRCELES, Juan Antonio (coords.): *La Facultad de Filosofía y Letras de Madrid en la Segunda República: arquitectura y Universidad durante los años 30*, Madrid, SECC: 42–57.

La Ciudad Universitaria (1943): Madrid, s.e.

Lourenço, Marta; Neto, Maria João (2011): *Património da Universidade de Lisboa. Ciência e Arte*, Lisboa, Tinta da China.

Marqués De Lozoya 81953): "La Ciudad Universitaria recinto de arte", *Revista Nacional de Educación*, 34: 31–33.

Medina Warmburg, Joaquín (2004): "Irredentos y conversos. Presencia e influencias alemanas: de la neutralidad a la postguerra española (1914–1943", in VVAA, *Modelos alemanes e italianos para España en los años de la postguerra*, Pamplona, Escuela Técnica Superior de Arquitectura Universidad de Navarra: 21–38.

Moscardó, Javier (1943): "El espíritu deportivo de las juventudes en la nueva Universidad", *Revista Nacional de Educación*, 34: 63–64.

Neto, Maria João; Soares, Clara (2016): "Os Edifícios da Cidade Universitária de Lisboa projetados por Pardal Monteiro. Um património artístico a conhecer", in Rivera Blanco, Javier (dir.), *Arquitectura Universitaria. Ciudades Patrimonio Mundial*, Universidad de Alcalá, Madrid: 269–289.

Pascoal, Ana; Neto, Maria João (2011): "A Cidade Universitária: ciência, espaço e função", in LOURENÇO, Marta; NETO, Maria João, *Património da Universidade de Lisboa. Ciência e Arte*, Lisboa, Tinta da China: 173–194.

Pedrazuela, Mario (2008): "El tajo sin retroceso. La vida académica bajo las bombas", in LÓPEZ-RÍOS MORENO, Santiago and GONZÁLEZ CÁRCELES, Juan Antonio (coords.): *La Facultad de Filosofía y Letras de Madrid en la Segunda República: arquitectura y Universidad durante los años 30*, Madrid, SECC: 610–627.

Pérez de Urbel, Fray Justo (1943): "La religión en nuestra Universidad", *Revista Nacional de Educación*, 34: 55–62.

Peset Reig, Mariano (1991): "La Ley de Ordenación Universitaria de 1943", in CARRERAS ARES, Juan José and RUIZ CARNICER, Miguel Ángel (eds.): *La universidad española bajo el régimen de Franco (1939–1975)*, Zaragoza, Institución Fernando el Católico: 125–158.

Rodríguez-López, Carolina (2002): *La Universidad de Madrid en el primer franquismo: ruptura y continuidad (1939–1951)*, Madrid, Dykinson.

Rodríguez-López, Carolina (2008a): "*Extirpar de raíz.* La depuración del personal docente universitario durante el franquismo. Los catedráticos de las facultades de Derecho", in FERNÁNDEZ-CREHUET LÓPEZ, Federico and ESPAÑA, Antonio Miguel (Hg.): *Franquismus und Salazarismus: Legitimation durch Diktatur*, Frankfurt am Main, Vittorio Klostermann: 61–99.

Rodríguez-López, Carolina (2008b): "Poder académico: los dirigentes universitarios en la España del primer franquismo (1939–1951). Una aproximación", in GHEDA Piero et al. (eds.): *La storia delle università alle soglie del XXI* secolo, Bolonia, CLUEB:199–210.

Rodríguez-López, Carolina (2015): *Paisajes de una guerra: la Ciudad Universitaria de Madrid,* Madrid, Servicio de Publicaciones de la Universidad Complutense de Madrid.

Rodríguez-López, Carolina et al. (2015): "De campos de Marte a palacios de Minerva: la reconstrucción de la Ciudad Universitaria de Madrid", in MOMOITIO, Iratxe and NÚÑEZ MONASTERIO Ana Teresa: *La reconstrucción del patrimonio después de una guerra*, Gernika, Fundación Museo de la Paz: 87–102.

Rodríguez-López, Carolina (2016): "*Estando muertos todavía hablan.* La Universidad de Madrid en el primer franquismo", *Ayer*, 101: 105–130.

Rodríguez-López, Carolina (2018): "Ruina y reconstrucción: La Ciudad Universitaria en la posguerra", in RODRÍGUEZ LÓPEZ, Carolina and MUÑOZ HERNÁNDEZ, Jara (eds): *Hacia el centenario: la Ciudad Universitaria de Madrid a sus 90 años*, Madrid, Ediciones Complutense: 283–314.

Ros, Samuel (1940): *A hombros de la Falange*, Madrid, Ediciones Patria.

Sánchez de Lerín, Teresa (2000): *Modesto López Otero. Vida y obra*, Madrid, Universidad Politécnica de Madrid.

s.a. (1942): "Una estatua ecuestre del Caudillo", *Revista Nacional de Educación*, 6: 95–96.

s.a. (1943): "La más bella Ciudad Universitaria del mundo, inaugurada", *Revista Nacional de Educación*, 34: 103–128.

s.a. (s.f.) *Una visita a la Ciudad Universitaria de Madrid*, Madrid.

Sanz Rodriguez, Pedro (1978): *Testimonio y recuerdos*, Barcelona, Planeta.

CHAPTER 5

Epistemic Communities and Science Makers in the Franco Regime: A Study of the Nuclear Energy Board

Albert Presas i Puig

'Technology serves politics
and politics serves technology'.
José M. Otero Navascués, 1969 (Otero (1969): 8)

Introduction

When the British economist Walter Bagehot published his essay 'Physics and Politics' in 1872,[1] he intended to demonstrate the effects of advances in physics and their transformation into technical inventions (railways, telegraphs) by the 'two old sciences' (politics and economy). Bagehot, member of the British historical institutionalist tradition, was particularly

[1] Bagehot (1872): 5.

A. Presas i Puig (✉)
Department of Humanities, Universitat Pompeu Fabra, Barcelona, Spain
e-mail: albert.presas@upf.edu

M. Janué i Miret, A. Presas i Puig (eds.), *Science, Culture and National Identity in Francoist Spain, 1939–1959*,
https://doi.org/10.1007/978-3-030-58646-1_5

concerned with why progress in science and technology in the final decades of the nineteenth century was not accompanied by similar progress in political bodies and public administration. He was interested in the effects of physics as a scientific discipline on the state's political activity. It was during this time that scientific activity became of definitive public interest; the state began to organise and fund scientific research with a clearly defined political purpose (e.g., on the grounds of what was understood as national identity, and economic and social progress). This meant defining science policies to secure conditions for scientific development, with the consequent incorporation of science into all areas of the productive economy (technical chemistry, electrotechnical industry, technical optics, telephony, etc.). At that time, the importance of knowledge in general, and science and technology in particular—both regarded as wealth generators—defined the interaction with institutionalised state power. This process involved an increase in bureaucratisation and, above all, the participation of experts in the design of public policies. Although the contexts differed, this process was common to many countries, irrespective of their political organisation, and it is within this tradition that we situate our article on Franco's Spain and the development of its nuclear programme.

One of the historiographic resources available to study the scientific and technological developments during the Franco regime is an analysis of the role played by its policy institutions. This text will focus on the constitution, development, and action of the Spanish Nuclear Energy Board (Junta de Energía Nuclear, JEN), created in 1951 to promote nuclear development and the consequent standardisation of Spanish science and technology. JEN was responsible for the development of the Spanish nuclear energy programme and a key strategic, institutional, and political element to understand most of the scientific, technological, and industrial development in Spain at that time. Further, it provided insight into key aspects of the Franco regime's economic and political history and international relations. While this text obviously cannot address all these aspects, we intend to create a proposal for their understanding. To facilitate subsequent comparative works, we begin with two theoretical propositions: first, the consideration of JEN as an epistemic community and, second, the actions of those whom we call 'science makers'. The historical development of the arguments presented is defined by Mitchell Ash's thesis on the relationship of interest and mutual convenience that is established between science and politics.[2]

Epistemic Communities

In the early 1990s, Peter Haas analysed the role of what he defined as epistemic communities in the definition and coordination of interstate policies.[3] As epistemic communities, Haas considered a network of professionals recognised as having expertise relevant to political implementation in a specific domain or subject area. According to Haas, due to increasing technification, policymakers had to make important decisions in an increasingly complex reality. Consequently, the advice was required from expert groups to make decisions in areas relevant to this new political reality (economy, society, energy, international relations, etc.). These expert groups formed what Haas defined as epistemic communities, in which the members share knowledge, as well as normative principles, causal beliefs, and a common understanding of group interests. Through their experience, experts disseminate specific values and knowledge, and can influence policy outcomes. The knowledge and expertise associated with such epistemic communities (self-)empower them to define norms, values, and strategies, and thereby influence government policies and outcomes.[4]

Although the idea of epistemic communities was developed to explain the role of experts in the management of international policies, the object of study is often the transfer of knowledge between different epistemic communities in the same country.[5] Considering that the scientific and technological policies in modern states are associated with scientific discourse, epistemic communities can influence the robustness of the options to be considered. However, the boundary between scientific and technological issues, on the one hand, and political issues, on the other, is permeable.[6] Haas analysed the mechanisms by which epistemic communities influence this process, arguing that expert communities control and channel information to decision-makers, and by those means determine patterns in state behaviour. Upon defining the existence and function of

[2] Ash (2002); also Ash in this volume.

[3] Haas (1992); Radaelli (1999). For the debate on epistemic communities, see Mai'a, Cross (2013).

[4] Criticisms on epistemic communities in Lidskog; Sundqvist (2002).

[5] Albert; Laberge (2007).

[6] Haas (1992): 11.

epistemic communities, Haas incorporated an element that deserves to be highlighted: ideas would be sterile without carriers.[7] The proposal of epistemic communities highlights the importance of these actors who can define and manage complex problems—particularly in those early stages of policy design when the uncertainty is maximal. In this conceptualisation of knowledge-generating processes, the experts who create knowledge are fundamental to political analysis.[8] Therefore, an epistemic community is a set of actors with professional recognition who make authoritative statements on issues considered politically and socially relevant. Moreover, the success of these actors depends not only on their epistemic resources but also on their skills at interacting with other political actors. If the epistemic community genuinely intends to influence political and social reality, it must generate this need for political advice and be able to interact positively with the different levels that define the state (politics, economy, industry, etc.).

The Science Maker, Generator of Ideas

The study of complex technological and scientific problems of advanced societies must consider the role of those actors who combine social skills with science and state administration goals. As advisers to government and industry, such individuals play a key role as generators of contemporary technological and scientific systems.[9] The development of scientific and technological research, particularly in the twentieth century, cannot be considered without these individuals who generate scientific policies and infrastructures, and, therefore, ultimately, provide knowledge.

In the late 1970s, Hugh Aitken focused on the constitution of scientific and technological systems and the role of actors who facilitate the exchange of information and knowledge required to maintain the complex technological system.[10] These actors share values, speak the 'same language', and act as translators between the different elements of the system. Aitken showed how the exchange of information requires individuals who can act as interfaces between the different levels. Therefore, the success of the system depends on these individuals' mediation and translation capacity.

[7] Haas (1992): 27.
[8] Radaelli (1995, 1997).
[9] See Needell (2000); Jo Nye (2004); Eckert (2006).
[10] Aitken (1976).

In our case, these actors serve as a driving force in the interaction between different elements and generate their own social network, which merges technology, society, business management, political action, economy, and so on. These actors we call science makers develop their role in a specific historical and social context. Their objectives are defined when they make a contingent interpretation of situations, institutional values, meanings, and power relations.[11] Like an entrepreneur, the science maker also has a special sense for perceiving an opportunity,[12] where opportunity is understood as a historical moment that is favourably valued to define a desirable and achievable future situation.[13]

Due to their creative nature, science makers frequently overcome the possible immediate deterministic institutional context. These contexts involve multiple actors with different skills. In this struggle between meanings, science makers ultimately impose the hegemony of their meaning, and it is then that the epistemic community they lead can impose itself. The recognition of science makers and, therefore, their epistemic community is based on a strong belief in his visionary ideas and communication strategy.

Science, Politics, and Mutual Mobilisation

We recall now the relationship already presented by Bagehot between science and politics. This has been a frequent topic among historians of science and technology, and there are several publications dedicated to the subject. Of all these publications, some of great relevance, we will highlight a text by Mitchell Ash for its potential and intrinsic interest in summarising a previous extensive work by the same author.[14] The title of the text is 'Wissenschaft und Politik als Ressourcen für einander', which can be translated as 'Science and Politics as Mutual Resources'.[15]

Mitchell Ash addresses the relationship between science and politics—not only how politics influences science, or even how science influences politics, but rather how the two serve and use each other as resources, and engage in mutual argument. The relationship between science and state,

[11] Jackson (2011).

[12] Shane; Venkataraman (2000).

[13] Stevenson; Jarillo (1990).

[14] See Ash in this volume.

[15] Ash (2002).

or science and politics, has often been understood as the subjection of science (in general, of knowledge) to the interests of politics and the state. The cases of dictatorial states (Nazi Germany, Soviet Union) are considered as examples. However, as Ash shows, this relationship cannot be considered unidirectionally. Scientists are also willing to get closer to the state to benefit from the resources it provides.[16] Ash presents a new form of discourse that addresses the relationship between science and politics in modern societies, proposing an extension of the concept of science beyond its cognitive content, as well as an extension of the concept of resource beyond its strictly economic and material meanings: mobilisable resources can be financial but also cognitive, artefactual, institutional, personal, rhetorical, or symbolic. On this basis, science is understood as a social subsystem that provides resources to other social subsystems. In turn, science appropriates or consumes resources from other subsystems of society. Considering science and politics as "mutually mobilisable resources", the initiative to bring the two entities together can also be taken by science and scientists, as they seek alliances with the state to define and conduct their agendas. In this sense, scientific development in modern societies involves the transformation or reorganisation of the mobilisable resources that science and politics provide for each other. Scientific autonomy and participation in political networks are by no means incompatible. Rather, the separation of the two functional systems (science and politics) is a significant division of the two fields to achieve common benefits.

Ash shows how these sets of resources are, in principle, bidirectionally mobilisable: scientists can mobilise and obtain resources from the political sphere for their interests in the same way as politicians can mobilise scientists and their skills or symbolic prestige for their interests. Alliances with institutional, but also ideological, bodies can be mobilised in this manner to provide resources to both parties. Throughout this process, scientists not only do not act as victims of certain political alliances or circumstances (for being forced by the state), but their recognised status as experts gives them a fundamental role as responsible acting subjects, which we can consider an epistemic community. As Ash indicates, this occurs in a complex social process of the increasing scientification of politics and increasing resources for science and technology. This involves both a scientification of politics and a politicisation of science. Through the new networks between science and politics, which are not established in a random or

[16] Ash (2002:32).

uncontrolled manner but were consciously built, increasingly scientific state institutional structures and research programmes have emerged from this effort by the state to achieve technocratic modernity. With these references and background, we will now address the case of nuclear science and technology in Spain during the Franco regime.

Nuclear Energy and JEN as an 'Epistemic Community'

Previous studies have argued that nuclear technology can provide important information on certain complex historical processes.[17] For this reason, the study of JEN as a driving force behind the Spanish nuclear programme allows us to analyse not only technology and science but also progress and economic development. Further, it helps us examine the political tensions and economic decisions that defined the process of social and economic modernisation in the final years of the Franco regime.[18] Considering the scope of this article, it is not possible to analyse all these possibilities. However, we will attempt to present the actions of certain actors and scientific groups that set priorities and strategies in energy, industrial, and technological development policies.

The Path to Nuclear Energy and the Creation of the Possibility

As in many other countries, the atomic explosions in Hiroshima and Nagasaki in August 1945 marked the first news that Spain had on the practical application of nuclear energy. The nuclear explosions in Japan were a demonstration of unlimited power and force: having the bomb meant having absolute strength. The fascination caused by this brutal manifestation of power had similar consequences in most countries, irrespective of their political systems. The expectations that arose from this first military application of nuclear energy led to numerous military and civilian discourses and rhetoric.[19] As in most countries,[20] the assessment of nuclear energy by the Spanish authorities had a military nature.[21] Although technical and economic difficulties were anticipated, the political benefits

[17] Anshelm (2010).

[18] See Rubio-Varas; De la Torre (2017).

[19] Eckert; Osietzki (1989), 248.

[20] See the materials developed in the HoNESt project (History of nuclear energy and society) (honest2020.eu).

[21] Mundo, VI, 276, 19.08.1945, p. 605.

of having nuclear weapons were considered by Spain—at the time internationally isolated due to its alliance with Nazi Germany and felt threatened by the new international order that had emerged from World War II.

Associated with a new form of large-scale scientific organisation known as Big Science, the atomic bomb illustrated the political and military importance of the new possibilities offered by science and technological development.[22] The rhetoric of the new Spanish authorities expressed the desire for an industrial complex that could provide military power for the defence of the country. This policy led to the creation of the National Institute of Industry (Instituto Nacional de Industria—INI).[23] Further, the subsequent commitment to nuclear energy was included in this rhetoric. With this, science was fully incorporated into the politics of the Franco regime.

The confirmation that Spain had uranium deposits was crucial to its decision to gain access to the new energy source.[24] From a strategic perspective, owning uranium meant owning an element with extraordinary exchange value, as, until then, only the United States and Great Britain had had access to uranium ore, which they sourced from the Congo.[25] The Spanish uranium deposits would thus prove an advantage to the Francoist authorities in their efforts to access the new technology.[26] Aware of the situation and in a hasty response, in October 1945, the government decreed its exclusive rights to exploit the uranium deposits on Spanish soil. The Commission for Nuclear Studies was developed in 1948 under the military authority of Carrero Blanco, who would become admiral in the Navy and was a trusted person of Franco.[27] The ambition to have access to nuclear energy was in line with the regime's industrialisation and economic development strategy. Considering the isolation imposed by the West on the Franco regime, industrial development policy was aimed at autonomously allowing national development.[28]

[22] *Galison*; Hevly (1992).

[23] San Román (1999).

[24] *El Noticiero Universal*, 10.08.1945, 3.

[25] Mongin (1997), 47.

[26] 'Prospecting for Uranium' on world deposits published by the United States Atomic Energy Commission (Washington, 1957) made no reference to Spanish deposits.

[27] Sabá (1991): 112.

[28] *Moya* (1975): 216. This policy is often described as autarchic, antagonistic to the one promoted by the new cadres of the Spanish administration at the end of the 1950s, which was considered an opening up and economic liberalisation. It must be said that, in general,

The Beginnings of Nuclear Research in Spain

After several conferences on Spain's nuclear possibilities, in late 1945, the Geological and Mining Institute began to explore possible uranium deposits throughout the Spanish territory.[29] Simultaneously, the Navy began to analyse all the available information on nuclear energy,[30] led by navy artillery engineer José María Otero Navascués (1907–1983), who would emerge as the main actor of Spanish nuclear development, particularly in the areas of scientific research and technological development. Otero's affiliation with the Navy explains why this department was tasked with overseeing this approximation work. Despite the favourable circumstances, it was, in fact, a fortuitous event that began the race for nuclear technology. In 1948, Otero had a casual meeting with Italian scientists who were interested in acquiring Spanish uranium ore in exchange for training Spanish scientists in their research centres; that is, exchanging uranium ore for knowledge. Otero convinced his military superiors of the convenience of such an agreement with the Italians.[31] From this point onwards, there was a flurry of hectic activity. In the summer of 1948, a Study Commission was established to focus on the training of personnel, the analysis of national uranium samples, and, in the medium term, the construction of a natural uranium and heavy water reactor. In September 1948, the legal and financial framework was established with the creation of the private company Studies and Patents for Special Alloys (Estudios y Patentes para Aleaciones Especiales—EPALE), which was later renamed the Board for Atomic Research (EPALE-JIA; Junta de Investigaciones Atómicas—JIA) and placed under the Presidency of the government, that is, of Carrero Blanco.[32] The tasks assigned to the new EPALE-JIA included personnel training, prospecting and mining, uranium extraction and subsequent treatment, research in physics and instrumentation, design of chemical and metallurgical plants, procurement of heavy water, construction of

the term 'economic autarchy' has been attributed to the Spanish case without comparative considerations with other countries. Very useful in this regard are iNash (1999) and Edgerton (2007). Both authors identify 'autarchic' practices not only in the fascist regimes, but also in countries such as Great Britain, France, and the United States, thereby specifying the term 'autarchy' and its use.

[29] BOE, 4 October 1945, 278, p. 2133.

[30] Otero (1957a): 216–23.

[31] Otero (1957a): 218; Villena (1984).

[32] *Journal of Applied Sciences*, Vol. 34, 1953, 458.

reactors, and production of isotopes for medicines.[33] At the end of 1951, Spain created the Nuclear Energy Board (JEN) based on those previous structures and institutions, and, with this, the first phase of the Spanish nuclear programme came to an end. The new JEN, already publicly recognised, would supervise all aspects of atomic energy and advise the government and companies interested in its development.[34] A small group of young scientists formed around Otero in the late 1940s and responsible for gathering information, can be considered the first Spanish group dedicated to nuclear issues.[35] Moreover, JEN would lead the first stages of the Spanish nuclear programme, thus beginning to take shape as an epistemic community.

José María Otero-Navascués, Science Maker of the Spanish Nuclear Programme

José María Otero Navascués (1907–1983) is among the most important figures in the scientific and technological development of Spain in the second half of the twentieth century, particularly for his contribution to the development of nuclear energy. Born in Madrid, he was a Navy artillery engineer and became a very powerful figure within the scientific and technological research system—sectors that he helped develop by following international management models. Otero was the promoter of JEN and, until his last days, its figure of reference. In the late 1920s, after completing his military training, Otero turned to physics. In Madrid he met the Swiss physicist, Paul Scherrer, and followed him to ETH Zurich to work with him, later moving to Berlin to work in optics. During this time, Otero visited the laboratories of the German technical optics company Zeiss, and its military subsidiary, Nedinsco, which provided him with first-hand information on how to relate academic research with industrial applications. These experiences gave Otero first-hand knowledge of what it meant to 'do science' in advanced countries, something unknown to Spanish science and industry at that time. After the Spanish Civil War (1936–1939), Otero, who was on the side of the coup army, was appointed secretary of the physics institute at the recently founded Spanish National Research Council (Consejo Superior de Investigaciones

[33] Otero (1957b): 3, 14–38.
[34] *Journal of Applied Sciences*, Vol. 22, 1951, 449–50, and Vol. 50, 1956, 269–71.
[35] Ordoñez; Sánchez Ron (1996).

Científicas—CSIC).[36] Gradually, Otero became the figure that defined strategies and priorities regarding physics and many related technical disciplines.[37] Otero used his advantage of having many international contacts to reinforce his position within Spanish science. Further, he became one of the most prominent figures in the Franco regime's strategy to overcome international isolation through the role of representation made by its scientists. Thus, Otero was a true defender of the Franco regime, which he represented on countless occasions in international forums.

Otero had a special relationship with the German scientific community, including physicists such as Werner Heisenberg and Karl Wirtz, with whom he established strong personal friendships. Wirtz was the bastion of the German nuclear programme[38] and played a fundamental role as an adviser to the Spanish authorities on nuclear energy issues and in the training programme for JEN personnel. Following Otero's strategy, and along with others who would move to the United States and Great Britain, the first young Spanish group was formed in Germany under the guidance of Heisenberg and Wirtz. These young scientists would later become responsible for Spanish nuclear research playing a significant role in technological research. This was the beginning of JEN's personnel training policy that defined much of its strategy on nuclear research. From its creation to its transformation into the Centre for Energy, Environment and Technology (CIEMAT) in 1986, JEN made great efforts for the development of nuclear technology. The most important achievements of its research on reactors were the assembly of the JEN 1 reactor in the context of the 'Atoms for Peace' programme and the development of the ARGOS, ARBI, and JEN II reactors and the CORAL fast reactor. Moreover, JEN guided nuclear industrial development, launching several activities, including ore extraction and refining, fuel production at a laboratory scale, isotope separation, production and extraction of plutonium, nuclear waste treatment, and nuclear instrumentation. In addition, the nuclear industry was the first to incorporate international norms and standards, a practice that was unknown in the Spanish industry at that time.[39] For the first time, Spain was among the group of 'qualified importers'; that is, technology-importing countries that could participate in a very high percentage of

[36] See Antolín's chapter in this volume.
[37] Caro et al. (1995): 62.
[38] Presas i Puig (2000).
[39] Presas i Puig (2007a).

nuclear technology developments.[40] Spanish nuclear research enjoyed a privileged position among all the countries of the world, as a result of which the regime was included in international scientific and technological forums. As the industrial implementation of this energy option was a capital-intensive process, it was present in international economic and funding forums, which slowly brought Spain closer to the international centres of political and economic decision-making.

Science, Politics, and Mutual Mobilisation: JEN as a Revitaliser of Spanish Physics

In August 1952 Otero gave a lecture on the situation of Spanish research at the time that JEN was established. Otero, who already had enormous influence at that time, complained about the state of Spanish research in general and basic research in particular, pointing out the prevailing scepticism towards research in the industry and government administration.[41] In his strategy as a science maker, Otero managed to make nuclear energy a priority for the Francoist authorities from a scientific, technological, and industrial perspective. JEN, as the organisation in charge of this task, monopolised the resources dedicated to research and technological development, to the detriment of other sectors. JEN's importance to the state can be observed by comparing its budgets with those of the CSIC. From 1948 to 1951, EPALE, JEN's precursor, had an annual budget of 18 million pesetas, which subsequently increased from 37 million in 1952 to 92 million in 1955. In 1957, it had 231 million, and in 1960, 293 million. In 1963, the budget reached 453 million.[42] A report by Wirtz for the German government reported a budget of 800 million for 1968.[43] In comparison, according to the corresponding Memoirs, the CSIC had a budget of almost 57 million in 1949, 67 million in 1951, 155 million in 1960, and 112 million in 1962. Therefore, for this final year, JEN had quadrupled its resources with respect to the CSIC.[44]

[40] Izquierdo (1998): 16.

[41] Presas i Puig (2000).

[42] Roca Rosell; Sánchez Ron (1990): 309.

[43] Wirtz (1968).

[44] Memoria CSIC 1949 (1951): 442; Memoria CSIC 1951 (1951): 466; Memoria CSIC 1952–54 (1958): 241; Memoria CSIC 1958 (1960): 424; Memoria CSIC 1960 (1963): 448; Memoria CSIC 1962 (1963): 468.

From this position of power and as part of his strategy to create a solid group that could act as an epistemic community, Otero designed a support policy for science faculties through financial assistance which enabled a basic infrastructure that did not yet exist in Spanish universities. This resulted in the first generation of physicists in the history of Spain who could dedicate their time exclusively to research.[45] The discretionary nature of the grants, which depended on Otero, gave him enormous power over research and the allocation of university positions. The researchers who held teaching positions in theoretical, atomic, and nuclear physics and in engineering related to nuclear technology all came from groups supported by JEN, that is, controlled by Otero.[46] In fact, Otero's actions in Spanish physics and technology were no exception. The scientific policies of the Franco government were a response to personal impositions or power relations, which impaired the implementation of efficient scientific policies.[47] All this was aggravated by, or a consequence of, the lack of a scientific tradition that could assist in the definition of objectives, as well as by the consequent absence of counter-figures that would require a debate on subsequent priorities and strategies.[48] This cannot be explained by the dictatorial character of the Franco regime alone, since other dictatorships underwent remarkable scientific and technical developments.[49] In our opinion, it was due to the lack of a recognised and solid tradition of 'doing science'. However, despite all the difficulties involving improvised strategies, physics was one of the scientific disciplines to most rapidly attain international standards, and once established, defined working methods and strategies for other developments and areas of knowledge.[50]

A crucial element in JEN's objectives was the Institute for Nuclear Studies (Instituto de Estudios Nucleares—IEN) dedicated to the training of personnel to oversee the development of this energy source, as well as related disciplines. The creation of the institute appeared in the 1964 nuclear energy law, and Otero, who was already president of JEN, coordinated the research and teaching related to nuclear energy, which was the culmination of his aspirations. The creation of the IEN was justified by the growing need for training in the new disciplines and the limited capacity

[45] See Pacheco's chapter in this volume.

[46] Velarde (2016): 178.

[47] See Antolín and Pacheco in this volume.

[48] Presas i Puig (2007a).

[49] Graham (2004). See also Rürup; Schieder (2000); Heilbron (2003); Kojevnikov (2004).

[50] Presas i Puig (2007a).

of Spanish universities up to that time in terms of their traditions, resources, and personnel. The model proposed for the IEN was based on MIT.[51] The board of the institute included representatives from JEN, universities, industry, and government administration, with Otero, once again, as the strong man of the group. The IEN played a crucial role both in the development of research and in the creation of groups and schools that would later institutionalise university research and subsequently advise the state on its nuclear programme.

Otero's Relationship with the Spanish Nuclear Industry

In his actions as a science maker and as a supporter of JEN as an epistemic community, Otero was aware of economic and industrial groups' reluctance and mistrust of Spanish nuclear research.[52] To decide what type of reactor would be used in future nuclear power plants, Otero designed a strategy to convince the Spanish authorities to opt for autonomous technology as much as possible. This was in response to the economic and industrial sectors' apparent desire to import all the necessary material and technology, without considering the possibility of developing its own. To import the technology and to disregard possible autonomous developments would spell the end for JEN.[53] In 1958, after overcoming most obstacles to the start of the nuclear programme, Otero continued complaining, and not without some bitterness, of the dependence on foreign countries and an inappropriate industrial strategy. His claim was that the industry did not trust research and that 'the country cannot be industrialised with borrowed equipment and borrowed ideas'.[54] In 1958, Otero warned about the lack of a scientific tradition in the country, and, simultaneously, rejected the excuse that there was a lack of subsidies: 'whether one can do research work or not is not just a question of money. It is an attitude. [...] There is undoubtedly a mistrust in the industry, especially in the manufacturing industry, which is almost proverbial, but it is necessary

[51] Sánchez Ron; Romero (2001).

[52] Letter of 8 July 1955 from Otero to Wirtz (LGAK, GLA Abt. 69/KfK INR, Zug Oct. 1995, Nr. 52).

[53] Letter of 26 April 1960 from Otero to Wirtz (LGAK, GLA Abt. 69/KfK INR, Zug Oct. 1995, Nr. 52).

[54] Otero (1957b): 35.

to stop this mistrust'.[55] Those responsible for the regime shared his views. General Suanzes, the president of the INI, showed similar concerns but emphasised a different aspect: 'Our basic industry has no incentive or desire to investigate on its own, nor does it particularly trust other centres in the country, which have no traditions or appropriate financial resources. This is perfectly logical and understandable, and the gradual restoration of external normality represents a setback for us in our adopted course. To do research on one's own account is very costly, and one's own research and that of others carried out in the country, if not supported by experience or a known name, is logically considered dangerous and also inconvenient'.[56] Within the INI, there were discussions among those in favour of importing the necessary technology, particularly from the United States,[57] and those who, like Otero, considered the necessity to develop their own technology as much as possible and to train their own personnel, thus preventing an excessive dependence on other countries that could hinder Spain's future development.[58]

Although JEN was the driving force, it was understood that efforts to develop nuclear energy production should be made by private companies. However, the electricity companies did not trust JEN as an adviser for the government due to its attempt to determine what industrial options would be followed. These companies also limited the funding for JEN's technological development programme until the government had ensured the viability of the nuclear business and that state actions (whether via JEN or INI) would not threaten their exclusive production and distribution of electricity of nuclear origin. This tension continued until the approval of the 1964 nuclear energy law, which already reflected the government's new economic policy. It redefined JEN's responsibilities and transferred the industrial planning, safety, and administrative authorisations to the Ministry of Industry. With this, JEN passed from the Presidency of the Government to the Ministry of Industry, as a delegated body without the authority to impose.[59] This redefinition of functions removed JEN's authority to intervene and was a consequence of the access to government

[55] Otero (1957b): 33–34.

[56] *Journal of Applied Sciences*, Vol. 43, March–April 1955, 159.

[57] Letter from Otero to Wirtz, 8 July 1955 (LGAK, GLA Abt. 69/KfK INR, Zug Oct. 1995, Nr. 52).

[58] For a summary of the nuclear development project, see Otero (1957a).

[59] Garrués-Irurzun; Rubio-Mondéjar (2018).

of the new 'technocratic' cadres,[60] who imposed a policy of economic and market liberation that granted maximum business guarantees to electricity companies. From 1964, and especially when the first nuclear power plants commenced operation, JEN lost strength to the electricity companies as a government adviser and its activities were limited to the training of personnel and provision of non-binding advice. The tensions that led to the state's new nuclear policy illustrate the aforementioned interactions and the transformation of meanings concerning political decisions on this energy source.

Despite these technical, political, and economic difficulties, JEN could continue its work in the development of nuclear technology. While JEN advanced in the field of nuclear energy, the Spanish industry commenced a modernisation process in the 1950s that, although it differed between areas, was highly significant in some cases. Under the supervision of JEN and with the support of the industry, the construction of nuclear plants was initiated, although the first plants, Zorita, Santa María de la Garoña, and Vandellós I, had little contribution from the Spanish industry. In 1969, the first national electricity plan was created and, with it, the decision to import six plants from Westinghouse: Almaraz I and II, Lemoniz I and II, and Ascó I and II. Another energy plan predicted the construction of five new plants. In both cases, there was a greater contribution from the Spanish industry, in some cases as much as 80%. The energy crisis of 1973/74 resulted in reconsideration of the energy field, which gave a new impetus to nuclear power. With Franco's death in 1975, democracy returned to Spain. The newly formed conservative government continued with the pre-existing plans and approved the construction of five new nuclear plants: Valdecaballeros I and II, Vandellós II, and Trillo I and II. In this case, the Spanish industry made a contribution of 90%. In the late 1970s, however, both national and international political pressures regarding specific areas of nuclear research became more intense, leading to specific projects being suspended. In 1983, the new socialist administration looked sceptically at the nuclear option and reviewed the old energy plans for the country. Consequently, new projects were not approved and projects that had not yet been completed were suspended. In 1992, the Spanish Congress, with a socialist government, decided that

[60] Technocrats refer to the reorganisation of the state administration, following technical and efficiency criteria. This should not be confused with the notion of technological promotion.

'no new nuclear power plants will start operating in the next decade'. With this, the history of the expansion of the Spanish nuclear programme came to an end.

Conclusion

The reaction of the Franco regime to nuclear energy is illustrated by the creation of JEN and corresponded to the initiatives of most surrounding countries. In this chapter, we have considered the creation and actions of JEN following the model of an epistemic community, since it aimed to lead the development of this energy option until its industrial implementation. Considering the national and international context and regardless of the particularities of Francoism, the strategy followed the model used in other countries regarding the challenge posed by this new energy source. This model changed international policy and had enormous economic importance in the Spain of the time and in its international relations; however, it also had vital consequences for the creation of a scientific and technological research network. In a country with no scientific or technological tradition and without industrial and academic networks to support new developments, this dynamic was begun by highly skilled actors with a special sense of opportunity. This is the role we attribute to José María Otero Navascués, whom we present as a knowledge production system generator and science maker. Following Ash's proposal, we understand science and its relationship with the state are understood as sets of interacting resources. The systems that support science and culture are not considered separate from scientific policy, but an integral part of politics in general, providing the latter with the necessary resources for its legitimisation. Based on this consideration of science and politics as resources for mutual mobilisation, Otero's actions assist in the analysis of the relationships between the industrial development policy of the regime (translated into strong economic and institutional support for nuclear energy) and his actions of offering the regime symbolic resources of modernity, technological and scientific development, and, as important as the these, connections with the most advanced scientific centres in the world. This was all done at a time when Francoism needed to integrate in the new international scene of the 1950s. With his extensive international activity, Otero not only promoted the regime in international organisations, but, through JEN as an epistemic community, contributed to the creation of a transnational network aimed at influencing transnational political actors,[61] thus ensuring that Spain participated in international forums in a subject of such relevance as energy. Considering JEN as an epistemic community helps to understand not only the actions of some of the actors who defined the Spanish nuclear development policy, but also the functional changes that resulted from changes in both national and international policies during the opening of the regime. In response to the new conditioning factors, economic strategies and international politics were modified during this process. As part of this new scientific political framework, JEN's original purpose to serve as the main actor in the nuclear programme changed when the political context drastically changed and, with it, its significance in the Franco regime's economic and technological development policy. At the time of its industrial implementation, nuclear energy acquired a new relevance in the new Spanish economic integration policy proposed by the new cadres of the administration, the so-called technocrats. The importation of technology, particularly American technology, to the detriment of the autonomous development proposed by Otero, would allow Spain to rapidly integrate in international credit markets, while establishing relations with exporting companies and thus opening up the possibility of sharing businesses in the country. All of this was designed to reinforce international acceptance of the Franco regime.

Due to the interdependence between politics, economics, and technology, the scientific expertise of an epistemic community is not the only source of knowledge on which to base political actions. In fact, the compartmentalisation of knowledge often prevents the use of expertise as a legitimising tool for topics that cover more than one political area. JEN is an example of this. The new state administration modified the status quo and incorporated new priorities (incorporation into international financing systems, becoming closer to Europe), opting once more to reaffirm and protect the interests of Spanish capital in the development of nuclear energy. Therefore, JEN and, with it, Otero, were reduced to mere companions and were no longer strategists. Despite this relocation of JEN and, therefore, of Otero's project, the actions of this science maker and this epistemic community greatly contributed to the establishment of a technological research and development system that was comparable to those of surrounding countries. Moreover, the regime greatly benefited

[61] Stone (2007).

from Spain's representation through JEN which allowed it to participate once more in international forums, not only scientific or technological but also economic and political.

References

Literature

Newspapers

El Noticiero Universal, 10.08.1945
Mundo, VI, 276, 19.08.1945.
Journal of Applied Sciences, vol. 34, 1953, 458.
Journal of Applied Sciences, vol. 22, 1951, 449–50, and vol. 50, 1956, 269–71.

Archive

Generallandesarchiv in Karlsruhe LGAK.

Aitken, Hugh G.J. (1976) *Syntony and Spark. The Origins of Radio,* New York, London: Wiley.

Albert, M.; Laberge, S. (2007) The legitimation and dissemination processes of the innovation system approach—The case of the Canadian and Quebec science and technology policy", *Science Technology and Human Values,* vol. 32, no. 2, 221–249.

Anshelm, Jonas (2010), "Among demons and wizards: the nuclear energy discourse in Sweden and the reenchantment of the world". *Bulletin of Science, Technology & Society,* vol. 30, no. 1, 43–53.

Ash, Mitchell (2002), Wissenschaft und Politik als Ressourcen für einander". In Rüdiger Vom Bruch, Brigitte Kaderas (Hrsg.), *Wissenschaften und Wissenschaftspolitik. Bestandsaufnahmen su Formationen, Brüchen und Kontinuitäten im Deutschland des 20. Jahrhunderts,* Stuttgart: Steiner.

Bagehot, Walter (1872, 1999), *Physics and Politics or Thoughts on the Application of the Principles of "Natural Selection" and "Inheritance" to Political Society,* Chicago: Ivan R. Dee Publisher

Caro, R., et al. (eds.) (1995), *Historia Nuclear de España,* Madrid: Sociedad Nuclear Española.

Eckert, M.; Osietzki, M. (1989), *Wissenschaft für Macht und Markt: Kernforschung und Mikroelektronik* in *der Bundesrepublik Deutschland,* München, Beck.

Eckert, Michael (2006), *The Dawn of Fluid Dynamics: A Discipline Between Science and Technology,* Weinheim: Wiley-VCH.

Edgerton, David (2007), *The Warfare State: Britain, 1920-1970*, Cambridge: Cambridge University Press.
Galison, Peter; Hevly, Bruce (eds.) (1992), *Big Science: The Growth of Large-Scale Research,* Stanford, Stanford University Press.
Garrués-Irurzun, Josean; Rubio-Mondéjar, Juan A. (2018), Between the Entrepreneurial State and the Regulatory State. The role of private interests in the first Spanish nuclear programme (c. 1951–1964). *FEG Working Paper Series* 01/18 (Faculty of Economics and Business, University of Granada, 2018).
Graham, Loren R. (2004), Science in Russia and the Soviet Union. A Short History, Cambridge: Cambridge University Press.
Haas, P.M. (1992), "Introduction: epistemic communities and international policy coordination". *International Organization*, vol. 46, no. 1, winter, 1–35.
Heilbron, J.L. (2003), History of Science, In J.L. Heilbron (Ed.), The Oxford Companion to the History of Modern Science, New York: Oxford University Press, 370–374.
Izquierdo, L. (1998). El *IEN* (IEE) y las ciencias y técnicas nucleares en *España. Revista SNE,* junio, *15–18.*
Jackson, Gregory (2011), "Actors and institutions". In Morgan Glenn et al. (Eds.), *Oxford Handbook of Comparative Institutional Analysis,* Oxford: Oxford University Press.
Kojevnikov, A. (2004), Stali's Great Science. The Times and Adventures of Soviet Physicists (London: Imperial College Press.
Lidskog, R.; Sundqvist, G. (2002), "The role of science in environmental regimes: the case of LRTAP", *European Journal of International Relations,* vol. 8, no. 1, 77–101.
Mai'a K.; Cross D. (2013), "Re-thinking epistemic communities twenty years later", *Review of International Studies*, vol. 39, no. 1, January, 137–160.
Memoria CSIC 1949, 1951, 1952–54, 1958, 1960, 1962, Madrid.
Mongin, D. (1997), *The French Atomic Bomb*, 1945–1958, Paris: LGDJ.
Moya, Carlos (*1975),* El poder económico en España, *1939–1970, Madrid: Tucar Ediciones.*
Nash, G.D. (1999), *The Federal Landscape: An Economic History of the Twentieth-Century West*, Tucson: University of Arizona Press.
Needell, A.A. (2000), *Science, Cold War and the American State: Lloyd V. Berkner and the Balance of Professional Ideals,* Amsterdam: Routledge.
Nye, Mary Jo (2004), *Blackett: Physics, War, and Politics in the Twentieth Century,* Cambridge, Mass.: Harvard Univ. Press.
Ordoñez, Javier; Sánchez Ron, J.M. (1996), "Nuclear energy in Spain: from Hiroshima to the sixties". In Paul Forman, and Sánchez Ron, José Manuel (Eds.), *Natural Military Establishments and the Advancement of Science and Technology. Studies in the 20th Century History* (Boston: Kluwer Academic Publishers), 185–213.

Otero Navascués, (1957a), "Programa español de energía atómica," DYNA, 4, abríl, pp. 216–23.
Otero Navascués, J.M. (1957b), "Hacia una industria nuclear". Energía Nuclear, vol. 1, July-September, 3, 14–38.
Otero Navascués, J.M. (1969), "Presentación". In B. Goldschmidt, Las rivalidades atómicas, 1939-1968, Madrid, JEN.
Presas i Puig, A. (2007a). Las ciencias físicas durante el primer franquismo. In M. A. Puig-Samper Mulero (Ed.), Tiempos de investigación: JAE-CSIC, cien años de ciencia en España, Madrid: CSIC, 299–303.
Presas i Puig, A. (2007b) The dream of a reactor: the DON Project. Methodological reflections on a technology development project during Francoism. Berlin: MPIWG.
Presas i Puig, A. (2000), "La correspondencia entre José M. Otero Navascués y Karl Wirtz, un episodio de las relaciones internacionales de la Junta de Energía Nuclear". *Arbor*, Vols. 659–660, November-December, 527–601.
Radaelli, C.M. (1995), "The role of knowledge in the policy process", *Journal of European Public Policy*, vol. 2, no. 2, 1995, 159–183.
Radaelli, C.M. (1997), *The Politics of Corporate Taxation in the EU: Knowledge and International Policy Agendas*, London: Routledge.
Radaelli, C.M. (1999), *Technocracy in the European Union*, London: Longman.
Roca Rosell, A.; Sánchez Ron, J.M. (1990) "Esteban Terradas. Ciencia y Tecnologia en la España contemporánea, Barcelona: INTA-Serbal.
Rubio-Varas, Mar; Joseba de la Torre (eds.) (2017), *The Economic History of Nuclear Energy in Spain*, London, Palgrave.
Rürup, Reinhard; Wolfgang Schieder (Hrsg) (2000), Geschichte der Kaiser-Wilhelm-Gesellschaft im Nationalsozialismus, Göttingen: Wallstein Verlag.
Sabá, K. (1991), "Spain's nuclear and non-proliferation policy". In H. Müller (Ed.), *How Western European Nuclear Policy Is Made. Deciding on the Atom*, London: Macmillan.
Sánchez Ron, José Manuel; Romero de Pablos, Ana (2001), Energía nuclear en España*: de la JEN al CIEMAT*, Madrid: Ministerio de Ciencia y Tecnologia.
San Román, Elena (1999) *Ejército e industria: el nacimiento del INI*. Barcelona, Crítica.
Shane, Scott; Venkataraman, S. (2000), "The promise of entrepreneurship as a field of research". *The Academy of Management Review*, vol. 25, no. 1, January, 217–226.
Stevenson, H.H.; Jarillo, J.C. (1990), A paradigm of entrepreneurship: entrepreneurial management, *Strategic Management Journal*, vol. 11, 17–27.
Stone, Diane (2007) "Transfer agents and global networks in the "transnationalisation" of policy". *Journal of European Public Policy*, vol. 11, no. 3, 545–566.
United States Atomic Energy Commission (1957), Prospecting for Uranium, Washington.

Velarde, Guillermo (2016). Projecto Islero (Córdoba: Guadalmazán, 2016).
Villena, L. (1984), "José María Otero Navascués (1907–1983)", *Pure and Applied Optics*, vol. 17, 1–12.
Wirtz, Karl (1968). The Atomic Energy in Spain. Confidential Report. 18 September 1968, Werner-Heisenberg-Archive of the Max Planck Institute for Physics in Munich.

CHAPTER 6

Science and Technology in the Nationalist Debate in Catalonia after the Civil War

Antoni Roca-Rosell

Introduction

This work addresses the ideological aspects of the relationship between science, technology, and nation in Catalonia in the early years of Francoism.[1] Given the dictatorship scenario, the public debate in Catalonia about its national identity disappeared and was instead withdrawn into the private sphere. Naturally, the debate continued in exile, and became perhaps even richer since it was integrated into the fight to recover democracy in Spain.

Catalonia represents a unique case as it was a stateless nation, as has been studied by several authors, such as Guibernau, while comparing its

[1] This paper is included in the projects HAR2016-75871-R and PRX18/00138.

A. Roca-Rosell (✉)
Universitat Politècnica de Catalunya – Barcelona Tech, Barcelona, Spain

Institute for Catalan Studies, Barcelona, Spain
e-mail: antoni.roca-rosell@upc.edu

M. Janué i Miret, A. Presas i Puig (eds.), *Science, Culture and National Identity in Francoist Spain, 1939–1959*,
https://doi.org/10.1007/978-3-030-58646-1_6

evolution with that of Scotland or Quebec.[2] While intellectuals play a relevant role in nationalism,[3] the role given to scientists and technicians must be highlighted in the Catalan case.[4] It seems clear that they were considered a prominent part of the 'potential elite' that Guibernau considers one of the elements of Catalan nationalism. As Ash points out in this same book, the concept of nation is highly variable. In Catalonia, it includes the Catalan language, history, tradition, and literature, combined with industry and productivity.[5]

The (complex) interrelationship between science, technology, and nation has occupied historians of science who have recognised that scientific activity and, of course, technology must be understood in a specific context, in terms of not only its dissemination and appropriation but also its production. This recognition, one of the foundations of the social history of science, has existed since the beginning of modern historiography, as illustrated by the works of the generation of George Sarton, Charles Singer, Hélène Metzger, or Aldo Mieli.[6] These scholars primarily addressed hegemonic nations, many of which led empires in which domination was strongly associated with the use of science and technology. There are references to 'French' or 'British' science, while national styles of technical activity are also considered.[7]

George Basalla (1967) proposed a model for the spread of 'Western' science throughout the world. While his article, written when he was a young Harvard graduate, was schematic—he said so himself—it paved the way for many studies that analysed the relationships between science and technology and local conditions, and the particularities of the circulation of ideas and practices, from starkly different perspectives. According to Anderson (2018), even though the Basalla model has since been surpassed or discarded, it represented a crucial turning point in the social history of science.

The case of secondary nations has been studied less from a scientific and technological perspective, but dissemination and appropriation are

[2] Guibernau, 2006.

[3] See the text of Ash in this volume. Also Guibernau, 2000.

[4] Roca-Rosell; Salavert-Fabiani, 2009.

[5] See, for example, Balcells, 1996.

[6] Fox, 2006.

[7] Among the classic references, see Crosland, 1977, Kranzberg (ed.), 1986, Pyenson, 2002, Gouzévitch et al. (ed.), 2004, Fox, 2012. The journal *Osiris* dedicated its 2009 issue to national identity: Harrison, Johnson (ed.), 2009.

considered relevant phases in the general history of science and technology. In Spain, the 'controversy of Spanish science' focused the debate on the scientific backwardness caused by religious obscurantism, for progressives, or disregard or marginalisation by the 'foreign' enemies of the country, for conservative sectors, in a new version of the 'Black Legend'.[8] On the other hand, Spain was included in the 'peripheral' countries, in a model proposed several years ago for the circulation and appropriation of science and technology, the intention of which was to go further in what they denounce as a 'diffusionist' model.[9]

Catalonia had a self-government within the Crown of Aragon until the end of the War of the Spanish Succession (1700–1714), when a unitary order was imposed in Spain, excluding cultural and political diversity. The movement to reclaim the Catalan identity was strengthened in the late nineteenth century.[10]

In the first third of the twentieth century, the rise of Catalan nationalism was associated with the promotion of science and technology. According to the architect and political leader Josep Puig i Cadafalch (1867–1956), science was an essential element of Catalanness.[11] His conception of science included technology, within the Catalans' practical and productive nature. In this context, the actions of the Catalanist movement in the early years of the twentieth century were significant: for example, technical education was promoted, including services and laboratories (the Industrial School of Barcelona—*Escola Industrial de Barcelona*, 1904), and an academic entity initially dedicated to history, literature, and art, the Institute for Catalan Studies—*Institut d'Estudis Catalans* (IEC, 1907), was established.[12] Simultaneously, private initiatives were developed, the most significant of which was from the Society of Jesus, which created a geophysical observatory in Roquetes in 1904, associated with the *Colegio Máximo of Tortosa*.[13] This college also promoted the establishment of a chemistry laboratory and a biology laboratory. These institu-

[8] José María López Piñero (1933–2010) proposed abandoning that controversy based on historical research, beginning in the imperial era of the sixteenth and seventeenth centuries. See López Piñero, 1979. Navarro Brotons and Eamon (2007) recently published a series of works on the 'black legend' and science.

[9] Gavroglu et al., 2008.

[10] Balcells, 1996.

[11] Roca-Rosell, 1988; Roca-Rosell; Salavert Fabiani, 2009.

[12] Roca-Rosell (coord.), 2008; Balcells; Pujol, 2007.

[13] García-Doncel; Roca-Rosell, 2007.

tions were transferred to Sarrià in 1915.[14] The chemistry laboratory soon became a school of chemical engineering, the Chemical Institute of Sarrià—*Instituto Químico de Sarrià* (IQS).[15]

After the collapse of the Bourbon monarchy, the Second Spanish Republic was proclaimed in 1931, and within this context, the new *Generalitat* of Catalonia was established as a self-governing body. At this time, Catalonia's system of science and technology could be described as promising.[16] The Spanish Civil War (1936–1939) had a more relevant scientific and technological component than is usually considered. In Catalonia, the medical and passive defence efforts, as well as the war industry, had remarkable scope.[17]

In this article, we propose discussing how science and technology were integrated into the Catalanist discourse after the Civil War, both among those in internal exile and among the diaspora.[18] Catalanism had gone underground, surviving with difficulty, in part, as some of its members had adhered to the Franco regime and now held high-ranking positions.[19] Abroad, some of the intellectuals in exile, while launching new scientific and technical careers, maintained their commitment to the Catalan identity, now in the context of the fight for the return of democracy to Spain. After winning the Civil War, the Franco regime imposed a dictatorship in which industrial development was promoted, and, according to studies of economic history, Spain can be considered an industrial country from the 1960s.[20]

[14] The municipality of Sarrià was integrated into the city of Barcelona in 1921.

[15] Puig Raposo; López García, 1994.

[16] Roca-Rosell, 2009.

[17] Català, Roca-Rosell, 2009. Valentines-Álvarez, 2018, analysed the role of industrial engineers.

[18] Català, Camarasa, 2009.

[19] See chapter 12 of Balcells, 1996.

[20] See, among others, Nadal et al., 1994, the study by Prados de la Escosura et al., 2012, and the recent review by Maluquer de Motes Bernet, 2014.

Post-War Internal Exile

The Spanish Civil War (1936–1939) culminated with the defeat of the Republicans.[21] The war caused considerable human and material damage, but also suppressed all kinds of progressivism, both through exile and repression, as well as imprisonment and execution, of many Republicans. Santiago López García has studied the loss of human capital in Spain, analysing the replacement of the Board of Extension of Studies—*Junta para Ampliación de Estudios* (JAE, 1907–1939) by the Higher Council for Scientific Research—*Consejo Superior de Investigaciones Científicas* (CSIC, created in 1939) in a complex process in which there was a strong ideologising of research, due to political control of both the personnel and even the content of the work. On the other hand, Francoism clearly incorporated technical research, which had been poorly promoted institutionally before the Civil War.[22] Jaume Claret i Miranda has studied the purge of professors from Spanish universities.[23] Of approximately 400 university professors submitted to purge, 141 suffered of reprisals, of whom 61 had gone into exile. The rest remained in Spain with varying degrees of repression.

In the case of technology in Catalonia, the School of Industrial Engineering in Barcelona, which was predominantly composed of conservative and Francoist teachers, was 'integrated' into a single school on a Spanish scale that became one of the three 'establishments' (until 1949).[24] The Industrial School of Barcelona was subjected to strict ideological control to erase its past as '*Universitat Nova*' (New University), as it had been dubbed in the previous era.[25] The IQS, which had been practically dismantled during the Republican era due to its association with the Society of Jesus, was re-established and commenced a new phase of operation, continuing its association with the chemical industry.

Official scientific research—included in the CSIC, which established a delegation in Barcelona—was then viewed with scepticism in the world of Catalanism.[26] The establishment of the delegation was the result of a

[21] Of the many references to the development of the Civil War, perhaps we can mention the classic Thomas, 1961, and Jackson, 1974.

[22] López García, 1997.

[23] Claret, 2004, 138. Also Claret, 2003.

[24] Lusa Monforte, Roca-Rosell, 2005.

[25] Roca-Rosell (coord.), 2008.

[26] Malet, 2009.

strategy to create research centres outside Madrid, making agreements with the universities, which was a departure from the JAE's previous policy. One of the purposes of establishing the CSIC in Barcelona was to openly replace the IEC, for its Catalan and republican commitment. The replacement even had to be 'physical': the CSIC delegation building was built and located next to the IEC headquarters until 1939.

Nevertheless, the IEC was not explicitly suppressed, and, as early as 1942, it resumed its activities in private meetings.[27] Many of its members were in exile or had died during the war period. Therefore, new members were appointed, whose Catalanist orientation (moderate and private) was not incompatible with their positions in, for example, the CSIC, as in the case of the geologist, palaeontologist, and priest Josep Ramon Bataller (1890–1962), who joined in 1942. Among the members residing in Barcelona, Eduard Fontserè i Riba (1870–1970),[28] a member of the Royal Academy of Sciences and Arts of Barcelona—*Real Academia de Ciencias y Artes de Barcelona* (RACAB), was particularly significant. He enjoyed a certain degree of impunity, probably as he was already 69 years old in 1939, and retired in 1940 without going through the purging process.[29] His longevity made him a living testimony to the scientific and technological boom of the pre-war period, although his case was exceptional. For example, his collaborator Manuel Álvarez Castrillón (1886–1957) was expelled from the university, tried, and imprisoned for 18 months.[30]

In 1939, the botanist Pius Font i Quer (1888–1964), who founded the Botanical Institute of Barcelona, was arrested and imprisoned.[31] A year and a half later, he was granted provisional freedom. He lost all his positions and spent the rest of his life working in publishing houses with outreach contributions. In 1942, he was appointed a member of the IEC.

The first publication (without legal deposit) of the IEC was released in 1947 and was a monograph by Frederic Duran Jordà (1905–1957), who had won an award that year for a study of the digestive mucosa.[32] As explained to the author (in around 1980) by Ramon Aramon, the

[27] Balcells, Pujol, 2007.
[28] Roca-Rosell et al., 2004.
[29] Claret, 2004.
[30] Roca-Rosell, 2007b.
[31] Artís; Camarasa, 1995.
[32] Frederic Duran Jordà (1947): *Histopatologia d'una nova capa d'epiteli semiescamós pla que cobreix les mucoses digestives*, Barcelona, IEC.

secretary general of the IEC since 1942, they decided to publish something that would not be subject to political persecution.[33]

The deployment of the CSIC in Catalonia created new professional opportunities, particularly in the laboratories of its applied science division, the *Patronato Juan de la Cierva*, and in fields such as geology, chemistry, and mathematics.[34] Some scientists had the opportunity to develop their careers in these centres, where ideological control was perhaps less severe than at the university, or at least that seems to be the case for the centres in Barcelona.[35]

An interesting case is that of Esteve Terradas Illa (1883–1950), a member of the IEC, but who assumed high-ranking positions in Francoist institutions and companies after his return to Spain in 1941.[36] Despite being a high-level technician of the regime, his personal life continued as normal compared to pre-war times, at least as far as language was concerned. In 1948, Fontserè wrote to Terradas, as a colleague at the IEC, to continue a publication that had commenced in the 1930s. Terradas replied, giving his opinion, including some nostalgic comments on the figure of Fontserè with whom he had socialised in the IEC and RACAB, and the correspondence continued in Catalan.[37] Thus, there was a clear contrast between the private and public spheres.

As observed earlier, the IEC attempted to continue its action by following in some way the culture of civil discourse. However, reality prevailed. Deprived of budget, forced into private and discreet activity, with great restrictions on Catalan as a language, and, consequently, without democratic freedoms, the IEC developed its career in internal exile.

In any case, as a result of this situation, the new generation of researchers—generally young and without a marked republican past—had to settle in institutions that were ideologically controlled by Francoism (the CSIC and the University), closed to the outside world which was seen as a threat; this experience of accommodation was, at the very least, unlikely to have been particularly stimulating from a civic perspective, as it impaired the civil discourse in science.

[33] Duran Jordà was a communist doctor who organised an early blood bank in 1936. See the biography and documentation on Duran Jordà in Bruguera-Cortada, 2012.

[34] Malet, 2009.

[35] Català; Camarasa, 2009.

[36] Roca-Rosell, 2005.

[37] Roca-Rosell; Sánchez Ron, 1990, 325.

Thomas Glick considered that the establishment of a civil discourse in science in the first years of the twentieth century was an essential element to explain its development in that period.[38] The civil discourse was manifested by the existence of few political and ideological interferences: Glick points out, among other cases, the 'truce' in the Darwinian debate and a general enthusiasm in the dissemination of relativity with few philosophical hostilities and less pronounced anti-Semitic attitudes compared to many other countries. On the other hand, the JAE developed its activity by overcoming the crisis of the monarchy, the establishment of the military dictatorship of Primo de Rivera, and the proclamation of the Second Republic. Francoism, as we have said, abrogated this civil discourse, with a new ideological use of science.[39]

Lino Camprubí has analysed engineering in the Francoist era.[40] He shows the backing of the regime in different areas, particularly public works and the agricultural industry. The technicians took charge of a plan to transform the country, which Francoism promoted for a capitalist modernisation that could also suffocate the demands of the population. Camprubí points out the accommodation and great opportunities that the engineers received at the time. However, engineers working in the productive private sector encountered more difficulties in the authoritarian policy of the regime, initially focused on autarchy.[41]

The importance afforded to engineers is explained by Franco's modernisation agenda. According to Malet (2019), the national-Catholic and fascist rhetoric served to find the ruling classes' support for this modernisation, classes that viewed it with distrust and hostility for their association with progressive movements. In this context, it should be noted that Catalan industrialists, in many cases integrated into the Catalanist movement, accepted—generally, without hesitation—the new established order, as it promoted their business interests. However, when Francoism made the first economic changes in the 1960s, some of these entrepreneurs attempted to return to their national commitment to Catalonia, as evidenced by the creation of the cultural entity Òmnium in 1961.[42]

[38] Glick, 1988 and 1993.

[39] On civil discourse including engineering, see Roca-Rosell, 2007a.

[40] Camprubí, 2014.

[41] This autarchic policy was progressively abandoned during the 1950s. See, for example, Maluquer de Motes Bernet, 2014.

[42] A biography of the main entrepreneur who promoted *Òmnium* has recently been published: Sinca, 2018.

According to González Blasco's (1980) sociological study, the researchers who were active during Francoism did not recognise any debt to the republican experience, from before the Civil War.[43] Furthermore, sectors of democratic ideology were a minority, although, in many cases, they had probably acquired this ideology thanks to stays abroad (which began to be important in the late 1950s). This context does not seem favourable to the idea that Catalanness and science and technology were particularly united elements; rather, the opposite was the case because science and technology acquired an official, Francoist, and Spanish image. Among scientists and technicians, as among other citizens, Catalanism moved into the private sphere.

The External Exile

In the external exile,[44] we find a continuity of the ideological place that science and technology had acquired in the republican period. Moreover, exiles, hosted by countries with democratic freedoms, such as Mexico, had the opportunity to publish and debate their positions. We propose analysing this continuity through the Catalan exile press with the help of some relevant publications. Fèrriz (2009) has compiled a catalogue of ten Catalan magazines published by Catalan exiles in Mexico, from which we can select the texts that are related to science and technology.[45] We choose *Quaderns de l'exili* and *La Nostra Revista*, two publications to which we had access.[46] The exile reflects the plurality of political and social options in the Catalanist world, from independence to communism.

Quaderns de l'exili appeared between 1943 and 1947 (26 issues), edited by Joan Sales (1912–1983) and Lluís Ferran de Pol (1911–1995), both of whom were exiled in Mexico but who returned to Spain in the late 1940s, where they worked as publishers, writers, and cultural activists. The articles of the *Quaderns de l'exili* that we have selected show a

[43] Presas i Puig (1998) points out that at a conference held in Germany in 1956, the secretary general of the CSIC presented the work of the CSIC as an evolution of what had begun around 1900, in contrast to the disavowals of the JAE experience by himself and his collaborators.

[44] An assessment of the Catalan, Valencian, and Balearic exiles in Català; Camarasa, 2009.

[45] Fèrriz, 2009.

[46] There is a facsimile edition of *Quaderns de l'exili: Mèxic, 1943–1947*, 1982 . It is possible to access some digitised numbers from *La Nostra Revista*, but we also obtained copies at the *Biblioteca del Pavelló de la República*, University of Barcelona.

concern for promoting science and technology in Catalonia (and in Valencia and the Balearic Islands) in order to promote independence. The scientific training has a strong focus on military training, but is also focused on renovating the university and keeping up to date with contemporary developments, particularly in the world of medicine, which in Catalonia started from a very relevant point in the time of the Republic and the Civil War. During World War II (1939–1945), the group promoting *Quaderns de l'exili* advocated that a Catalan military battalion should be organised to combat fascism, and, consequently, to secure allies in favour of the formation of a Catalan state.

The issues of *Quaderns de l'exili* had a regular structure based on almost fixed sections. One of these was 'Les armes i la història' (Weaponry and history), with its first text signed by the Menorcan Colonel Vicenç Guarner (1893–1981) on the 'glorious democratic, military, and civic tradition of Catalonia'. Other texts in the same section addressed specific episodes of Catalan soldiers' activity and of their presence during World War II in the United States Army and the British Royal Air Force. An article signed by August Pi-Sunyer (1879–1965) was published in March–April 1944 on Josep Trueta's (1897–1977) contributions to the treatment of war wounds. In the same line of argument, in the issue of May–June 1945, Joan Sales devoted the section on weaponry and history to disavowing the left-wing 'pacifists', in response to an article by the writer Joan Oliver 'Pere Quart' (1899–1986) in the magazine *Germanor* in Santiago de Chile which questioned the need for a Catalan army.

Another section, 'El treball i les ciències' (Work and science), which is less belligerent, is dedicated to the dissemination of current scientific news, mainly medical. For example, in the issues of January–March and October–December 1946, the medical doctor Albert Folch i Pi (1905–1993),[47] who established his career in Mexico, wrote an extensive article on vitamins and their role in diet and certain diseases.

On the other hand, the June 1946 issue of *Quaderns de l'exili* is devoted almost exclusively to the university and its role in Catalan culture. The editorial contains the history of the University of Barcelona, from its foundation to its suppression by the Bourbons after 1714, The new university that the authors conceive, not necessarily located in Barcelona, should be established by criteria of utility to the culture and development of

[47] Bruguera-Cortada, 2019.

Catalonia. In his note, Joan Sales attacks the 'floralistic'[48] conception of Catalan culture, which, according to him, had been prolonged during the exile. The militant of the Workers' Party of Marxist Unification—*Partido Obrero de Unificación Marxista* (POUM)—Jordi Arquer i Saltor (1907–1981) made a contribution in another article entitled 'Els estralls de l'autodidactisme en el moviment obrer' (The negative consequences of self-learning in the workers' movement). He states that the lack of training had been a limitation for the revolutionary process and advocates an educational system that does not marginalise workers.

Another orientation was shared by the members of the *Partit Socialista Unificat de Catalunya*, Unified Socialist Party of Catalonia (PSUC), a party born in July 1936 as a fusion of communist and socialist groups in the context of the outbreak of the Civil War. It joined the Third International and maintained cooperative relations (and friction) with the Spanish Communist Party. In general, the PSUC had a favourable attitude towards the freedom of Catalonia in the context of a federation of the peoples of Spain. Moreover, in the publications of the PSUC, mainly after 1945, it is stated that science and technology are the foundation of Marxism and of the revolution.[49]

José Pardo Tomàs, Àlvar Martínez Vidal, and Enrique Perdiguero introduced the concept of 'medical Catalanism' several years ago to interpret the commitment of many doctors to the vindication of Catalan identity from the late nineteenth century.[50] This medical Catalanism included the idea that medical practice should be a collective task in which research plays a relevant role. One example is the creation of a hospital in Toulouse by several doctors associated with the PSUC (and with the Spanish Communist Party) to treat evacuees from the Nazi concentration camps in 1945 and, simultaneously, care for the wounded of the *maquis*—the anti-Franco guerrilla movement—and the Spanish refugees residing in the city of Occitania.[51] One of the hospital promoters, which was called 'Varsovia' as it was on Varsovie Street, was the PSUC medical doctor Josep Bonifaci (1893–1989), who had already participated in the renovation of the

[48] *Floralistica*, that is, following the tradition of the Floral Games, Catalan poetry competitions that were resumed in the mid-nineteenth century that Sales and his colleagues considered to be completely outdated.

[49] Roca-Rosell, 1986.

[50] Pardo; Martínez-Vidal; Perdiguero, 2006.

[51] Martínez-Vidal (coord.), 2010.

healthcare system in Catalonia. The Varsovia Hospital, still existent,[52] should be considered an extension of the Catalan medical tradition.

The industrial engineer Eduard Barba Gosé (1901–1973), a PSUC militant exiled in the Dominican Republic, published a summary of the history of technology in Catalonia in *La Nostra Revista* (1948).[53] His objective was to fill a gap which, according to him, comprised the lack of studies on the history of technology in Catalonia.[54] Further, he based his interest in the history of technology on the important role it plays in the history of Catalonia, given the close relationship between technology, economy, and politics.

Barba's aim was to vindicate the role of Catalonia in the creation of the 'peninsular' manufacturing system and to claim the invention or introduction of some relevant technologies. He divided his discussion into branches of technology, within which in many cases his study dates to the Middle Ages or earlier. The discussion highlights unique contributions in textiles, such as the velvet looms of Jacint Barrau (1810–1884) and Ferran Alsina (1861–1907), or the stretched threads of Ferran Casablancas (1874–1960), technologies that had a significant international reach. Regarding the chemical industry, he mentions the contributions to alcohol distillation of the Majorcan Ramon Llull (1232–1316) and Arnau de Vilanova (1240–1311) and its uses as a medicine and a drink (with aromatic herbs). The participation of Josep Roura i Estrada (1797–1860) and Jaume Arbós i Tor (1824–1882) in the gaslighting production technique, which pioneered this industry in Catalonia, is also highlighted. Regarding metallurgy, Barba recalls the Catalan forge, the hegemonic metallurgical system used until the nineteenth century.[55] He then explains that the first railway was the Barcelona-Mataró line (1848), which was promoted by the businessman Miquel Biada (1789–1848). He explains the history of the mechanical workshops of the early nineteenth century that culminated in the company "Maquinista y Terrestre" of Barcelona in 1855, among other mechanical companies that reached, according to Barba, a distinguished position in the 'manufacturing production system' of their time. As for individual contributions to mechanics, Barba refers to the contributions of

[52] It is called *Hospital Joseph Ducuing*.

[53] Barba, 1948.

[54] Probably, Barba made use of the historical references in Pi i Sunyer, 1983.

[55] Barba refers to the Catalan forges as blast furnaces but, in fact, they were bloomeries in which the direct casting method was used. See, for example, Lusa Monforte, Roca-Rosell, 2019.

Francesc Santponç i Roca (1756–1820) and Francesc Salvà i Campillo (1751–1828), with minor inaccuracies, given that the general evaluation is correct.[56] Something similar occurs when he mentions the naval industry,[57] railways, and automobile construction,[58] subjects to which he dedicates the final three parts of the section. He also points out that the first photograph in Spain—a daguerreotype—was taken in 1839 in Barcelona. To conclude, he mentions electricity and the role of Catalonia in the electrification of Spain (Barba considers that the 1888 Barcelona Exposition was the moment that promoted this), and adds another reference to the contributions to telegraphy of Salvà i Campillo.

One of the most explicit texts on the relationship between Catalan identity and science and technology was written by the chemist and pharmacist Alfons Boix i Vallicrosa (1899–1965), a member of *Esquerra Republicana de Catalunya* (Republican Left of Catalonia).[59] He published an article titled 'Science and Catalanness' in *Lletres*, vol. II, No. 5, January 1945 (reproduced in *La Nostra Revista*, vol. I, No. 10, October 1946). Boix Vallicrosa (1945) explains that physics has changed our view of the world and the universe in the twentieth century. He considers that quantum physics has proposed a new view of the atom and radiation, blurring the boundaries, he says, between matter and energy. Furthermore, the transmutation of the elements ('transelementation', says Boix Vallicrosa) is conceived in the new theoretical framework, materialising the alchemists' dream. Regarding the resistance of religion to new theories, such as those proposed by Galileo and Darwin, the successors of the ancient apologists, literal interpreters of the Holy Scriptures, now adhere to the laws of the new quantum mechanics, such as Heisenberg's uncertainty principle, to maintain their dogmatic positions. It seems that science has become an

[56] According to Barba, the Santponç's steam engine was imitated outside Spain, but there is no confirmation of this. On the other hand, Salvà was a pioneer of telegraphy, but not of submarine cables, as Barba says. However, he correctly identifies Santponç as the pioneer of steam technology in Catalonia and his collaboration with Salvà in a new machine for scutching hemp and flax as a relevant technical contribution. See Sánchez Miñana, 2007.

[57] In addition to commenting on the medieval tradition of shipbuilding in the different Catalan shipyards, there is an interesting error when mentioning 'Isaac Monturiol' as the pioneer of submarine navigation, thus linking two true pioneers in Spain—the Catalan Narcís Monturiol (1819–1885) and the Cartagena native Isaac Peral (1851–1895).

[58] He cites Hispano-Suiza as a pioneering company in the automotive sector.

[59] This organisation is a social-democratic and catalanist party created in 1931, which evolved into independentism.

ally: according to what they say, the probabilistic nature[60] of the physical world, which challenged mechanicism, will end up confirming religious truths. Furthermore, the intellectuals of the Third Reich attempted to provide a philosophical basis for Nazism. The probabilistic nature of the world supported the denial of the role of individuals. Similarly, developments in biology have been instrumentalised to endorse racist and xenophobic policies.

In the face of this confusion, many thinkers, says Boix Vallicrosa, attempt to 'bring order' with precautionary ideas and with the conviction that the new science does not deny progress. Bertrand Russell (1872–1970), quoted by Boix Vallicrosa, raises the need to reformulate the basic concepts of science. Max Planck himself, he tells us, believes that the development of physics will lead to a new conception of causality.

It is in this part, commenting on the need for 'consideration and balance', that Boix Vallicrosa states that the Catalan tradition has faced problems with these attitudes of consideration and balance. This leads him to consider the relationship between science and people. Pure science, despite not intending to do so, is 'the first instrument of technical progress', but scientific development is not only a factor driving material progress, since knowledge contributes to raising the moral dignity of people; hence the importance of science as an 'educational factor'. Conversely, according to Boix Vallicrosa, people can shape science. As it is universal, he continues, it makes no sense to speak of a national science, but people can influence and bring nuances to scientific development. He mentions, for example, the contributions of the Hindu Chandrasekhara Venkata Raman (1888–1970) and his school that have given Orientalist overtones to physics.

Boix Vallicrosa says that new physics needs a balanced view, and this can come, according to him, from Catalan scientists, in the same way that thinkers like Ramon Llull had achieved this in the past. Against extreme positions of uncertainty, Boix Vallicrosa claims to remain in the 'human zone', which, according to him, is the Catalan zone of '*seny*' (good sense) and balance.

Boix Vallicrosa then proposes the possible contribution of the Catalan 'mentality' for the resolution of the crisis in science of his time, a vindication of the Catalans' role in the history of science. Barba stated something similar with regard to technology.

[60] Boix says '*estadística*', statistics.

This—peculiar—relationship between Catalanness and science also appears in an editorial of *Quaderns de l'exili*, in issue 15 of September–October 1945. Although this editorial is not signed, we dare to attribute it to Joan Sales (1945). The title 'De Miguel Servet a la bomba atòmica' (From Miguel Servet to the atomic bomb) arises from a biography of the Aragonese scientist (from the Catalan region La Franja,[61] the author highlights) Miguel Servet (1509–1553), a posthumous work by Jaume Aiguader, who died in Mexico in 1943. Servet was burned at the stake in Geneva for his scientific activities, which included the discovery of blood circulation. The author of the review indicates that Servet said that matter is 'condensed light', which made him a predecessor (!) to the developments in nuclear physics of the first half of the twentieth century. In this part, the author of the review introduces the notion of an atom that was now understood differently than in ancient times, for which he provides a quote from Louis de Broglie (1892–1987). On the one hand, Becquerel, the Curies, and Rutherford, continues the author of the review, uncovered the 'small atomic disintegrations', that is, radioactivity. On the other, Bohr, Planck, Chadwick, Joliot, and others showed that the atom is a kind of planetary system. However, the nucleus turned out to be composed of protons[62] and neutrons, the latter discovered in 1932. A long quote from Louis de Broglie, from a 1939 work, leads to the reflection that there were no Catalan names associated with the great undertaking of the 'disintegration of the atom' that de Broglie compared to the discovery of America. According to the author of the review, it would seem that Catalans are focused on 'floral games and culture', a criticism that would have been disallowed, says the author of the review, if it were not for the explosion of the atomic bomb, which demonstrated that 'this science is very respectable'. The bomb, he claims, is nothing more than a 'minor' consequence of a series of 'philosophical, mathematical, and physical musings' that would have seemed irrelevant and prosaic for being too far from poetry. However, he affirms that there is much more poetry in an atom than in all the 'green and contemplative dormouse *carrinclonas*' imaginary. From this, he concludes that it is necessary to retain young people with scientific talent who are forced to emigrate.

It can be concluded that the Catalanist group located abroad showed a great interest in the world of science and technology, closely following

[61] The '*Franja*' is the Aragon territory of Catalan culture.

[62] It is erroneously said that there are also electrons in the nucleus.

their interrelationship, and, in a certain way, with hope. They thought about deepening the experience of teaching and research that had been developed during the Republican period and, in this way, preparing for the return of democracy to Spain, a return that some saw as associated with independence. For a time, they had relied on a military solution that they were forced to abandon as the Allies did not intervene at the end of World War II. A general characteristic, as we have seen, was to consider that scientific and technological modernisation had been an essential element of the Catalan identity and that it should be so in the future.

Conclusion

The breakdown in the civil discourse of science caused by Francoism in Spain resulted in an environment of mistrust in the professors and researchers who continued or commenced their careers after the Civil War (except, of course, for those followers of the regime). The CSIC's decision to establish itself in Catalonia created new opportunities, although limited for decades. The Catalanist world that remained in Spain had to adapt to the new rarefied environment and, in many cases, was a victim of repression.

The world of exiles dealt with the situation of science and technology in Catalonia from different perspectives. It seems clear that the debate was a continuation of the one that had commenced before the war and that had continued during the war. Both during and after the war, the need to promote research, and make the most of talent, came under stress. There was agreement from many perspectives on the importance of pure science, both for the advancement of knowledge and for the promotion of technology and economic development. Many authors considered historical reflection to be the foundation of general reflection. In one case, that of Barba Gosé, he explicitly thought that the history of technology was a very relevant instrument for understanding the present and preparing for the future. Moreover, there was an agreement on the need of science and technology for development, even in a peripheral region.

The exile of many researchers, the private (and discreet) activity of the IEC, and the development of official scientific and technical research contributed to the weakening or breaking of the association between science and Catalanness in Catalonia. Francoist rhetoric, fully integrated into national Catholicism and rancid Spanishism, paved the way for the scientific and technical research that was required for the industrialisation process that the regime was promoting. In this hostile context, the connection between Catalanism and scientific progress could only create a crisis. In fact, the generation of researchers that emerged during the Franco regime was characterised by its distance from politics. In this way, the Franco regime fought the "potential elite" of Catalanism.

With the triumph of Francoism in 1939, a significant part of the Catalan (and Spanish) scientific community went into exile, and, in many cases, these front-line people had a second chance in Mexico, Venezuela, the United States, France, and Great Britain. These exiled scientists continued to associate the Catalan identity with the promotion of science and technology. In Spain, Francoism sought to make a clean sweep of the previous scientific and technical system, and set out to establish a new reactionary ideology of science, in which Franco intended to create a science and technology that was 'authentically' Spanish, and simultaneously, at the service of the country's industrialisation. Catalan scientists and technicians had to submit to this context, generally adopting 'apolitical' positions and maintaining their connection with the Catalan identity only in the private sphere.

References

Anderson, Warwick (2018): "Remembering the Spread of Western Science", *Historical Records of Australian Science*, 29(2), 73–81.

Artis, Mireia; Camarasa, Josep M. (1995): "Pius Font i Quer (Lleida, 1888, Barcelona, 1964). La maduresa de la botànica catalana", in: Camarasa, J.M.; Roca-Rosell, A. (coord.) *Ciència i tècnica als Països Catalans*, Barcelona, Fundació Catalana per a la Recerca, 1245–1276.

Bruguera-Cortada, Miquel (2012): "Frederic Duran Jordà", *Galeria de metges catalans*, http://www.galeriametges.cat/galeria-fitxa.php?icod=EK [consulted in March 2020].

Bruguera-Cortada, Miquel (2019): "Albert Folch i Pi", *Galeria de metges catalans*, http://www.galeriametges.cat/galeria-fitxa.php?icod=EFLF [consulted in March 2020].

Balcells, A. (1996): *Catalan Nationalism*, London, Macmillan.

Balcells, Albert; Pujol, Enric (2002/2007): *Història de l'Institut d'Estudis Catalans*, Barcelona, Institut d'Estudis Catalans, 2 vols.

Barba i Gosé, Eduard (1948): "Apunts per a una Història de la Tecnologia Catalana", *La Nostra Revista*, vol. III, num. 35–36, november-december, 356–360.

Basalla, George (1967): "The Spread of Western Science", Science, Vol. 156, Issue 3775, 611–622.

Boix i Vallicrosa, Alfons (1945): "Ciència i catalanitat", *Lletres*, vol. II, num. 5, January. Reproduced in *La Nostra Revista*, vol. I, num. 10, october 1946.

Camarasa, J. M.; Català, J. I. (2009): "La diàspora i l'exili interior", in Parés, R.; Vernet, J. (dir.) *La Ciència en la història dels Països Catalans*, Barcelona, València, Institut d'Estudis Catalans, Universitat de València, vol. 3, 829–852.

Camprubí, Lino (2014): *Engineers and the making of the Francoist regime*, Cambridge, Massachusetts, The MIT Press. Updated Spanish version, Barcelona, Crítica, 2017.

Català, J. I.; Roca-Rosell, A. (2009): "La Guerra civil (1936–1939) i la ciència", in Parés, R.; Vernet, J. (dir), vol. 3, 809–828.

Claret i Miranda, Jaume (2003): *La Repressió franquista a la universitat catalana: La Universitat de Barcelona autònoma, de la segona república al primer franquisme*, Barcelona, Institut Universitari d'Història Jaume Vicens i Vives; Vic, Eumo.

Claret i Miranda, Jaume (2004): *La Repressió franquista a la universitat española*. Tesis doctoral, Barcelona, Universitat Pompeu Fabra. Institut Universitari d'Història Jaume Vicens i Vives.

Crosland, Maurice (1977): "History of Science in a National Context", *The British Journal for the History of Science*, vol. 10, 95–113.

Fèrriz Roure, Teresa (2009): *Escriptors i revistes catalanes a l'exili. Índex de deu publicacions editades a Mèxic (1939–1981)*, Barcelona, Editorial UOC.

Fox, Robert (2006): "Fashioning the Discipline: History of Science in the European Intellectual Tradition", *Minerva* 44 (4), 410–432.

Fox, Robert (2012): *The Savant and the State. Science and Cultural Politics in Nineteenth-Century France*, Johns Hopkins University Press.

García Doncel, Manuel; Roca-Rosell, Antoni (2007): *Observatorio del Ebro. Cien años de historia (1904–2004)*, Roquetes, Tarragona, Observatori de l'Ebre, Edicions el Mèdol.

Gavroglu, Kostas, et al (2008): "Science and technology in the European periphery: some historiographical reflections", *History of Science*, vol. 46, 153–175.

Glick, Thomas F. (1988): «Introduction», in: Glick, Thomas F.: *Einstein in Spain*, Princeton University Press, 3–16.

Glick, Thomas F. (1993): «Ciencia, política y discurso civil en la España de Alfonso XIII», *Espacio, Tiempo y Forma*, Serie V, Hª Contemporánea, t. 6, 1993, 81–98. https://dialnet.unirioja.es/servlet/articulo?codigo=148023.

González Blasco, Pedro (1980) *El Investigador científico en España*, Madrid, Centro de Investigaciones Sociológicas.

Gouzévitch, Irina; Grelon, André; Karvar, Anousheh (coord.) (2004): *La Formation des ingénieurs en perspective: modèles de référence et réseaux de médiation XVII–XX siècles*, Rennes, Presses universitaires de Rennes.

Guibernau, Montserrat (2000): "Nationalism and Intellectuals in Nations without States: the Catalan Case", *Political Studies*, vol. 48, 989–1005.

Guibernau, Montserrat (2006): "National identity, devolution and secession in Canada, Britain and Spain", *Nations and Nationalism*, vol. 12 (1), 51–76.

Harrison, Carol E.; Johnson, Ann (ed.) (2009): *National Identity. The Role of Science and Technology*, The University of Chicago Press. Issue of *Osiris*, second series, 24.

Jackson, Gabriel (1974): *A Concise history of the Spanish Civil War*, London, Thames and Hudson.

Kranzberg, Melvin (ed.) (1986): *Technological education technological style*, San Francisco, San Francisco Press.

López García, Santiago (1997): "El 'Patronato Juan de la Cierva' (1939–1960). I Parte: Las Instituciones Precedentes", *Arbor*, CLVII, 619, 201–238.

López Piñero, José María (1979): *Ciencia y técnica en la sociedad española de los siglos XVI y XVII*, Barcelona, Labor.

Lusa Monforte, Guillermo; Roca-Rosell, Antoni (2005): "Historia de la ingeniería industrial. La Escuela de Barcelona 1851–2001", *Documentos de la Escuela de Ingenieros Industriales de Barcelona*, vol. 15, 13–95.

Lusa Monforte, Guillermo; Roca-Rosell, Antoni (2019): *La "farga" de l'ETSEIB: un testimoni singular de l'enginyeria industrial*, Barcelona, ETSEIB. (Patrimoni ETSEIB).

Malet, Antoni (2009): "La Guerra civil i les institucions científiques catalanes: el cas de la recerca matemàtica (1907–1967)", en Parés; Vernet (dir.), 719–759.

Malet, Antoni (2019): "Science and power: Francoist Spain (1939–1975) as a case study", *Centaurus*, 61, 111–132.

Maluquer De Motes Bernet, Jordi (2014) *La economía española en perspectiva histórica. Siglos XVIII–XXI*, Barcelona, Pasado & Presente.

Martínez Vidal, Àlvar (coord.) (2010): *Exili, medicina i filantropia: L'Hospital Varsòvia de Tolosa de Llenguadoc (1944–1950)*, Catarroja, Barcelona, Afers.

Nadal, J.; Carreras, A.; Sudrià, C. (comps.) (1994): *La economía española en el siglo xx. Una perspectiva histórica*, Barcelona, Ariel.

Navarro Brotons, Víctor; Eamon, William (eds) (2007): *Más allá de la leyenda negra: España y la revolución científica = Beyond the black legend: Spain and the scientific revolution*, València, Instituto de Historia de la Ciencia y Documentación López Piñero.

Pardo Tomàs, José; Martínez-Vidal, Àlvar; Perdiguero, Enrique (2006): "Per la Ciència i per la Pàtria": medical catalanism (1898–1936)", Sixth ESSHC. Amsterdam, 22–26 March 2006. Network 'Health'. Session 'Health and Nations', 11 pages. http://digital.csic.es/bitstream/10261/7884/3/Medicalcatalanism.pdf.

Pi i Sunyer, Carles (1983): *L'Aptitud econòmica de Catalunya*. Edition by Ricard Vinyes, Barcelona, La Magrana/Diputació de Barcelona. It includes the two volumes published in 1927 and 1929.

Prados De La Escosura, Leandro; et al (2012): "Economic Reforms and Growth in Franco's Spain", *Revista de Historia Económica = Journal of Iberian and Latin American Economic History*, Año n° 30, Número 1, 45–90.

Presas i Puig, Albert (1998): "Nota histórica: una conferencia de José María Albareda ante las autoridades académicas alemanas", *Arbor*, n° 631–632, julio-agosto, 343–357.

Puig Raposo, Núria; López García, Santiago M. (1994): "Chemists, engineers and entrepreneurs. The Chemical Institut of Sarrià's impact on Spanish industry (1916–1992)", *History & Technology*, vol. 11, 345–359.

Pyenson, Lewis (2002): "An End to National Science: The Meaning and the Extension of Local Knowledge", *History of Science*, vol. 40, 2002, 251–290.

Quaderns de l'exili: Mèxic, 1943–1947 (1982), Barcelona, Estudis Nacionalistes. Edición facsímil.

Roca-Rosell, Antoni (1986): «La necessitat de la ciència, la necessitat de la llibertat», in: *La nostra utopia*, Barcelona, Planeta, 69–89. (70–90 of the Spanish edition).

Roca-Rosell, A. (1988): "Ciencia y sociedad en la época de la Mancomunitat de Catalunya (1914–1923)", in Sánchez Ron, José Manuel (editor): *Ciencia y sociedad en España*, Madrid, Ediciones el arquero/CSIC, 223–252.

Roca-Rosell, Antoni (2005): «Professionalism and Technocracy: Esteve Terradas and Science Policy in the Early Years of the Franco Regime», *Minerva*, 43, 147–162.

Roca-Rosell, Antoni (2007a): "El discurso civil en torno a la ciencia y la técnica", en Suárez Cortina, M.; Salavert Fabiani, V. (ed.) *El regeneracionismo en España*, València, Universitat de València, 241–159.

Roca-Rosell, Antoni (2007b): "Manuel Álvarez Castrillón (1886–1957), Orígens de la Meteorologia matemàtica", *XIII Jornades de Meteorologia Eduard Fontserè*, Barcelona, Associació Catalana de Metereologia, 2007, 69–76.

Roca-Rosell, Antoni (coord.) (2008): *L'Escola Industrial de Barcelona. Cent anys d'ensenyament tècnic i d'arquitectura*, Barcelona, Diputació de Barcelona, Ajuntament de Barcelona, Consorci de l'Escola Industrial de Barcelona.

Roca-Rosell, Antoni (2009): "La República, la Generalitat i la renovació de les institucions científiques", en Parés; Vernet (dir.), 785–808.

Roca-Rosell, Antoni; Sánchez Ron, José Manuel (1990): *Esteban Terradas (1883–1950). Ciencia y Técnica en la España contemporánea*, Barcelona, Madrid, El Serbal, INTA.

Roca-Rosell, Antoni; Batlló Ortiz, Josep; Arús Dumenjó, Joan (2004): *Biografia del doctor Eduard Fontserè i Riba (1870–1970)*, Barcelona, Associació Catalana de Meteorologia.

Roca-Rosell, Antoni; Salavert Fabiani, Vicent L. (2009): "Catalanisme, valencianisme i ciència en el canvi de segle", in: Parés; Vernet (dir.), 523–569.

[Sales, Joan?] (1945): "De Miquel Servet a la bomba atómica", *Quaderns de l'exili*, num. 15, september–october, i–ii+x.
Sánchez Miñana, Jesús (2007): *Vida y obra del doctor Salvà i Campillo (1751–1828)*, Madrid, FECYT. https://www.fecyt.es/es/publicacion/vida-y-obra-del-doctor-salva-y-campillo.
Sinca, Genís (2018): *El Cavaller Floïd: biografia de Joan Baptista Cendrós*, Barcelona, Proa.
Thomas, Hugh (1961): The Spanish Civil War, London, Eyre and Spottiswoode Publishers Ltd. (There are many editions).
Valentines-Álvarez, Jaume (2018): "Seeing like a factory: Technocratic nationalism in Catalonia, 1930–1939", *History and Technology*, vol. 34 (3–4), 235–258.

PART III

Women's Space in the Science and Culture of the Regime

CHAPTER 7

In the Land of Men: Women in Applied Sciences at the CSIC

Fernando García Naharro

Introduction

"Just as Christ, as man, is subject to God (…), so too the woman is subject to the man." These were the words of Pope Pius XII in September 1941, a time when the Holy Father perceived an "unhealthy atmosphere of renascent paganism". Faced with this situation, and in his will to restore the essentially different human male and female nature, the pope recalled what St. Paul taught to his disciples of Ephesus: "Let women be subject to their husbands, as to the Lord: because the husband is the head of the wife, as Christ is the head of the Church. (…) As the Church is subject to Christ, so also let the wives be to their husbands in all things".[1] Alongside these concerns, and after receiving the gift of several religious books published by the Spanish National Research Council (CSIC, for its initials in

[1] Pio XII (1945): 286–287.

F. García Naharro (✉)
Europa-Universität Flensburg, Flensburg, Germany
e-mail: fernando.garcia-naharro@uni-flensburg.de

M. Janué i Miret, A. Presas i Puig (eds.), *Science, Culture and National Identity in Francoist Spain, 1939–1959*,
https://doi.org/10.1007/978-3-030-58646-1_7

Spanish), in 1943, Pope Pius XII praised José Ibáñez Martín, the Minister of National Education in Spain, for the service the Council provided "to counteract the pernicious influence that those who sow bad seeds unfortunately have had on the Spanish intellectual terrain".[2] With this, however, the pontiff blessed not only the programme of Catholic cultural revival entrusted to the CSIC since its founding in 1939,[3] but also the political orientation of the Spanish dictatorship towards the reactionary immobilism of the Catholic Church.

In consideration of this, the present chapter seeks a detailed and intimate understanding of the performative function of this frame of reference. Hence, I aim to contribute to the works collected in this book by providing an analysis of the political influence that these ideas had over women in science in Francoist Spain. In so doing, I focus on the official discourse of the regime that identified masculinity as an intrinsic feature of the brand new Spain.[4] Framed by this premise, I argue that the dictatorship also ensured a privileged position to the male subject-knower (the *modest witness*[5]) within the Spanish scientific field—a field wherein I shall nonetheless intend to situate women. To this end, I restrict the scope of my study to the reality of the CSIC, the main scientific institution in the Francoist state.[6] However, within a dictatorial context strongly biased towards applied sciences, I will pay particular attention to the branches of technical research that, according to José Ibáñez Martín, were not receiving proper attention until the creation of the Council.[7]

In consequence, and drawing on research on gender and science and women in science in Spain,[8] I shall put female scientists at the forefront of my inquiry: What kind of women had sufficient legitimacy and independent status to qualify as *modest witnesses* under Franco? Therefore, by analysing the socio-professional profile of female researchers at the

[2] CSIC (1944): 58.

[3] CSIC (1940).

[4] Box, Zira (2019):125–146.

[5] Haraway, Dona (2004): 223–250.

[6] Malet, Antoni (2007): 211–256; Puig-Samper Mulero, Miguel Ángel (ed.) (2007); Delgado, Lorenzo; López, Santiago M. (eds.) (2019).

[7] CSIC (1948): 147.

[8] Pérez Sedeño, Eulalia; Alcalá Cortijo, Paloma (coord.) (2001); Becerra Conde, Gloria; Ortíz Gómez, Teresa (coord.) (1996); Barral Morán, María José; Magallón Portolés, Carmen; Miqueo Miqueo, Consuelo; Sánchez González, María Dolores (coord.) (1999); Magallón Portolés, Carmen (2004); Santesmases, María Jesús (2000); Romero De Pablos, Ana; Santesmases, María Jesús (coord.) (2017).

CSIC—those who were publishing in scientific journals, holding research positions or accessing training programmes abroad—I aim to provide valuable information about those female scientists who worked in fields such as engineering and applied sciences in Franco's Spain.

Science: The Official Discourse of Science Under Franco

From the very beginning, the Franco regime sought to find in science the legitimising foundations that would sustain its project. It was, then, the very concept of science that was at stake. Performing as an argumentative substrate linked to an entire (alleged) scientific tradition *of the purest national character*,[9] the image of science embedded in the official discourse of the regime[10] was purportedly united and harmonious, rooted in Catholicism and (supposedly) free,[11] albeit always subordinated to "the living realities that the State has set forth".[12] In this regard, José Ibáñez Martin asserted that to exclude research from its commitment to Spain would be "to divinise science, to exempt it from a required tribute to the national heritage and to manufacture anarchic pedestals or territories free from an obligation imposed on others. Duties and responsibilities that are everyone's concern, even more demanding for cultivators of intelligence".[13]

However, who were those savants? Who could bear that label within a dictatorial regime greatly interested in policing scientific boundaries? According to José Ibáñez Martín, this category was open only to certain kinds of people, namely the "chosen aristocracy of the men of tomorrow", those who were properly educated within scientific and intellectual circles. A class of chosen men, he argued, "who have emerged purified from the crucible of the red revolution and the bloody war".[14] As men defined by this honourable image, these scientists performed in the official discourse as the antithesis of those *others* who walked on erroneous paths, namely the scientists then identified with the Second Spanish Republic

[9] Antolín Hofrichter, Andrés (2021).

[10] Pictured in this chapter by the authoritative voice of José Ibáñez Martín, the Minister of National Education (1939–1951) and the founding father and first president of the CSIC (1939–1967).

[11] García Naharro, Fernando (2017): 45–89.

[12] CSIC (1942): 2.

[13] CSIC (1942): 39.

[14] CSIC (1942): 37–38.

(1931–1936). Dubbed as "the hierophants of impiety and anti-patriotism",[15] "outcast intelligentsia"[16] or "anti-Spanish representatives",[17] these abject Spanish researchers were very often present in the Francoist discourse during the first half of the 1940s, always acting like the "constitutive outside"[18] to the body of the legitimate scientists.

In addition, at around the same time Spain portrayed itself in the official discourse as absorbed in its own affairs and with "nothing to learn from those Spaniards who (…) proclaiming themselves 'free', actually wanted to make Spain a cultural vassal of minor secondary figures from abroad".[19] However, with time and changes in the international context this revenge discourse would gradually lose ground, paving the way for a brand new motto: the "spring-like and fragrant peace of Spain".[20] Moreover, this proposition marked the so-called *Third Stage of Spanish Science*,[21] which—according to José Ibáñez Martín—guaranteed "the permanent and universal unity of Science, to whose passionate and constructive tasks repel the circumstantial historical discrimination into winners and losers".[22] Thus, in 1951 José Ibáñez Martín denied the reality of exile and the purging that he justified just a few years earlier. However, while he saluted the labour of those *competent and modest savants*—employing Menéndez Pelayo's terms—of the Council, José Ibáñez Martín recognised, too, another compelling reality: the scientific exodus in Spain.[23]

Vir Modestus. Shaping the "Legitimate" Scientist

Even though modern science demanded an increasing number of professionals,[24] science in Francoist Spain was still conceived as a select, vocational and unprofitable activity.[25] This *scientific asceticism* was at the core of the professional *ethos* of CSIC personnel. For instance, in the mid-

[15] CSIC (1942): 30.
[16] CSIC (1943): 51.
[17] CSIC (1944): 48.
[18] Butler, Judith (1993): 3.
[19] CSIC (1944): 48.
[20] Sesma Landrín, Nicolás (2006): 45.
[21] CSIC (1951): 68.
[22] CSIC (1951): 66.
[23] CSIC (1952): 88.
[24] CSIC (1960).
[25] Albareda, José María (1951): 13.

1950s the CSIC had "fellows (mostly PhDs) earning 250 pesetas per month; collaborators (university and institute professors) earning 400 pesetas; and section chiefs earning a maximum gratuity of 1000 pesetas per month (the same amount received at the National Institute of Physics and Chemistry before 1936)".[26] Likewise, in the early 1960s these remunerations were still deemed insufficient by the Association of Researchers and Collaborators of the CSIC.[27]

In addition to this idea of austerity, the model of the scientist was then built out of the premise of a solitary man performing a self-reflective work free from disturbance. Thus, this activity was carried out within the private context of discovery that José Ibáñez Martín situated "in the calm half-darkness of the laboratory and in the quiet place of the library."[28] However, the patterns of action in which this scientific *ethos* was embedded also constitute such a crystallised form of social organisation which, therefore, is a means of regulating the scientific community.[29] In this regard, this scientific method portrays itself as shaped by the values of the Christian rhetoric: Humility, constriction, asceticism and personal sacrifice would put these scientists in contact with nature, *the nourishing mother*,[30] whom they must treat with respect and objectivity, avoiding any kind of intervention.[31]

The scientist would therefore be the discoverer of the secrets of Divine Wisdom, a modest man characterised by attributes such as moderation, discipline and selfless work.[32] The *vir modestus* was constructed upon deep-seated cultural biases that also carried sexual and gender meanings that devalued women and overvalued men.[33] In this regard, while its attributes appear to be founded on reality, they are actually built out of the original male/female opposition: *rational/emotional, rational/irrational, objective/subjective, universal and impartial/specific and personal, culture/ nature, big/small, top/bottom, public/private* and so on.[34] Hence, from

[26] AGUN/JIM/592/Memoria Proyecto Presupuesto 1956–1957.
[27] Letter from February 5, 1960 (AGUN/JIM/590/S-21).
[28] CSIC (1951): 77.
[29] Shapin, Steven; Schaffer, Simon (2005): 44.
[30] Pérez Sedeño, Eulalia (2011): 105–106.
[31] Lora Tamayo, Manuel (1969): 29.
[32] Albareda, José María (1951): 15–76.
[33] Haraway, Donna (2004): 230–232; Harding, Sandra (1997): 186–200.
[34] Lloyd, Genevieve (1984); Howkesworth, Mary E. (1989); Bourdieu, Pierre (2007): 107–128.

these classification systems emerge guidelines for action that establish equivalences between physical and social space, thus facilitating entry for those who adopt the manners that are both recognised and recognisable within the scientific field.

In consequence, while this image of the scientist portrays itself as universal, it actually supports a particular way of being that is sexually characterised—and even legally sanctioned—and "tailor-made for men".[35] A masculine scientific modesty was linked to trustworthiness, a model the dictatorship reinforced strategically. With this in mind, I will now consider the importance of basic social institutions such as the family, church, state and school in strengthening male domination within the conditions and precepts of National Catholicism in Franco's Spain. This in turn shaped the rest of the practices involved in the governing of bodies as well as the practices of indoctrination and social control during the different stages of the Francoist period.[36]

Modest Witness? Science and Gender in Dispute

If God had reserved for women "the pains of childbirth, the work of breastfeeding, and the early education of children",[37] during the first Francoism, Spain reserved for women the mandate of domestic subjection. Revolving around the home, the dictatorship articulated its discourse of femininity, which regulated everything—ranging from clothing to conscience—in accordance with "the high mission of spiritual and social order that, in the perfect Christian life, corresponds to the female sex, which is responsible for defending traditional family values".[38] In this regard, these values were in open opposition to those belonging to the emancipated woman—then dubbed as a product of "masculinising feminism"—or the woman that turned into an "arid intellectual product".[39] Both female ways of being were then seen as opposed to the truly feminine values,[40] those embodied in the figure of the woman as *mother and wife*. A female mode that was taught at school, routinised on the street and shielded by law. Thus, while women were prevented by law from accessing certain profes-

[35] Bourdieu, Pierre (2000): 75–83.
[36] Cayuela Sánchez, Salvador (2014).
[37] Pio XII (1945): 291.
[38] Roca I Girona, Jordi (1996): 81.
[39] Di Febo, Giuliana (2003): 35; 28.
[40] Roca I Girona, Jordi (1996): 143–148.

sions, by *common sense*, people could, with complete impunity, brand women as bad mothers and wives—as the director of the "Balmes" Institute at the CSIC did—for merely holding scientific positions.[41]

Notably, if the Second Spanish Republic was a political regime that defended—normatively and legally—the equal rights of men and women,[42] the Franco regime returned, however, to a patriarchal order wherein women were controlled through the reproductive, educational and labour policies of the dictatorship. Indeed, rather than being considered as autonomous selves, under Franco women were dependent individuals[43] that, due to their role in the family, were seen as guarantors of those sacred values revolving around motherhood, home and care: The ideal of the "Angel of the Home".

To this end, and employing norms and decrees (but also socially shared perceptions[44]), the dictatorship made it difficult for women to feel comfortable even in allegedly neutral spheres such as those of high culture. Consider, for instance, the case of Carmen Conde; in May 1959, José Ibáñez Martín called on the Council to promptly resolve her situation in the CSIC, and treat her "like the rest of the personnel and, therefore, assimilating her to their condition".[45] Likewise, as the physicist of the CSIC María Egüés recalled, José Casares—a former member of The Board for Advanced Scientific Studies and Research (JAE, for its initials in Spanish) and later the first director of the CSIC's Institute of Physics and Chemistry[46]—did not want women at his Institute. "He tolerated Piedad (de la Cierva) because she had worked there for a long time and he knew her family."[47]

In the same vein, the Spanish pioneer of chromatography, María Josefa Molera, also experienced gender discrimination in her youth. After obtaining her degree in chemistry, Julio Casares Gil rejected her as a PhD student just because she was a woman.[48] Even the renowned biochemist

[41] Alcalá, Paloma; Magallón, Carmen (2008): 160.

[42] *Gaceta de la República*, no. 35 (04/02/1937): 635–636.

[43] Gallego Méndez, Mª Teresa (1999): 214–215.

[44] Nash, Mary (2000): 627–646.

[45] Letter from José Ibáñez Martín to the vice president of the CSIC José Royo (May 23, 1959) (AGUN/JIM/590/R-7).

[46] Suay Matallana, Ignacio (2012): 127–132.

[47] Moya De Guerra, Elvira (2002): 564.

[48] Carrascosa, Alfonso V (2012): 383–384.

Alberto Sols[49] remembered that, when Margarita Salas went to his laboratory in search of a site for her doctoral research, he thought: "Bah, a girl. I will give her a research topic that isn't very interesting. If she doesn't succeed, it won't matter". In this regard, the biochemist and molecular biologist Margarita Salas argued that such statements "could reflect the mentality that existed in 1961 regarding what could be expected from the scientific work of a woman".[50]

Unlike her experience as a student of chemical sciences in Madrid (1955–1960) where Margarita Salas experienced no sexual discrimination, working with Alberto Sols she never felt she was treated as a peer. Even during the 1960s and early 1970s, after having completed her research training with Alberto Sols and Severo Ochoa, Margarita Salas continued to be seen in Spain as "Eladio Viñuela's wife". Although her marriage with an outstanding scientist (like Salas, Viñuela was a pioneer in the field of molecular biology) could have overshadowed her contributions as a researcher, it also provided her with a convenient social cover that enabled her to develop professional interactions that, otherwise, may have been seen as unacceptable for a woman in a man's world.[51]

Undoubtedly, in Franco's Spain where the paid and extra-domestic work of married women was dangerous for the stability of the family (and an affront to the male role),[52] for female scientists to count with a partner and in surroundings conducive to their interest in acquiring knowledge made things easier. Similarly, this proximity to (male) scientists granted these women an essential social capital in a time and environment dominated by strong sexist prejudices. In this regard and framed by this premise, I shall present some information in the following pages about those women who had sufficient and legitimate independent status to qualify as scientists under Franco. Some female scientists, for instance, had "the privilege of travelling to train abroad", something that, as María Jesús Santesmases, Montserrat Cabré i Pairet and Teresa Ortíz Gómez argued, then turned into an exclusive "way of gaining access to professionalisation and academic prestige".[53]

[49] According to Margarita Salas, Alberto Sols was "the best biochemist there was at that time in Spain". She was able to work with him as a result of a recommendation from Severo Ochoa (Interview with Margarita Salas conducted in her office at the Centre for Molecular Biology (CBM) on October 24, 2016).

[50] Salas Falgueras, Margarita (2011): 20.

[51] Santesmases, María Jesús (2000): 114.

[52] Roca I Girona, Jordi (2005): 87–91.

[53] Santesmases, María Jesús; Cabré I Pairet, Montserrat; Ortiz Gómez, Teresa (2017): 397.

Women in Applied Sciences During Franco's Dictatorship

"Science and Technology need their men: their self-sacrificing, dedicated men." Thus began the introduction of *Colaboradores e investigadores del Consejo Superior de Investigaciones Científicas* (1956), the third volume of the General Publications of the CSIC. The aim of this book was to take stock of the changes that occurred over fifteen years in this institution; a period of time wherein several job positions at the Council—ranging from "scientific collaborator" (1945) to "scientific researcher" (1947)—first appeared. Therefore, this volume represented an index of the scientific collaborators and researchers who worked in the CSIC until 1955.

Flipping through the pages of this book,[54] we can see that regarding the Applied Sciences branches of the Council, for instance, only four of the thirty-two people belonging to the "Alfonso el Sabio" Board of Mathematics, Physics and Chemistry were women. One of those was the chemist Mercedes Cubero Robles, disciple and collaborator of Luis Brú Villaseca (himself a disciple of the distinguished Spanish physicist Julio Palacios), a woman who was married to Agustín Vioque Pizarro, researcher at the Institute of Fats at the Council.[55] However, the other three women belonged to the "Antonio de Gregorio Rocasolano" Institute of Physical Chemistry: Isabel Martín Tordesillas, María Domínguez Astudillo and María Josefa Molera Mayo. While the chemist Domínguez Astudillo was a member of the Teresian Association in Spain who would direct the Radiobiology Laboratory of the CSIC in the 1950s,[56] the chemist Molera Mayo—the daughter of a commandant of the Nationalist faction who died during the Spanish Civil War—would direct the section of chemical kinetics in the 1960s. Moreover, in 1957 Molera Mayo was married to Joaquín Hernáez Marín, who was also a researcher at the CSIC.[57]

Likewise, regarding the "Juan de la Cierva" Technical Research Board, only two of the twenty people listed were women. Among the female researchers was Olga García Riquelme, outstanding member of Miguel Catalán's team (the Spanish scientist who discovered the "multiplets") at

[54] CSIC (1956).

[55] Marquez Delgado, Rafael (2004): 117; ABC Sevilla, 11/10/1969.

[56] (http://www.institucionteresiana.org/es/mujeres-destacadas/maria-astudillo); Paz Velazquez, Flavia (2002): 56.

[57] Carrascosa, Alfonso.V. (2012): 384.

the "Daza de Valdés" Institute of Optics,[58] the centre wherein Maria Teresa Vigón Sánchez was the director of the Photochemistry Division.[59] As the daughter of General Juan Vigón—Minister of the Air Force and, later, chief of the Nuclear Energy Board (JEN, for its initials in Spanish)—Maria Teresa Vigón enjoyed close ties in fields of power and a favourable family environment, which both she and her sister María Aránzazu Vigón employed pursuing a successful scientific career in Franco's Spain. In this regard, since 1949, María Teresa Vigón was also one of the few women holding a "Scientific Researcher" position at the Council. Therefore, some of the premises already mentioned seem to be true: These women's profiles were distinguished by close links to men in the Spanish scientific field and even to members of the Francoist elite. Along with this essential social capital within a field dominated by men, those women too were legitimated scientists: they were not purged or did not go into exile after the Spanish Civil War (1936–1939).

Furthermore, like many of their male colleagues at the CSIC, they would not dismiss the chance to intern abroad and obtain legitimation for themselves within the international scientific field. However, a quick glance at the volume *Becarios del Consejo Superior de Investigaciones Científicas* (1957) reveals that the majority of internships awarded then belonged to men.[60] For instance, looking at the same boards of trustees mentioned earlier, between 1940 and 1956, among the 128 researchers awarded at the "Alfonso el Sabio" Board with internships abroad, only eight were female researchers (6.2% of the board's fellowship grant holders). Curiously, these were women with names we are familiar with: while Mercedes Cubero Robles spent six months at the University of Graz (Austria) in 1952, María Josefa Molera Mayo got an internship in 1950 for a year working at the Institute of Physical Chemistry in Oxford (England). A little earlier, in 1949, Domínguez Astudillo spent five months at the Radiochemical Laboratory in Cambridge (England).

Some years earlier, in 1946, Amelia D´Ocon Asensi—PhD student and wife of Manuel Lora-Tamayo[61]—spent a month in Switzerland and England. Likewise, Baldomera Velázquez Sánchez, who preferred a non-

[58] Sánchez Ron, José Manuel (1999): 319–327; Sánchez Ron, José Manuel (1994): 360; 345–394.

[59] Romero De Pablos, Ana (2017): 335–344.

[60] CSIC (1957).

[61] Lora-Tamayo, Manuel (1993): 74.

English-speaking destination, got a grant in 1955 to spend one year and seven months at the Institut Français du Caoutchouc in Paris (France). However, most of these female scientists travelled to Anglo-Saxon nations in the post-war period. In this regard, after studying in the United States with a scholarship from Bryn Mawr College,[62] Margarita Bernis Madrazo spent three weeks in London in 1951 perfecting her knowledge of electronics. At around the same time, Antonia Martín-Tesorero Álvarez—who was married to José García Santesmases, Professor of Physics at the University of Madrid[63]—was studying calculations of optical systems in Boston (USA) while, in 1956, María Luisa Canut Ruiz, wife of José Luis Amorós Portolés (a former researcher of the Council), spent six months at University College London (England).

For its part, between 1940 and 1956, the "Juan de la Cierva" Technical Research Board awarded fellowship grants to 103 researchers of which only seven were women (6.8% of the board's fellows), including some of the female scientists already mentioned such as María Teresa Vigón Sánchez. In November 1947, she got a grant to study abroad for ten months at the Technical School of Zürich (Switzerland) with Professor Eggert. Another such case is Olga García Riquelme, who, in 1952, obtained a grant to work for eighteen months at the University of Lund (Sweden) with Professor Edlén. Some years earlier, in 1948, Piedad de la Cierva Viudez got an internship at the National Bureau of Standards in Washington (USA), even if it was not her first training experience abroad—in 1936, she was already working with Professor Georg von Hevesy at the Universitetets Institut for Teoretisk Fysik in Copenhagen, thanks to a scholarship funded by the JAE.[64] The daughter of a well-connected family,[65] Piedad de la Cierva was also the first woman at the Institute of Physics and Chemistry of the CSIC due to both her career accomplishments and her personal ties to major scientific figures of the Francoist era.[66]

While in 1956 María Teresa Sardiña Gallego opted to spend four months at the Sorbonne (Paris) with Professor Mathieu, other women chose instead to find their way in Germany. On November 1, 1954, Victoria Redondo Mena obtained a grant to work for fifteen months at the

[62] Bryn Mawr College (1950): *Report of the President to the Board of Directors (1949–1950)*, December: 27.

[63] ABC (10/12/1950).

[64] Magallón Portolés, Carmen (2004): 240–242.

[65] Romero De Pablos, Ana (2017): 325–326.

[66] Alva Rodriguez, Inmaculada (2016): 6; Pimentel, Juan (2020): 291–342.

Optisches Institut der Technische Universität Berlin with Professor Slevogt, while María del Carmen Sánchez Conde left in 1955 to work with Professors Neuhaus and Konopicky at the Bonn Institutes of Mineralogy and Refractory Research. In 1956, and despite the misgivings of a colleague,[67] Juana Bellanato Fontecha got an internship to work for nine months at the Physical Chemistry Institute of the University of Freiburg with Professor Reinhard Mecke. Two years earlier, Bellanato had defended her doctoral thesis directed by the rehabilitated professor Miguel Catalán, who had agreed to do so upon the proposal of José Barceló Matutano,[68] researcher at the Council and Bellanato's former teacher.

Nevertheless, if we expand the scope of study to include all of the Boards of the Pure and Applied Sciences within the Council,[69] we can see that, according to official data, internship grants were awarded to 514 men and 49 women over a period of sixteen years, representing female scientist as only 8.7% of the fellows. The sum is insignificant indeed, but worthy of consideration compared with the fact that of the 113 individual grants awarded to women by the JAE between 1908 and 1934, only 18 (16%) were designated for studies of a scientific nature.[70] On the other hand, these data present some interesting findings. Among these female research fellows, more than half (57%) worked at the Institutes of Natural and Agricultural Sciences. These centres employed a greater number of researchers who specialised in more *feminised*[71] areas of knowledge. Therefore, the dynamics particular to the scientific field come into play not only at the national level but also at the international level.

It becomes apparent that within the scope of what we call "science", there are more women in some fields than in others. Fields such as engineering or the technological and applied branches of science are prime examples of this segregation. In this regard, women within these fields would be known as "pioneers", "exceptions" or "token" women in science. This is because, traditionally, gaining access to these fields was difficult for women. In many cases, those who achieved such access were those

[67] Juana Bellanato Fontecha o la voluntad de ser científica (http://161.111.98.22/web/guest/juana-bellanato) (Accessed 04.03.2020).

[68] "Recuerdos de Juana Bellanato y medalla Miguel Catalán" (https://www.tendencias21.net/fisica/recuerdos-de-juana-bellanato-y-medalla-miguel-catalan_a32.html) (Accessed 04.03.2020).

[69] García Naharro, Fernando (2017): 120–131.

[70] Alcalá, Paloma; Magallón, Carmen (2008): 149.

[71] Santesmases, María Jesús (2000): 85.

who had relatives, husbands or male friends already working in these fields. Because of this, the historian Margaret W. Rossiter termed these scientific terrains difficult for women to access as "peripheral" fields.[72] Building upon this premise, I will finish this paper looking at the women who published scientific articles in journals belonging to these areas of science in Franco's Spain.

Women in "Peripheral" Fields. Female Authors in the *Revista de Ciencia Aplicada*

Some features discussed earlier deal with the importance of the sociopolitical assumptions involved in making individualist attributions of authority, something crucial for understanding who had a legitimated voice in the field. In addition to that, I argue that these ideas could be embodied in (and could be understood through) the centrally authorised scientific journals produced under the regime.[73] In this regard, the scientific journals of the CSIC were then considered the official *locus* of scientific authority in Spain. Built out of specific names, these journals were an active instrument in shaping the Spanish scientific community, as well as a vehicle for revealing the alleged "consecration and inner vitality" of the Council to the world.[74] Although the publications do not seem to have been as much of a determining factor as seniority was in the internal promotion of researchers,[75] in them, however, we see the names of those individuals labelled as *modest witnesses* in Franco's Spain.[76] In order to show the catalogue of approved authors who deserve to appear in their publications, I will look at the women who published in the *Revista de Ciencia Aplicada*, "the central journal" of the "Juan de la Cierva" Board of the CSIC, whose first issue was published at the end of 1947.

Looking a little closer reveals that scientific journals can speak to the larger issues. In this regard, if we look at the more than 280 authors who signed—as authors or co-authors—articles in the *Revista de Ciencia Aplicada* during the twenty-year period between 1947 and 1967, we see that at least twenty-seven women appear among the signatories (9.6% of

[72] Rossiter, Margaret W. (1997): 171.
[73] García Naharro (2020a).
[74] CSIC (1943): 47.
[75] Santesmases, María Jesús (2000): 137–138.
[76] García Naharro (2020b).

the authors and co-authors). Despite all of these female authors belonging to official scientific institutions, their professional profiles contrast with those of most of the male signatories, who were active engineers linked to the CSIC, Spanish or foreign universities, big companies and institutions wherein they held positions of responsibility.

The first article signed by a female researcher appeared in volume 27 of the journal in 1952. The author was María Alicia Crespí, who was identified in the article as a PhD in Chemical Sciences but with no institutional affiliation available.[77] However, Crespí was a woman raised in an intellectual environment who worked in "Piritas Españolas" of the National Institute of Industry and moved to JEN in March 1957.[78] Awarded in 1970 with a grant from the Juan March Foundation,[79] María Alicia Crespí would later become one of the first women to hold a professorship at a Superior Technical School in Spain, as chair of Electrical Engineering at the School of Architecture in Madrid.[80]

Nevertheless, female authors in the journal normally signed their works co-authored with higher-ranking colleagues. Thus, for instance, in 1953 Fermina González Pellissó, labelled as a chemistry graduate and researcher at the Institute of Fats, published an article co-authored with the chief of the Chemistry Division, José María Rodríguez de la Borbolla, and the researcher Carlos Gómez Herrera. It was the ninth article published with the same generic title in different journals but the first that appeared in this publication with Fermina as co-author. Likewise, in 1954 Felisa Núñez Cubero published the article "*Imanes permanentes*"[81] along with her doctoral adviser Salvador Velayos. Despite his background as a disciple of the renowned physicist Blas Cabrera at the National Institute of Physics and Chemistry of the JAE, Salvador Velayos was then working at the Council, in the Division of Magnetism of the "Alonso de Santa Cruz" Institute of Physics.[82] In 1956, the sole author of "*La curva de desimanación de materiales para imanes*" Felisa Núñez nonetheless expressed her "gratitude to Professor S. Varela for the great interest with which he

[77] *Revista de Ciencia Aplicada*, No 27, Year VI, issue 4 (July–August, 1952) pp. 331–346.

[78] *La Voz de Galicia* (04/11/2012).

[79] Grant *España*, 1970. *Estudios Técnicos e Industriales* (http://recursos.march.es/web/prensa/anales/1966-1970/28-Indice-onomastico-1966-1970.pdf).

[80] *ABC* (28/04/1977): 22; BOE (02/07/1977).

[81] *Revista de Ciencia Aplicada*, No 40, Year VIII, Issue 5 (September–October, 1954) pp. 385–392.

[82] Baltá Calleja, Francisco J. (2009): 176. Otero Carvajal, Luis Enrique (2006): 155.

directs my work, and to the Spanish National Research Council, which has made it possible to carry out this work".[83] With these words, Felisa Nuñez Cubero returned kindness to the institution that gave her the chance for receiving training abroad. In the late 1950s, she got an internship grant to spend a year in England, working with the physicist Leslie Fleetwood Bates at the University of Nottingham.[84] At around the same time, she was Lecturer of Physics at the University of Madrid,[85] where she would later become the first to hold a professorship at the Engineering School of Madrid.[86]

However, it was in 1955 that an article signed by a female researcher holding a position of responsibility first appeared: the author was the chemist Piedad de la Cierva, who was the head of the Chemistry Laboratory at the Naval Laboratory (LTIEMA). Her article reported the work of the team that had been awarded by the "Juan de la Cierva" Technical Research Board in 1954.[87] This pioneering female scientist, who has been mentioned throughout these pages, was also the first to publish in the journal a paper co-authored with another woman. The co-author was Francisca de Andrés, a member of her research team at the Naval Laboratory and her former PhD student.[88] De Andrés and de la Cierva signed the paper "Propiedades físico-químicas de arcillas españolas", the opening article of volume 63 of the *Revista de Ciencia Aplicada*. As they noted, however, this article was the third of a series of papers presented at international conferences and meetings.[89]

Nevertheless, it was not until the 1960s that other female authors holding high-ranking positions published articles in the journal. In this regard, one of those women was the chemist María Josefa Molera, labelled in the article as the head of the Chemical Kinetics Division at Rocasolano Institute of the Council.[90] This female scientist, who suffered gender dis-

[83] *Revista de Ciencia Aplicada*, No 48, Year X, Issue I (January–February, 1956) pp. 30–33.

[84] CSIC (1961): 70.

[85] *Revista de Ciencia Aplicada*, No 73, Year XIV, Issue 2 (March–April, 1960) pp. 97–106.

[86] "Las mujeres en la Universidad Politécnica de Madrid" (Exhibition *(1910–2010) Mujer y Universidad, un siglo de vida "La Mujer en la Universidad, la Ingeniería, la Arquitectura y el Deporte* (November, 2011).

[87] *Revista de Ciencia Aplicada*, No 45, Year IX, Issue 4 (July–August, 1955) p. 289.

[88] Alva Rodriguez, Inmaculada (2016): 9.

[89] *Revista de Ciencia Aplicada*, No. 63, Year XII, Issue 4 (July–August, 1958) pp. 289–292.

[90] *Revista de Ciencia Aplicada*, No. 112, Year XX, Issue 5 (September–October, 1966) pp. 385–394.

crimination in her youth, was by then a veteran in her field. Furthermore, like some of the women already mentioned, she had achieved a certain kind of power and prestige within the CSIC, the major scientific institution in Franco's Spain that, however, as we have seen, was also characterised by an intrinsic unequal distribution of power between male and female researchers.[91]

Conclusions

In sum, during the Franco regime only a few women could adapt to both the dynamics of the scientific field and the professional segregation that prevailed in the CSIC.[92] However, in assessing individuals, categories such as social class, systems of thought and gender should never be underestimated as explanatory variables in scientific biographies throughout history. For instance, most of the women who passed through the different divisions of the National Institute of Physics and Chemistry during the Second Spanish Republic belonged to "the enlightened middle class, linked to the republican circles".[93] Therefore, having family support and being raised in environments conducive to an interest in science are traits that have traditionally distinguished the biographies of women scientists in Spain. This is even more true in the context of Franco's dictatorship, where the importance of such determinants became more pronounced. Immersed as they were in a heavily ideologised atmosphere, women found themselves facing a regime that proclaimed motherhood as women's paramount value and the domestic sphere its realm of action.

This system of beliefs also informed the "emphatically Catholic"[94] official image of science under Franco. Running alongside these concerns, I have argued that the Franco regime supported the traditional ethos of the *vir modestus* with the social and political implications that it carried within: beliefs about the nature of the experts and their knowledge claims, premises that then featured importantly in the construction of those who had a legitimate voice in the Spanish scientific field. However, despite all of these assumptions and expectations, some female scientists also knew how to find a place there. These women had brilliant records and careers; many of

[91] Alcalá, Paloma; Magallón, Carmen (2008): 161.
[92] Alcalá Cortijo, Paloma (1996): 61–76.
[93] Magallón Portolés, Carmen (2010): 346.
[94] CSIC (1942): 48.

them had also been trained in select institutions and were raised in affluent and Christian environments where the demand for academic excellence was the norm. Furthermore, and importantly, all of them had favourable social connections and possessed social capital that they deployed, as we have seen, in networks emerging from recommendations and from social and conjugal relations with (male) scientists of their time. This social capital proved crucial and very advantageous in their navigation of an adverse environment and an authoritarian and dictatorial era dominated by ideological principles and sexist prejudices.

Primary Sources

General archives of the Universidad de Navarra (AGUN): José Ibáñez Martín (JIM) Fund.

Memories of the General Secretary of the Consejo Superior de Investigaciones Científicas (CSIC).

Revista de Ciencia Aplicada (Patronato Juan de la Cierva/Consejo Superior de Investigaciones Científicas).

Interview with Margarita Salas (1938–2019) conducted in her office at the Centre for Molecular Biology (CBM, for its initials in Spanish) on October 24, 2016.

References

Albareda, José María (1951): *Consideraciones sobre la investigación científica*, Madrid, CSIC.

Alcalá, Paloma; Magallón, Carmen (2008): "Avances, rupturas y retrocesos: mujeres en las ciencias experimentales en España (1907–2005)" en Romero De Pablos, Ana y Santesmases, María Jesús (coords.), *Cien años de política científica en España*, Bilbao, Fundación BBVA: 141–169.

Alcalá Cortijo, Paloma (1996): "Españolas en el CSIC. Presencia y status de las mujeres en la investigación científica española: el CSIC 1940–1993" en Ortiz Gómez, Teresa; Becerra Conde, Gloria, (eds.), *Mujeres de ciencias. Mujeres, feminismos y ciencias naturales, experimentales y tecnológicas*, Granada, Universidad de Granada: 61–76.

Alva Rodriguez, Inmaculada (2016): "Piedad de la Cierva: una sorprendente trayectoria profesional durante la Segunda República y el Franquismo", *Arbor*, 192 (779): a322.

Antolín Hofrichter, Andrés (2021): "La modernidad ajena": Orden simbólico y política científica en el CSIC" en Jauné-Miret, Marició; Presas I Puig, Albert (eds.), *Nacionalización e internacionalización de la cultura y la ciencia en el primer Franquismo (1939–1959).*

Baltá Calleja, Francisco J. (2009): "Sinopsis sobre la investigación en Física del CSIC en el edificio *Rockefeller*: periodo 1940–1975" en González Ibáñez, Carlos; Santamaría García, Antonio (eds.), *Física y química en la Colina de los Chopos. 75 años de investigación en el edificio Rockefeller del CSIC (1932–2007)*, Madrid, CSIC: 175–180.

Barral Morán, María José; Magallón Portolés, Carmen; Miqueo Miqueo, Consuelo; Sánchez González, María Dolores (coord.) (1999): *Interacciones ciencia y género. Discursos y prácticas científicas de mujeres*, Barcelona, Icaria editorial.

Becerra Conde, Gloria; Ortíz Gómez, Teresa (coord.) (1996): *Mujeres de ciencias: mujer, feminismo y ciencias naturales, experimentales y tecnológicas*, Granada, Instituto de Estudios de la Mujer/Universidad de Granada.

Bourdieu, Pierre (2007): *El sentido práctico*, Buenos Aires, Siglo XXI.

Bourdieu, Pierre (2000): *La dominación masculina*, Barcelona, Anagrama.

Box, Zira (2019): "Los atributos de la nación. Género y clase en la España franquista" en Encarnación Lemus; Manuel Peña (eds.), *Alianzas y propaganda durante el primer franquismo*, Barcelona, Planeta: 125–146.

Butler, Judith (1993): *Bodies that matter. On the discursive limits of "sex"*, Routledge, New York/London.

Carrascosa, Alfonso V (2012): "Molera Mayo, Maria Josefa" en *Diccionario Biográfico Español*, Tomo XXXV, Mestre y Bosch-Montoya: 383–384.

Cayuela Sánchez, Salvador (2014): *Por la grandeza de la patria. La biopolítica en la España de Franco (1939–1975)*, Madrid, Fondo de Cultura Económica.

CSIC (1940): *Estructura y norma de la investigación Nacional*, Madrid, CSIC.

CSIC (1942): *Memoria de la Secretaría General (1940–1941)*, Madrid, CSIC.

CSIC (1943): *Memoria de la Secretaría General (1942)*, Madrid, CSIC.

CSIC (1944): *Memoria de la Secretaría General (1943)*, Madrid, CSIC.

CSIC (1948): *Memoria de la Secretaría General (1946–47)*, Madrid, CSIC.

CSIC (1951): *Memoria de la Secretaría General (1950)*, Madrid, CSIC.

CSIC (1952): *Memoria de la Secretaría General (1951)*, Madrid, CSIC.

CSIC (1956): *Colaboradores e investigadores del Consejo Superior de Investigaciones Científicas*, Madrid, CSIC.

CSIC (1957): *Becarios del Consejo Superior de Investigaciones Científicas*, Madrid, CSIC.

CSIC (1960): *Organización y presupuestos de la investigación en el mundo*, Madrid, CSIC.

CSIC (1961): *Becarios del Consejo Superior de Investigaciones Científicas 1957–1960*, Madrid, CSIC.

Delgado, Lorenzo; López, Santiago M. (eds.) (2019): *Ciencia en transición. El lastre franquista ante el reto de la modernización*, Madrid, Silex.

Di Febo, Giuliana (2003): "Nuevo Estado, nacionalcatolicismo y género" en Nielfa Gloria (ed.), *Mujeres y hombres en la España franquista: sociedad, economía, política y cultura*, Madrid, Editorial Complutense: 19–44.

Gallego Méndez, Mª Teresa (1999): "Mujeres en el franquismo o la desmesura de lo privado" en Fagoaga Concha (coord), *1898–1998. Un siglo avanzando hacia la igualdad de las mujeres*, Madrid, Dirección General de la Mujer: 209–221.

García Naharro, Fernando (2017): *El papel de la ciencia. Publicaciones científicas y técnicas durante el franquismo (1939–1966)*, Madrid, UCM.

García Naharro, Fernando (2020a): "Reshaping Scientific Journals. Applied Science and its audience under Franco (1939–1966)", *Journal of Iberian and Latin American Studies* (https://doi.org/10.1080/14701847.2020.1826180).

García Naharro, Fernando (2020b): "El "audible silencio" durante el franquismo. Artículos científicos y técnicos publicados en revistas y lenguas extranjeras por investigadores del CSIC (1939–1964)", *Llull*, Vol. 43, Nº 87: 227–246.

Haraway, Dona (2004): "Modest_ Witness@Second_Millennium", in *The Haraway Reader*, New York/London, Routledge: 223–250.

Harding, Sandra (1997): "Women´s Standpoints on Nature. What makes them possible?", *Osiris*, Vol 12: 186–200.

Howkesworth, Mary E. (1989): "Knowers, knowing, known: feminist theory and claims of truth" *Signs*, Spring 1989; 14, 3: 533–557.

Lora Tamayo, Manuel (1969): *Un clima para la ciencia*, Madrid, Gredos.

Lora-Tamayo, Manuel (1993): *Lo que yo he conocido. Recuerdos de un viejo catedrático que fue ministro*, Cádiz, Federico Joly y Cia.

Lloyd, Genevieve (1984): *The Man of Reason. "Male" and "Female"* in *Western philosophy*, Minneapolis, University of Minnesota Press.

Magallón Portolés, Carmen (2004): *Pioneras españolas en las ciencias: las mujeres del Instituto Nacional de Física y Química*, Madrid, CSIC.

Magallón Portolés, Carmen (2010): "Las mujeres que abrieron los espacios de las ciencias experimentales para las mujeres, en la España del primer tercio del siglo XX", *Arenal*, 17:2; julio-diciembre: 319–347.

Malet, Antoni (2007): "Las primeras décadas del CSIC: investigación y ciencia para el franquismo" en Ana Romero de Pablos y María Jesús Santesmases, *Cien años de política científica en España*, Bilbao, Fundación BBVA, pp. 211–256.

Marquez Delgado, Rafael (2004): "La física en la antigua Facultad de Ciencias de la Universidad de Sevilla" en Castillo Martos, Manuel; Ternero Rodríguez, Miguel (Coord.), *La ciencia en la historia de la universidad española. 92 años de Química en Sevilla*, Sevilla, US: 101–124.

Moya De Guerra, Elvira (2002): "Mujeres en Ciencia y Tecnologías Físicas en el CSIC", *Arbor* CLXXII, 679–680 (Julio-Agosto): 549–577.

Nash, Mary (2000): "Maternidad, maternología y reforma eugénica en España, 1900–1939", en DUBY Georges y Perrot, Michelle (dir.), *Historia de las mujeres. El siglo XX*, Buenos Aires, Taurus: 627–646.

Ortiz Gómez, Teresa; Becerra Conde, Gloria, (eds.), *Mujeres de ciencias. Mujeres, feminismos y ciencias naturales, experimentales y tecnológicas*, Granada, Universidad de Granada.

Otero Carvajal, Luis Enrique (2006): *La Destrucción de la ciencia en España. Depuración Universitaria en el franquismo*, Madrid, UCM.

Paz Velazquez, Flavia (2002): *Sal de tu tierra*, Cuadernos biográficos Pedro Poveda, 7, Madrid, Narcea.

Pérez Sedeño, Eulalia; Alcalá Cortijo, Paloma (coord.) (2001): *Ciencia y género*, Madrid, Editorial Complutense.

Pérez Sedeño, Eulalia (2011): "El sexo de las metáforas" *Arbor*, Vol. 187–747 enero-febrero: 99–108.

Pimentel, Juan (2020): *Fantasmas de la ciencia española*, Madrid, Marcial Pons.

Pio XII (1945): *La familia cristiana. Discursos del Padre Santo a los recién casados 1939–1945*, San Sebastián, Editorial Pax.

Puig-Samper Mulero, Miguel Ángel (ed.) (2007): *Tiempos de investigación, JAE-CSIC, cien años de ciencia en España*, Madrid, CSIC, 2007.

Roca I Girona, Jordi (1996): *De la pureza a la maternidad. La construcción del género femenino en la postguerra española*, Madrid, Ministerio de Educación y Cultura.

Roca I Girona, Jordi (2005): "Los (no) lugares de las mujeres durante el franquismo: el trabajo femenino en el ambito publico y privado", *Geronimo de Uztariz*, 21: 81–99.

Romero De Pablos, Ana; Santesmases, María Jesús (coord.) (2017): *Historia biográficas: género y científicas en España*, Dossier en *Arenal*, 24:2; julio-diciembre.

Romero De Pablos, Ana (2017): "Mujeres científicas en la dictadura de Franco. Trayectorias investigadoras de Piedad de la Cierva y María Aránzazu Vigón", *ARENAL*, 24:2; julio-diciembre: 319–348.

Rossiter, Margaret W. (1997): "Which science? Which women?" *Osiris*, Vol. 12: 169–185.

Salas Falgueras, Margarita (2011): "Mujer y ciencia" en Mateos Cachorro, Ana y Perote Alejandre, Alfonso (coord.), *Genes, ciencia y dieta. Lecciones sobre evolución humana*, Madrid, Instituto Tomás Pascual/CENIEH: 13–24.

Sánchez Ron, José Manuel (1994): *Miguel Catalán. Su obra y su mundo*, Madrid, Fundación Ramón Menéndez Pidal/CSIC.

Sánchez Ron, José Manuel (1999): *Cincel, martillo y piedra. Historia de la ciencia en España (siglos XIX y XX)*, Madrid, Taurus.

Santesmases, María Jesús (2000): *Mujeres científicas en España (1940–1970): profesionalización y modernización social*, Madrid, Instituto de la Mujer.

Santesmases, María Jesús; Cabré I Pairet, Montserrat; Ortiz Gómez, Teresa (2017): "Feminismos biográficos: aportaciones desde la historia de la ciencia", *ARENAL*, 24:2; julio-diciembre: 379–404.
Sesma Landrín, Nicolás (2006): "Franquismo ¿Estado de derecho? Notas sobre la renovación del lenguaje político de la dictadura durante los años 60", *Pasado y memoria. Revista de Historia Contemporánea*, 5: 45–58.
Shapin, Steven; Schaffer, Simon (2005): *El Leviathan y la bomba de vacío. Hobbes, Boyle y la vida experimental*, Bernal, Universidad Nacional de Quilmes.
Suay Matallana, Ignacio (2012): "Ciència, política i diplomacia: José Casares Gil (1866–1961)" en *Actes de la IX Jornada sobre la Història de la Ciència i l'Ensenyament*, Barcelona, SCHCT-IEC: 127–132.

CHAPTER 8

A Field Open to Women: Censorship of Children's and Youth Literature Under Franco Through Women Readers

Ramón Tena-Fernández and José Soto-Vázquez

Introduction: Objectives and Methodology

In this article, we address women censors of children's and youth literature (CYL) in Spain to shed light on their names and professional profiles and understand how the process of censoring these works was enacted under Franco. We have four objectives. First, given that censors were

This work is part of the activities of the Children's and Youth Literature Research Group of the Junta de Extremadura (SEJ036). Co-financed by FONDOS FEDER. Operational programme FEDER and Junta de Extremadura. File number GR18026. It also benefited from participation in the international symposium *Cultura y ciencia nacionales en el primer franquismo [National culture and science in early Francoism] (1939–1959)*, UPF, Barcelona, 13–14 December, 2018. FFI-HAR Funds 2016-75559 (AEI/ERDF, EU).

R. Tena-Fernández (✉) • J. Soto-Vázquez (✉)
Universidad de Extremadura, Badajoz, Spain
e-mail: rtena@unex.es; jsoto@unex.es

M. Janué i Miret, A. Presas i Puig (eds.), *Science, Culture and National Identity in Francoist Spain, 1939–1959*,
https://doi.org/10.1007/978-3-030-58646-1_8

designated with a "reader" number, we seek to determine whether a single number was used to refer to different readers specialising in a particular literary genre or type of audience and to determine the gender of CYL censors, with a special focus on Reader 22. Second, we demonstrate how censorship activity was influenced by the specific legal regulations established in 1955 for the management of children's and youth publications. In addition, we intend to document the specific training of women readers to provide an initial biographical approach. Finally, we consider the issue of gender in the management of children's texts during Francoism.

Methodologically, the analysis is based on contrasting primary sources from the General Archive of the Administration and comparing censorship legislation enacted during the Franco regime. Thus, the main sources on which this research is based are censorship reports from different time periods and for different literary genres; galley proofs and versions of texts generated during the process; the personal records of censors; legal proceedings against punished authors and publishers; and research published by censors. Because an integral view of the censorship process will help demonstrate how it was perceived in the cultural realm, we have conducted interviews with authors and illustrators who were censored (images were among the most closely reviewed elements of books); editors of newspapers and books for children and youth; and columnists from this era. To this end, we will review the scientific literature with the intention of determining what the work of censors was like in Spain. Subsequently, we will seek to describe the censorship process that associated specific numbers with specialised readers, a process that shifted (at least with regard to children's and youth literature) following the specific legislation of 1955. Finally, we will present the profile of Reader 22 and the women who signed most of the assessment reports under this pseudonym, including María Isabel Niño Mas.

Difficulties in the Study of Censorship

Many gaps remain in the research on Francoist censorship. This lack of knowledge does not stem from a lack of researchers or relevant bibliographic records but from the difficulty of accessing the necessary information. With regard to the figure of the censor, the response of the General Archive of the Administration is decisive. The files do not differentiate between children's and adult literature; nor is it feasible to conduct a specialised consultation on the basis of literary genres, types of works, or the censor in charge of their evaluation. To date, there is no official public list of censors who worked for the state; nor is there a record book of the reports processed. To access information about the censors, it is necessary to present documentary proof that they are deceased and that 20 years have passed since their death. This is difficult considering that no list of the censors' names exists and that, upon accessing the files they judged, one finds that they largely worked under reference numbers. Despite this, a number of specialists have laid the groundwork for future research.[1]

Censoring Women Censors

Cisquella et al.[2] describe a group of state workers composed mostly of officials. Members of this staff were officially called "readers" and, during the Franco period, consistently included 25 to 30 evaluators under the head reader, who was subordinate to the head of Editorial Planning at the Ministry of Information and Tourism (1951–1975).

However, no single model existed. Payment for the completion of each report is stipulated in the Order of 7 March 1952, which reorganised the Reading Service of the General Office of Information. It notes that specialist (non-permanent) readers were paid 100 pesetas per unit (200 book pages). However, for texts in French, Italian or "regional languages" (Basque, Galician or Catalan), payment was 150 pesetas for every 100 pages. Most lucrative were English texts proposed for translation and those with complex content, whose readers received a minimum of 200 pesetas per 100 pages. Finally, works in German, which paid 300 pesetas per 100 pages, were the most coveted by reviewers. Ruiz Bautista[3] and Peña[4] clarified this situation through an analysis of the announcement of a selection process that was published in 1942 to fill the posts of six censors. Posts were open to candidates who met any of the following requirements:

1. *Hold a degree on any subject area.*
2. *Have published, or present to the tribunal, a work (whether completed or not) of scientific research or literary criticism.*

[1] Abellán (1984): 110; Cisquella, Erviti and Sorolla (1977); Martínez (2003); Sinova (2006); Bautista (2008); Larraz Elorriaga (2014); Cerrillo and Sotomayor (2016b) or Box (2019).

[2] Cisquella, Erviti and Sorolla (1977).

[3] Ruiz Bautista (2008): 55–57.

[4] Peña (2019): 147–174.

3. *Be able to translate some language.*
4. *Be a member of the Old Guard or the Requetés prior to 18 July 1936.*
5. *Be a member of the military (of any position), a provisional member or a reservist.*
6. *Be a priest (of the regular or secular clergy).*
7. *Be a party militant believed to possess sufficient merit for this due to services provided to Spain and the Catholic Church.*

These requirements indicate that future censors were to have a strong ideological component and ties to the Franco regime. According to Manuel Peña,[5] if an individual passed the selection process, *the exercise of his functions should be carried out with the greatest secrecy and without the slightest hint of corruption, at least in theory.* However, we cannot be certain that these criteria were maintained throughout the Franco era, although they attest to the rigour of the early years.

Based on Ruiz Bautista,[6] we know that the censors had to carry out their work with the greatest of secrecy. If their identity were discovered, they would be dismissed and, hence, they could not be too discreet. Bibliographic sources confirm that this anonymity was exercised impeccably because, decades later, we still do not know who worked there, although some female readers or readers' wives have been identified.[7]

Regarding children's and youth literature, of particular interest is the study of Sotomayor et al.,[8] which, after the analysis of hundreds of reports, confirms that women were mostly involved in the evaluation of books for children. Among those whose identities are known are Montserrat Sarto, M. C. Serna, María África Ibarra and Isabel Niño, whose judgements are said to have been much stricter than those of the head censor, Francisco Aguirre, and even those of the ecclesiastical censor, particularly in the case of Isabel Niño.[9] Most of the women readers had ties to the Catholic Reading Office of Santa Teresa de Jesús (which we will discuss later) and usually were not involved in evaluating adult literature.

Although under Franco, female employment was understood as a danger to the social function of women, who were associated with

[5] Ibid.: 154.
[6] Ibid.
[7] Cisquella, Erviti and Sorolla (2006): 51.
[8] Sotomayor et al. (2016): 64–66.
[9] Ibid.: 64–65.

motherhood and childcare,[10] the case of censorship was different. Society had a classificatory capacity, functioning as a prism through which the meanings—always multiple and changing—assigned to gender difference and interwoven with class divisions proved to be inseparable from the construction of the national ideal.[11] The administration believed that a woman's perspective was necessary for the screening of children's content, and this belief paved the way for the work that would later be performed by other women in both homes and schools.

An informative exercise in comparison reveals researchers' frequent allusions to Isabel Niño as one of the strictest readers. In fact, an article published by Peio H. Riaño in the journal *Público*[12] states that Niño oversaw *Los niños tontos* by Ana María Matute and argued that "*it is a pity that the* 'tremendismo' *applied to children prevails in most of them [the stories]. They are true nightmares, as are the drawings; in very bad taste, modernist as they are attempting to be*". Niño analyses all of the short stories, noting that some are incomprehensible and sad, using the adjectives "depressing" or "cruel". With these assessments, she concludes that "*children must be treated with more respect. Your publication was entirely REJECTED*". However, 13 days later, the reader F. Aguirre, who had the final say, allowed the book to be published.[13] This situation, in addition to demonstrating the arbitrariness of censorship and Niño's strictness as a reader, invites us to consider the training and aspirations that allowed women censors to write the reports that were required of them.

Previously, we enumerated the criteria for the selection process;[14] however, these cannot be assumed to be the only criteria that were used, since censorship was subject to various ministerial changes throughout the dictatorship (Vice Ministry of Public Education, 1941–1945; Ministry of National Education, 1945–1951; and Ministry of Information and Tourism, 1951–1975). Herein lies one of the reasons for the arbitrariness of the judgements: different profiles for readers were prioritised depending on who was in charge of the administration. We have no record of the academic differences between readers of children's literature and censors of adult books. The consulted literature does not indicate whether an

[10] Blasco Herranz (2008): 241.
[11] Zira Box (2019): 130.
[12] Riaño (2010).
[13] Sotomayor et al. (2016): 64–65.
[14] Ruiz Bautista (2008).

education in child development was needed to "correct" titles for children or whether, simply based on a sexist mentality, censoring books for children was understood to be a woman's job and, hence, they were considered suited to the post. However, men dominated the hierarchies of these sectors, which leads us to wonder what other positions were held by women censors. Some names have been revealed to date, such as those already cited by Sotomayor et al.[15] and Cerrillo Torremocha and Sanz Tejeda; however, all that is indicated regarding these women censors is their participation in some religious offices,[16] associations or social entities, and not their professions or the agencies and institutions where they worked.

If we look at the characteristics of this work and seek to understand the work environment, wages, possible prestige of serving as a state reader and even the possibilities for professional advancement, we must point out the lack of specialised studies on these topics.[17] However, we can derive relevant data from studies such as that of Sinova,[18] which provides data on the working environment of press readers: they were "under severe pressure" and were constantly monitored by superiors. Furthermore, we know that these facts are difficult to document because the sanctions for these employees were always kept secret and, hence, to investigate their career paths, it is necessary to access their personal records. Therein lies the difficulty of these types of studies. The absence in the General Archive of the Administration (AGA) of a list of personnel serving as government readers has forced investigators to extract evidence, interpreting virtually illegible signatures in the attempt to identify some names. These are the reasons why the authors cited here argue that it is imperative to determine whose hands Spanish literature was in under Franco.

[15] Sotomayor et al. (2016).
[16] Cerrillo Torremocha and Sanz Tejeda (2016).
[17] Larraz Elorriaga (2014).
[18] Sinova (2006): 154.

Reader 22's Association with Children's and Youth Literature

The ministries in charge of censorship usually employed between 25 and 30 reviewers. This fact, provided by Cisquella et al.,[19] coincided with the number of pseudonyms we counted in the files that they evaluated. Immediately after the end of the Civil War, the censors' edits were made in small files that did not provide much information and did not include many points to analyse. However, the earliest reports did note the name of the censor, which facilitated knowledge about this initial group. However, this lasted for just over two years. As time went by, the system became structured, and so did the reports that were to be completed by the evaluator. The Ministry of the Interior, which housed the Press and Propaganda Service, was replaced in 1941 by the new Vice Ministry of Public Education (1941–1945) and, subsequently, readers worked with a kind of dossier that laid out the aspects to be assessed: morals, religion, literary quality, dogma and politics.

Censors no longer wrote their last name on the reports; instead, each was assigned an identification number. The fact that pseudonyms consisting of numbers below 30 were used on most reports has led many critics to argue that there were about thirty reviewers on the payroll, although there is evidence of a different model. The former claim emanates from primary sources; in a comparison of more than 600 censorship reports for both children's literature and adult literature, we found several books examined by censors who signed with the numbers 31 (exp. 6147-69), 34 (exp. 2224-54), 35 (exp. 6149-69) and 36 (exp. 6150-69). Additionally, together with this small group, other reports have been found signed with the numbers 63 (exp. 9940-73) and 67 (exp. 148-73).

We also have the list of readers established by Abellán[20] from before the archive became the object of purges and when information on its readers was not protected from queries. The names he provides point to a team formed in 1954 comprising 20 "administrative censorship personnel" and 40 readers, including both specialists and permanent ones, which fits with our figures.

The third piece of evidence is drawn from the staff responsible for hiring readers. There is a letter dated 1967 signed by the Director General

[19] Cisquella et al. (1977).

[20] Abellán (1984).

for Information addressed to the Assistant Secretary of the Department recommending the hiring of Antonio Iglesias as a book censor on the grounds that several terminations had occurred simultaneously in January, and the book inspection section needed new reviewers. However, if we examine the reports for that year, we see that all the numbers remained active, a detail that leads us to believe that the pseudonyms were not associated with a particular reader but with a thematic specialty. Bear in mind that exactly one year had elapsed between the terminations and the new appointments and that along with these voluntary terminations that needed to be covered, there were likely censors who left due to death, illness or the end of a contractual relationship. It is easy to imagine that the 30 readers cited in previous studies were insufficient to review all of the literature in the country when in years such as 1967 the team performed its work with just over 20 censors; this number seems even less sufficient if we consider that, in 1964, approximately 5294 works were published.[21] The temporal aspect must also be added to the description of this situation because rarely did more than three days elapse between a work being received for censorship and the head reader assigning a reader number for its evaluation. The speed at which the reports were completed, the enormous amount of work that was assigned and the lack of suspension of any of the numbers lead us to believe that behind each reader number lay a type of specialty and more than one name.

Given that there was a hiring distinction made between payroll readers and specialists, it would not be absurd to hypothesise that the latter served as extra help for the permanent staff. Fernando Larraz[22] clarifies that the permanent staff were ministry officials who were on duty and took orders from the head of the section, while specialists, who did not have fixed hours and charged on the basis of services provided, formed an arbitrarily chosen group for which there was no legal basis for regulation.

With the intention of proving that each reader number was assigned a specialty and was used by several censors, we have developed a statistical analysis that evaluates the characteristics of the work performed by each reviewer based on more than 600 censorship reports, from which we have extracted the "identifying information" described in the entry register (recipient, subject matter, gender, editorial nature or price) and the incidents and verdict used in the "resolution" section. In this regard, one of

[21] Díaz Plaja (1994): 16.
[22] Larraz Elorriaga (2014): 87.

the objectives we established for our study is to determine with certainty whether the galley proofs of children's and youth books that publishers sent to the ministry were assigned a specific reader number used for examining titles for children.

Discovering whether there was a typology of censors by literary genres is crucial for understanding censorship in the Franco regime. The Ministry of Information and Tourism created the Children's Press Advisory Board through a ministerial order of 21 January 1952. Its purpose was to prepare reports on the nature and content of children's publications and to establish guidelines for action and control. The board cannot be regarded as temporary since it gained greater responsibilities over time; furthermore, in 1954, it was recognised as having authority over non-periodical publications and books and became known as the Advisory Board for Children's Publications.

Two facts demonstrate the importance of CYL in relation to censorship. The first is the Decree of 24 June 1955 and the Press and Printing Act of 1966. Together with the decree, an order of the Ministry of Information and Tourism was approved that managed children's publications, thus establishing a specific regulation to which all references to CYL were required to adhere. The review criteria for censorship had never before been published, and this legislation indicated for the first time the criteria to be followed for children's works (religion, morality, nation/politics, literary values and psychological/educational aspects).

The second fact that attests to the differentiation of children's literature from adult literature is that, following the adoption of the new Press and Printing Act (1966), it was established that censorship would change from a "compulsory consultation" to a "voluntary consultation". However, within the framework of this state campaign regarding freedom of the press and literature, the Statute on Children's and Youth Publications (1967) was adopted subject to the previous law; it states that children's titles would continue to be submitted to the censorship offices before publication.

Based on our previous research[23] and that of Hanna Martens in 2016,[24] we assume that Reader 22 was the main reviewer of children's and youth literature. Martens analysed the censorship of Charles Perrault's stories translated from the French and distinguished two distinct phases in the 39

[23] Tena and Soto (2019): 51–70.
[24] Martens (2016).

reports considered. In the first phase (1939–1960), a variety of criteria are observed in the numbers the administration assigned to each reviewer, and there appears to be no uniformity. However, in the second phase (1961–1975), greater homogeneity is observed, because all the reports bear the same number: Reader 22. However, given the discretionary nature of the judgements issued under this pseudonym, Martens clarifies that Reader 22 was probably not always the same person.

Women Readers of Children's and Youth Literature

We have divided our sample of censorship reports according to the reader number that the publisher of the book indicated when registering the work for consultation. However, not all reports were useful: several were completed without any assessment (they only indicated the verdict), others were reissues and a few lacked identifying data regarding their censor. The study was therefore carried out using 506 reports, of which 322 were for adult literature and 184 were for children's literature. We organised the latter reports based on the number of the reader who evaluated them. However, the sample of children's titles was reduced to 164 because 20 reports were poorly conserved and thus were not reliable sources for determining authorship. These caveats aside, the results are frankly revealing because, from a payroll of nearly 40 reader numbers, the total is very well distributed, and each number accounts for a small percentage of the total, which shows that children's books were distributed equitably among censors. However, there is one clear exception: 63% of the evaluations of children's books were conducted by Reader 22, who judged 104 of the 164 titles analysed. The remainder went to reader numbers with very sporadic entries for this type of literature, such as 1, 2, 9, 10, 16, 21, 30 and F (Table 8.1).

Indeed, the percentage of works that was subject to the criteria of Reader 22 is what leads us to think that this number does not correspond to the identity of a single person but instead reflects a group dedicated to children's texts. We believe that the number of books requiring analysis would be far too great to be evaluated by a single reviewer who assiduously issued judgements the day after receiving the task. In the interest of determining the readers' names, we interpreted some signatures in the files they signed; among them, two names stand out for their frequency:

Table 8.1 Data extracted from the evaluated files from between 1951 and 1975

Reader number	*No. of works analysed*	*Percentage*
Reader 1	3	1.83
Reader 2	3	1.83
Reader 9	5	3.05
Reader 10	11	6.71
Reader 16	2	1.22
Reader 21	10	6.10
Reader 22	104	63.41
Reader 30	22	13.41
Reader F	4	2.44

María Isabel Niño and María África Ibarra. In 1953, María Isabel Niño reviewed *Aladino y la lámpara maravillosa* (Aladdin and the Magic Lamp) (file no. 1953), *Blancanieves y los siete enanitos* (Snow White and the Seven Dwarves) (file no. 3487-53), *El pastorcillo Cristian* (The Little Shepherd Christian) (file no. 4022-53) and *Alí Babá y los 40 ladrones* (Ali Baba and the 40 Thieves) (file no. 430-54). This last story is evidence that links the two readers: because before Isabel Niño judged this book (1954), the reader Ibarra assessed it (1953). Moreover, illustrating her reputation, and as we pointed out earlier, it was Niño who imposed suppressions on the text, while her predecessor simply accepted it.

We cannot ignore the fact that shortly after, as a result of their friendship, the two readers jointly published the manual *Bibliotecas infantiles. Instalación y funcionamiento* (Children's libraries: Installation and functioning)[25] regarding library management, on which they were both specialists. However, although Niño's name appears most frequently, followed by the number 21, Ibarra's signature also appears on a broad list of works, all completed in 1953 under the number 22: *El mejor regalo* (The Best Gift) (file no. 6742-53), *Tito y Lía van al Polo* (Tito and Lia go to the North Pole) (file no. 365-53), *Wendy y Peter Pan* (Wendy and Peter Pan) (file no. 403-55), *Los cromos encantados* (The Enchanted Trading Cards) (file no. 6725-53) and *Piel de asno* (Donkey Skin) (exp. 3485-53).

[25] Niño Mas and Ibarra y Oroz (1956).

The Background of the Censor María Isabel Niño Más

Although the link between Reader 22 and CYL is clear, and the identities of some names have been corroborated, we continue to overlook the origins of these women's involvement with censorship. In response to this issue, we have performed an unprecedented analysis of the personal record of the reader whom the specialised bibliography indicates was the most critical one in the entire book inspection section: María Isabel Niño Más (Benavente, 1900–Madrid, 1969).

Her name is among those that appear most frequently on all censorship reports, and it is not surprising to observe that she was far stricter than the head reader, who sought to empathise with authors and often "disregarded" Niño's negative reports and authorised the circulation and sale of the title.

Consulting Niño's personal file (file no. 42-05051) allows us to understand her academic background, parallel jobs and trajectory in the ministries that were in charge of censorship. However, it must be clarified that this information, due to its personal nature, was part of the regime's internal and secret documentation and, hence, these files not only included the service sheet, but also the reader's background, political ideals and ties during the Civil War and were therefore considered reserved material for internal consultation only. This is why Sinova[26] stressed the difficulty of studying this topic: a close inspection of the signatures on the files indicated that, in most cases, only the last name was shown and, sometimes, a full name cannot be surmised without risking error. In fact, there are some allusions to Niño in *Censura y LIJ en el siglo XX* (Censorship and CYL in the Twentieth Century);[27] however, as Sinova warned, even in that work there are multiple uncertainties that sometimes allude incorrectly to the identity of María África Niño and confuse the names of Niño and María África Ibarra y Oroz. These data correspond to two different people, as we determined at the Ministry of Justice and corroborated with the death certificates of the two women, which enabled our access to their personal records.

These records demonstrate that Niño and Ibarra, contrary to what might be assumed, possessed broad academic training. Ibarra, who was

[26] Sinova (2006).
[27] Sotomayor et al. (2016).

the daughter of the historian Eduardo Ibarra y Rodríguez, earned a doctorate in philosophy and letters.[28] In her career, she translated some historical documents from English. During the Civil War, she served at the Library of the Universidad de Valencia,[29] attaining the post of director of the Regional Archive of Valencia. Later, in 1959, she went to the Library of the Royal Academy of History,[30] which she eventually led.[31] Additionally, she was part of the Reading Office of Santa Teresa de Jesús from its origins in 1942 and published several works in addition to the one cited.[32] Her relationship with Niño may have begun much earlier; in 1928, both she and Felipa Niño (María's sister), who became vice director of the National Archaeological Museum,[33] were part of a group of intellectuals associated with the Centre for Historical Studies in which Sánchez Albornoz and José Maria Lacarra were involved. This led both sisters to defend the outstanding librarian, bibliographer and lexicographer María Moliner in the face of her record purge after the end of the Spanish Civil War.[34] Ibarra died in 1996.[35]

Niño, on the other hand, graduated with a degree in philosophy and letters and pursued doctoral studies, which she combined with extensive work experience that began before she became part of the book inspection services. The ministry maintains an accreditation issued by Hilario Juan Arnau, secretary of the Professional and Domestic School for Women, that certifies that Niño was appointed honorary assistant on 20 January 1921 by the proposal of a professor of history of the decorative and industrial arts at the school and served until the end of the 1921–1922 term. Shortly afterwards, she joined the body of archivist functionaries of the *Delegación de Hacienda de Zamora* (Tax Office of Zamora) (1922), where she remained for four years. Her promotion to supernumerary in 1926 permitted her to transfer to the General Archive of Alcalá de Henares a year later and be appointed as a second-level official, a merit that led to her promotion in a matter of three years. In June 1930, after serving in the Treasury Archive and the General Archive of the Administration, she

[28] Peiró and Pasamar (2002): 329.
[29] Cabeza (2000): 164.
[30] Peiró and Pasamar (2002): 329; *BOE* (1959): 8411.
[31] *Boletín de la Real Academia de la Historia* (1973): 227–228.
[32] Ibarra y Oroz (1949).
[33] Zozaya (1992): 111.
[34] De la Fuente (2011).
[35] Peiró (1996): 194.

moved to the National Library of Spain (BNE, for its initials in Spanish) under similar employment conditions. In 1931, she was promoted again. Her time at the BNE would represent the most rapid ascent of her professional career because it only took her six years for her to reach the seventh category (within her rank as a second-level official) and thus double the initial salary she had earned with the state. However, this was only the beginning of her ascent; after the end of the Civil War, she soon gained fifth category status. Successively and with an uninterrupted pace, she would go on to achieve a promotion by ministerial order to second category functionary. This role was unusual for a woman during the dictatorship, particularly if we observe that Niño obtained up to 14 promotions and that her career was entirely uninterrupted. Manuel Carrión,[36] in his "*Semblanza cronológica*", states that Niño was a true institution among the body of functionaries and was quite a figure at the National Library. She was a person who succeeded in bringing together in one life "the demands of a soul of the Lord" with the highest professional demands. Therefore, Niño appears as a religious woman who was diligent in her work, conservative-minded and loyal to the Franco regime, encompassing all the qualities required to ascend in a BNE directed by Miguel Artigas, who rewarded loyalists.

Also at the BNE, only personnel loyal to the Franco regime were selected.[37] At the headquarters and in department offices such as those occupied by Niño, workers avoided allowing their managers to present books in exhibits and display cabinets that, although authorised by censorship, were suspicious or would encourage criticism. It was equally important to quench ideas contrary to those of the Franco regime and to encourage the reading of works of propaganda.

Niño was one of the individuals responsible for promoting reading related to the regime and for helping to silence discourse that departed from its ideals. However, in order to attain these levels of power, she had to ascend positions and acquire notoriety; her first milestone was the temporary post she took in 1945 as head of the music section, which launched her name to the forefront of the BNE. She combined this work with her participation in the Office of Santa Teresa de Jesús, where, in 1944, she

[36] Carrión Gútiez (1969): 9–10.
[37] De la Fuente (2011); Egoscozábal and Mediavilla (2012).

joined its founding members Paulina Junquera and África Ibarra,[38] the latter of whom was also a censor of children's literature.

The Office of Santa Teresa was created by Catholic Action's Superior Council of Women,[39] and among its objectives was the defence of morality and religious sentiment through children's texts in the attempt to guide young female readers. The office assessed works that had already passed official censorship and submitted them to a second evaluation, publishing its own *Catálogo crítico de libros para niños* (critical catalogue of books for children) with multiple selections of evaluations carried out in various years (1961, 1963, 1967, 1972 and 1975).

Prior to these catalogues, Niño[40] had described the office's process in detail: the model of the cataloguing sheets for their reviews, the surveys of librarians and children's readers and the weekly peer readings among office members, together with the initial work carried out in the *Catálogos críticos de libros para niños* by Catholic Action's Superior Council of Women. In early April 1945, Niño, having newly arrived at the Office of the Head of the Music Section, sent a letter to her superior in which she stated the following:

> *That you are the only boss who does not charge gratification for said headship and having created several headships that will be assigned by contest, according to the Office communique dated 1 of the current month. I beg you to grant the Head of Indies or Theatre or any of the other newly created headship without this claim signifying my resignation from my position in the Music Section.*[41]

This appeal would bear results years later: first, in 1952, she was appointed as a specialist reader and began to work in the ministry; second, far from leaving the BNE to devote herself to reading, her earlier letter proved effective, and in 1958, she became Head of the Section on Music and Archives of the Spoken Word.[42] However, although no connection to censorship is mentioned in her obituary, irrefutable proof has been found

[38] Carrión Gútiez (1969): 9–10.
[39] García Padrino (2018): 41–45.
[40] Niño y Mas (1958): 595–598.
[41] S.A. (n.d.): file 42-05051.
[42] Following this new appointment, she would create the article "Brief historical review of the Section on Music and Archives of the Spoken Word" as a supplement to the *Revista de Archivos, Bibliotecas y Museos*, LXXIII, 1966.

in the notification with the title of specialist reader signed by the Director General of Information.[43]

Niño's connection with children's literature would increase, as demonstrated by the fact that, in 1960, after years as shadow director of the Office of St. Teresa of Jesus, she became the institution's official director.[44] Thus, she not only imposed corrections as part of the official state censorship service but was now also the leader of the re-censorship service outside the ministry; she carried out official censorship through institutions and an unofficial censorship through the recommended catalogues of CYL recommended by the Reading Office of Santa Teresa de Jesús. Therefore, at this point, she could silence works or authors she considered undesirable because, unlike many other censors, she had two methods at her disposal with which to obscure or enhance the literary productions of certain authors. Additionally, starting in 1962, Niño was part of the Catholic Commission for Children, serving on the jury of the awards granted by the Spanish Catholic Commission for Children (CCEI, for its initials in Spanish). During these years, she organised annual exhibitions of books for children and youth at the National Library together with the National Reading Service and the Reading Office of Santa Teresa de Jesús.[45] Niño came to be so prominent in literary criticism of CYL that in 1971, the National Library posthumously named the first library for researchers of CYL in Spain, created through the Ministry of Information and Tourism, after her. In 1986, this library was included in the general collections.[46]

The reports that the administration collected on Niño's professional career reveal that she was indeed recognised as a pre-eminent specialist in literature for young people under Franco. The proof is found in a piece she wrote just two years before her death, in which she states as follows:

> *That I would like to carry out work corresponding to overtime in the children's sections of the Public Libraries as a specialist in the selection and formation of children's libraries, duly endorsed by having led the Reading Office of Santa Teresa de Jesús of Catholic Action's Superior Council of Women for 28 years.*

[43] S.A. (s. f.): file 42-05051.
[44] Carrión Gútiez (1969): 9–10.
[45] S.A. (1964): 69.
[46] According to Díaz Plaja (1994), it closed in 1989: 16–17.

This request was granted by the inspector after he was informed by the director of the Public Libraries of Madrid that Niño was undoubtedly the most suitable person due to her specialisation in selecting children's works. Through children's books, she forged an authority that extended to any cultural practice relating to children. As an example, Carolina Toral (head of the children's section of the publishing house Editorial Escelicer), Niño's colleague in the Reading Office of Santa Teresa de Jesús, closed the foreword to her *Literatura infantil Española* (Spanish children's literature)[47] with these words:

> *Those who desire a serious moral-religious guarantee in children's works can, and should, consult the various Catálogos Críticos de Literatura Infantil, which have been published by the General Directorate of Archives and Libraries, 1954; the Association of Archivists, Librarians and Archaeologists, in 1952; and Catholic Action's Superior Council of Women, in 1948; all of them drafted and selected by the Reading Office of Santa Teresa de Jesús, the only entity in Spain that is dedicated to these activities.*

Conclusions

There was no single homogeneous period of censorship under Franco. After a few early years during which readers signed their files and for which the attribution of specific readings to specialised censors is not clear, an initial normative attempt can be observed from 1941 onwards with the creation of the Vice Secretary of Public Education and would last until 1945, although with fluctuations in the numbers of readers and the attribution of readings. It was not until the publication in 1955 of the Decree on the Management of Children's and Youth Publications that a broader number of works for children's audiences was reviewed by Reader 22, among others. Furthermore, at that time, a more comprehensive and specific model of reports was established for this genre, and all works of this type were subject to censorship even after the enactment of the Press and Printing Act of 1966. Similar results are suggested by a statistical analysis of all the analysed works, which showed that 63% of children's texts were evaluated by Reader 22. This pseudonym, according to hints in some of the records, covered a group of women that included María África Ibarra, M. C. Serna, Montserrat Sarto and María Isabel Niño Mas, all of whom

[47] Toral Peñaranda (1957): 14.

were linked to the Reading Offices of Santa Teresa de Jesús and Catholic Action's women's group. Women—who were barred from other disciplines and arts[48] and who did not occupy the highest posts in the ministry or the censoring institutions—did occupy prominent roles in the censorship of children's literature, which demonstrates that the regime conceived of children's care as an area reserved for women. This literary censorship was carried out, to a large extent, by female officials and women who were associated with libraries and the world of children's publishers and who were supporters of the regime and defenders of the precepts of the Catholic Church. María Isabel Niño Mas would perform a double censorship of texts: officially, as a member of the censorship body, and critically, as an evaluator of children's and youth publications in the Reading Office of Santa Teresa after 1945.

References

Abellán, Manuel L. (1978): "Censura y práctica censoria", *Sistema*, 22: 29–52.

Abellán, Manuel L. (1980): *Censura y creación literaria es España (1939–1976)*, Barcelona, Ediciones Península.

Abellán, Manuel L. (1984): "Literatura, censura y moral en el primer franquismo", *Papers: Revista de sociología*, 21: 153–172.

Blasco Herranz, Inmaculada (2005): ""Sección Femenina" y "Acción Católica": la movilización de las mujeres durante el franquismo", Gerónimo de Uztariz: 55–66.

Blasco Herranz, Inmaculada (2008): "Mujeres y "cuestión social" en el catolicismo social español: los significados de la "obrera", *Arenal: Revista de historia de mujeres*, XV (2): 237–268.

Box, Zira (2019): "Los atributos de la nación. Género y clase en la España franquista", en Lemus, Encarnación; Peña, Manuel (eds.): Alianzas y propaganda durante el primer franquismo, Barcelona, Ariel: 125–146.

Cabeza Sánchez-Albornoz, Mª Cruz (2000): *La biblioteca universitaria de Valencia*, Valencia, Universitat de València.

Carrión Gútiez, M. (1969): "El viaje de Isabel Niño (1960–1969)", *Boletín de la Dirección General de Archivos y Bibliotecas*, XVIII, CX: 9–10.

Cerrillo Torremocha, Pedro C.; Sánchez Ortiz, César (coords.) (2016): *Prohibido leer. La Censura en la literatura infantil y juvenil contemporánea*, Cuenca, Ediciones de la Universidad de Castilla La Mancha.

Cerrillo Torremocha, Pedro C.; Sanz Tejeda, Arantxa (2016a): "Lectura, Iglesia y sociedad", *Revista de Estudios Socioeducativas (RESED)*, 4: 27–36.

[48] Manuel Peña (2019).

Cerrillo Torremocha, Pedro C.; Sotomayor, Mª Victoria (eds.) (2016b): *Censuras y LIJ en el siglo XX*, Cuenca, Ediciones de la Universidad de Castilla- La Mancha.

Cisquella, Georgina; Erviti, José Luis; Sorolla, J. A. (1977): *Diez años de represión cultural. La censura de libros durante la ley de prensa (1966–1976)*, Barcelona, Anagrama.

Cisquella, G; Erviti, J. L.; Sorolla, J. A. (2006): *La represión cultural en el franquismo: diez años de censura de libros durante la Ley de Prensa, 1966–1976*, Barcelona, Anagrama.

De La Fuente, Inmaculada (2011): *El exilio interior. La vida de María Moliner*, Barcelona, Turner.

Diaz Plaja, Aurora (1994): "Tres décadas del libro infantil y juvenil", *Educación y biblioteca*, 48: 15–17.

Díaz-Andreu, Margarita (2009): "Niño Mas, Felipa", en Díaz-Andreu, M.; Mora, G.; Cortadella, J. (eds.): *Diccionario Histórico de la Arqueología en España (siglos XV–XX)*, Madrid, Marcial Pons: 483–484.

Egoscozábal Carrasco, Pilar; Mediavilla Herreros, María Luisa (2012): "La bibliotecaria Luisa Cuesta Gutiérrez (1892–1962)", *Revista general de información y documentación*, 22: 169–187.

García Padrino, Jaime (2018): *Historia crítica de la Literatura Infantil y Juvenil en la España actual (1939–2015)*, Madrid, Marcial Pons.

Ibarra Y Oroz, Mª África (1949): "Libros y niños", *Bordón*, 1: 33–39.

Jefatura Del Estado (1938): "Ley por la que se modifica la de 30 de enero de 1938, que organizó la Administración Central del Estado", *Boletín Oficial del Estado*, 183: 3216–3217.

Jefatura Del Estado (1952): "Orden de 21 de enero de 1952 por la que se crea la Junta Asesora de la Prensa Infantil", *Boletín Oficial del Estado*, 32: 475.

Jefatura Del Estado (1966): "Ley 14/1966 de 18 de marzo, de Prensa e Imprenta", *Boletín Oficial del Estado*, 67: 3310–3315.

Larraz Elorriaga, Fernando (2014): *Letricidio español. Censura y novela durante el franquismo*, Gijón, TREA.

Martens, Hanna (2016): *Tradición y censura en las traducciones de literatura infantil y juvenil en la cultura franquista: Los cuentos de Perrault en español hasta 1975*, Tesis doctoral, Cáceres, Universidad de Extremadura.

Martínez-Michel, Paula (2003): *Censura y represión intelectual en la España franquista: el caso de Alfonso Sastre*, Hondarribia, Argitaletxe Hiru.

Ministerio De Información Y Turismo (1952): "Orden de 5 de febrero de 1952 por la que se crea la Junta Asesora de la Prensa Infantil", *Boletín Oficial del Estado*, 52: 805.

Ministerio De Información Y Turismo (1956): "Orden acordada en Consejo de Ministros de 24 de junio de 1955 por la que se desarrolla el Decreto sobre ordenación de las publicaciones infantiles y juveniles", *Boletín Oficial del Estado*, 33: 841–845.

Ministerio De Información y Turismo (1967): "Decreto 195/1967, de 19 de enero, por el que se aprueba el Estatuto de Publicaciones Infantiles y Juveniles", *Boletín Oficial del Estado*, 37: 1964–1967.

Niño Mas, Mª Isabel; Ibarra y Oroz, Mª África (1956): *Bibliotecas infantiles. Instalación y funcionamiento*, Madrid, Dirección General de Archivos y Bibliotecas.

Niño Mas, Mª Isabel (1958): "El Gabinete de Lectura Santa Teresa de Jesús al servicio de los niños", *El libro español*, 11: 595–598.

Peña, Manuel (2019): "Censuras y censores en el primer franquismo" en Lemus, Encarnación; Peña, Manuel (eds): Alianzas y propaganda durante el primer franquismo, Barcelona, Ariel: 147–174.

Peiró Martín, Ignacio; Pasamar Alzuria, Gonzalo (2002): *Historiadores españoles contemporáneos*, Madrid, Akal.

Peiró, I. (1996): *Boletín de la Real Academia de la Historia*, CXCIII: 194.

Real Academia De La Historia (1973): "Memoria de las actividades de la biblioteca durante el año 1972", *Boletín de la Real Academia de la Historia*, CLXX, Cuaderno 1: 227–228.

Riaño, Peio H. (2010): "Matute, la "inmoral"". *Diario Público*. Sección Cultural, https://www.publico.es/culturas/matute-inmoral.html (consultado 27 de agosto de 2018).

Ruiz Bautista, Eduardo (coord.) (2008): *Tiempo de censura: la represión editorial durante el franquismo*, Gijón, TREA.

S. A. (1964): "Inauguración de la exposición de libros infantiles", *ABC*: 69, http://hemeroteca.abc.es/nav/Navigate.exe/hemeroteca/madrid/abc/1964/10/15/069.html (consultado 19 de septiembre de 2018).

S. A. (s. f.): Expediente personal de Isabel Niño (exp. 42-05051), Alcalá de Henares, Archivo General de la Administración.

Sinova, Justino (2006): *La censura de prensa durante el franquismo*, Madrid, Espasa Calpe.

Sotomayor, Mª Victoria; Cerrillo Torremocha, Pedro C.; Sánchez, César; Cañamares, Cristina (2016): "La censura de libros para niños: 1939–1976", en Cerrillo, Pedro C; Sotomayor, Mª Victoria (eds): *Censuras y LIJ en el siglo XX*, Cuenca, Ediciones de la Universidad de Castilla La Mancha: 53–122.

Tena Fernández, Ramón; Soto Vázquez, José (2019): "Las revisiones de los *Cuentos de hadas* de Andersen durante la Guerra Civil y la censura franquista", en Margarida, Ana (ED.): *Tendências contemporáneas da investigaçao em literatura para a infância e juventude*, Oporto, Universidad de Aveiro: 51–70.

Toral Peñaranda, Carolina (1957): *Literatura infantil española (apuntes para su historia)*, Vol. I, Madrid, Editorial Coculsa.

Zozaya, Juan (1992): "Necrológicas. Ante la muerte de Felipa Niño y Mas (1902–1991)", *Boletín de Museo Arqueológico Nacional*, X: 111.

CHAPTER 9

The Contribution of the Female Section to the Hispanic Community of Nations

Vanessa Tessada Sepúlveda

Introduction

In May 1951, the Female Section successfully held the First Hispanic-American and Filipino Women's Congress in Madrid, bringing together approximately 250 Spanish and Hispanic American women for discussions and proposals on the role of women in society. The success of this unique transatlantic women's meeting can be explained by the support and diplomatic alignments of the Spanish Ministry of Foreign Affairs. The ministry, in the context of seeking strategies for international insertion in a scenario that excluded Spain from new international organisations and post-war

This article summarises part of the author's doctoral research '*Las estrategias de proyección internacional de la Sección Femenina española hacia América Latina y su recepción en Chile (1937–1977)*' (The strategies of international projection of the Spanish Female Section in Latin America and Its Reception in Chile (1937–1977)). University of Valladolid. 2017.

V. Tessada Sepúlveda (✉)
Universidad Autónoma de Chile, Santiago, Chile

M. Janué i Miret, A. Presas i Puig (eds.), *Science, Culture and National Identity in Francoist Spain, 1939–1959*,
https://doi.org/10.1007/978-3-030-58646-1_9

economic aids, used Hispanist ideology to create a space that would facilitate its positioning and international prestige.

The First Hispanic-American and Filipino Women's Congress was part of several international activities subsidised by the Spanish government to make Spain the spiritual guide and cultural focus of Hispanic-American development.[1] The Congress succeeded in opening a space, until then rare, for women to reflect on their condition, position, and mission in the post-war world. Its two most outstanding achievements were, on the one hand, the visibility of social, political, and cultural problems from a perspective that placed women as subjects, and, on the other, the attempt to establish the Female Section as an international leader in these issues. Consequently, a series of permanent exchange programmes between Spain and America emerged, which led to the creation of a space for the transfer of knowledge, models, and ideologies. In this chapter, we will delve into the Hispanic American Women's Cultural Circles and the system of scholarships and exchanges, although the programme also included other elements, such as the travels of the leaders, the awarding of decorations, congresses, and so on.

This space of exchange was able to survive, as it was associated with the longing for international insertion that marked the Francoist diplomacy after World War II.[2] The 'Hispanic Community of Nations' was the rhetorical construct used for that purpose, and it focused on the re-establishment of cultural ties with the former Spanish colonies based on religion, language, and a common past, that is, on Hispanidad. Spain would provide cultural, educational, economic, and technical cooperation, acting as a spiritual bridge with Europe.[3]

From its Second Council (1938), the Female Section began to coordinate its relations with the outside world. The position of Director General of the Foreign Service was created at that time, dependent on the Foreign Service of the Falange Española Tradicionalista (FET) and the Juntas de Ofensiva Nacional Sindicalista (JONS) during the Civil War, and which counted among its main tasks the creation of Female Sections abroad and the acceptance of aid sent in support of the rebel side. This first initiative to establish connections did not endure after the end of the Civil War and

[1] Delgado (1988).
[2] Delgado (2003).
[3] Del Arenal (1994): 40; Martín Artajo (1953).

disappeared when the Falange was banned abroad around 1939.[4] The end of World War II brought new opportunities and the Female Section began to cooperate directly with the Ministry of Foreign Affairs, as evidenced by the memoirs of its international activities.[5] This collaborative work allowed the female Falange to act as an agent of Spanish cultural diplomatic relations and be integrated into the strategies for creating the Hispanic Community of Nations—to which it contributed, as we will see, with a 'model of Hispanic woman' that brought together women from both sides of the Atlantic and a series of long-term bonding actions with Latin America.

The Model of the Hispanic Woman and the Hispanic Community of Nations

The historiography of the Female Section has increased significantly since Rosario Sánchez reported the few available studies on the institution.[6] However, the participation of the Falangists in the diplomatic policies of the State and in foreign spaces has only caught the attention of female researchers. Recognition of this work can be found in the memoirs of Pilar Primo de Rivera,[7] the National Delegate of the female Falange, and in pro-Falangist texts such as those of Luis Suarez.[8] Nevertheless, despite its scarcity, the current historiography shows us three thematic lines: the relationship established with European fascisms during the interwar period and World War II,[9] the colonial action in Africa during the 1960s,[10] and the travels of the Choirs and Dances.[11] This research aims at filling a historiographic gap by analysing the role of the Female Section as an agent of these policies, while addressing the gender consequences of the cultural diplomatic policies of early Francoism.

The Female Section had a journey of approximately forty years, during which it incessantly pursued—and without much success according to

[4] Tessada (2017a).
[5] Tessada (2017b): 125, 129, 132.
[6] Sánchez (1993).
[7] Primo de Rivera (1983).
[8] Suarez (1993).
[9] Blanco-Camblor (2005); Morant I Ariño (2013).
[10] Bengochea (2012); Nerín (2007); Morales and Vieitez (2014).
[11] Casero-García (2000); Stehrenberger (2012); Amador (2003).

S. Rodríguez[12]—the framing of the Spanish woman. Once it became the female arm of the single party at the end of the Civil War, it turned its efforts to the education of women. For this, it had a series of educational services, of propaganda and training, which would serve as a transmission belt for the female model that National Syndicalism defended. Traditionally, this model is associated with motherhood and marriage, and values such as subservience, sacrifice, and self-renunciation, ideas that emerged in the line of thought of José Antonio Primo de Rivera, founder of the Spanish Falange. These conservative and anti-feminist slogans were related to Catholic postulates on the essence of femininity in motherhood, reproduction, and the family.

However, with their actions, the female Falangist militants broke with that initial gender mould, pushing the limits of the generic gender order. From the Second Republic, it became evident that the action boundaries for women had shifted, and it is precisely there that the female Falangists discovered the founding story of their model of femininity. The values extracted from that period nurtured what was called the Falangist style and way of being, which was characterised by discipline, austerity, self-renunciation, service, courage, selfless dedication, joy, and sympathy.

For B. Barrera, only the contradiction between the 'Falangist model of the woman' and the 'model of the Falangist woman'[13] was apparent, as the continuity of Falangist femininity was provided by a normative emotional identity centred on a style and way of being. This imprint was reflected in any task that the woman developed, either in the public or domestic space.[14] Along the same line of thought, Cenarro proposes that the Falangists managed to revise the model, pushed by the multiplicity of discourses on the role of women and the tensions that this generated.[15] The historical nature of this construction and the ability to negotiate and revise the discourse in Spain allowed the Falangists to attempt to internationalise their female model. However, that model had to be contextualised both to the strong Hispanic and national Catholic component that the Spanish international exposure of the time required and to the Latin American sociocultural realities.

[12] Rodríguez (2005).
[13] Barrachina (1991).
[14] Barrera (2019): 254.
[15] Cenarro (2017): 96.

When Alberto Martín Artajo was appointed Minister of Foreign Affairs in 1945, the Spanish diplomatic guidelines changed direction for two related reasons. First, the end of the World War had left Spain in a position of isolation from the international scene. Second, it was necessary to dissipate the Falangist imprint that had prevailed in foreign policy until the early 1940s. The international context led the regime and the new Minister to generate strategies for international recognition. While a de-fascist make-up was promoted in the country (Constitutive Law of the Cortes in 1942, prohibition of the fascist salute in 1945, Charter of the Spanish People in 1945, Law of Succession in 1947), Latin America was being conceptualised abroad as an appropriate space for the construction of international privilege.[16]

The Hispanidad, as a common and connecting theme for this space, ideologically supported the foundation of what Martín Artajo called the 'Hispanic Community of Nations'. This 'Community' would be responsible for producing the rhetoric that would help Spain to legitimise the articulation with Latin America. This is how he expressed it in 1949:

> The great truth of our present time in the landscape of international relations is the failure and asphyxiation of isolated countries, of States that romantically abandon their own geography and their exclusive sphere. On the contrary, the concept of the community of nations historically associated by a common spirit and interest is becoming stronger as a fruitful idea for supremacy and defence.[17]

In practical terms, this 'Community' was created within the actions of various institutions in Spain and America that were dependent on or promoted by the Ministry of Foreign Affairs. These initiatives included the Institute of Hispanic Culture in Madrid (1947),[18] promotion of local counterparts in the main Latin American cities, publication of periodicals such as *Mundo Hispánico* and *Cuadernos Hispanoamericanos*, university courses and conferences dedicated to Ibero-America,[19] and the opening of the *Colegio Mayor Hispanoamericano de Nuestra Señora de Guadalupe* (1947). They were also complemented with discussions about the

[16] See Delgado (2003); Del Arenal (1994).
[17] Lacalle (1957): 107.
[18] Cañellas (2014).
[19] Delgado (1988): 172.

possibility of a common nationality (the Hispanic supranationality),[20] treaties of recognition of studies, the creation of a common monetary fund, and easier customs facilities,[21] among other projects. All these initiatives aimed at overcoming international isolation.[22]

The Female Section obtained a place in this cultural diplomatic plan when the Choirs and Dances of Spain were invited to tour Latin America. The positive result of this cooperation with the Ministry of Foreign Affairs offered opportunities for international projection, unimaginable even for Pilar Primo de Rivera:

> [I]n that year [1947], paradoxically to the doors that, on the one hand, closed the world to us, all the possibilities were opened to us in Hispano-America (sic). What could interest us the most, the Spanish World, wanting to find itself free from strange influences (sic), turned its eyes to Spain, and more specifically to the Falange. And our Foreign Service, which had hardly ever anything to do, although we did not provide a budget, suddenly found itself at the centre of our activities.[23]

The first trip of the Choirs and Dances of Spain was to Argentina (1948), which was organised in exchange for the visit of Eva Duarte de Perón in June of the previous year. The success of this first venture in terms of propaganda and politics gave Pilar Primo de Rivera negotiating power with the government and reinforced the cultural work of the Female Section. Further, it demonstrated the usefulness of the apolitical imprint of the political-cultural initiatives of the female Falangism.[24]

The benefits of this first trip were evident in the short term for the institution. Another transatlantic tour was organised in 1949 comprising nine countries (Venezuela, Dominican Republic, Peru, Chile, Ecuador,

[20] Martín Artajo (1956).

[21] Royal Academy of History, *Archivo Asociación Nueva Andadura. 'Veinticuatro países iberoamericanos se reúnen en Bogotá'*, *Teresa, revista para todas las mujeres*, No. 58, October 1958, p. 12.

[22] In 1946, the UN imposed a political veto on Spain, requesting the departure of the ambassadors. However, the *Chargé d'Affaires* continued the relations. The consequences of the veto lasted until 1949. In 1953, the concordat with the Vatican and the Pact of Madrid with the United States were signed. In 1955, Spain became a member of the UN.

[23] Royal Academy of History, *Archivo Asociación Nueva Andadura*, Blue Series, Folder 1B '*Historia de la Sección Femenina por Pilar Primo de Rivera*'. Part 5: mainly refers to the projection of the Female Section abroad, from 1945 to 1951, p. 161.

[24] Casero-García (2000): 51.

Panama, Colombia, Haiti, and Puerto Rico), with performances in more than twenty cities. All this took place under the strategic eye of the National Delegate who saw possibilities of the Female Section in terms of political representation, as revealed in her discourse at the beginning of this tour:

> The Female Section is once again on its way to America, to serve, as last year. [...] We are doing this because America has to remember that it is connected to Spain by a common destiny, and in these times when any other representations of Spain (sic) could arouse suspicion, our Choirs and Dances emerge, with all the Spanish traditions, and, when carried by you, the female youth of the Falange, they acquire new vigour and a new grace.[25]

This potential for representation was also perceived at the government level. Carlos Cañal (*Chargé d'Affaires* in Chile) evidences this in the positive evaluation he sent to Carrero Blanco after the 1949 tour:

> My duty would not have been fulfilled [...] if I did not point out, especially to Your Excellency, the extraordinary effectiveness of the reported visit of the Choirs and Dances to this country [Chile], for propaganda and promotion of our Spain. [...] Undoubtedly, from a political 'utilitarianism' perspective, this visit provided a significant fruit.[26]

In this manner, supported by the Ministry of Foreign Affairs, the Female Section obtained the necessary political and budget support. This gave life to a joint space with Latin America that would last until 1977, when the institution disappeared following the publication of the Royal Decree-Law that put an end to the Movement.

After the trips of the Choirs and Dances, the new initiative comprised the organisation of the 1951 Congress mentioned earlier. This idea was valued among the contacts obtained by the Falangist Command during their tours of the folklorists. The first trip to Argentina would already have warned of the propagandist potential of these visits.[27] Therefore, from the

[25] General Archive of the Administration, *Delegación Nacional Sección Femenina Departamento Coordinación*, Caja 5802, '*Mensaje de Pilar a los Coros y Danzas*', 1949.

[26] Archive of the Ministry of Foreign Affairs, R 2948 Exp. 71. '*Informe de Carlos Cañal a Luis Carrero Blanco*', 16 January 1950.

[27] General Archive of the Administration, *Delegación Nacional Sección Femenina Departamento Coordinación* Caja 5802, '*Carpeta Coros y Danzas. Organización viaje Argentina*', 1948.

second tour emerged a report announcing the positive acceptance of the idea among prominent conservative women who belonged to feminine groups related to Hispanism or with close relations to the local Institutes of Hispanic Culture.[28] In Spain, this project was supported by the director of the Institute of Hispanic Culture, Alfredo Sánchez Bella, who participated in the preparatory meetings.[29] In this manner, the Congress was part of the Francoist attempts to win the loyalty of the leading sectors and the Hispanic-American elites to support Francoism and propaganda channels,[30] besides incorporating educational promotion, or intellectual renovation.[31]

This First Hispanic-American and Filipino Women's Congress condensed the exaltation of Hispanism of this period. The meeting was held to commemorate the fifth centenary of the birth of Queen Isabel *La Católica*, which allowed it to be dedicated to this model of a strong and exemplary woman, the architect of the 'discovery of America'. This association was also indicated in the use of the caravel, a boat used by Christopher Columbus in his discovery of America, as a symbol of the Congress. Piedad Colón de Valdés, a descendant of Christopher Columbus, was even invited as an honorary member.[32] The success of this meeting can be measured by its vast attendance. A total of 229 communications were presented, 159 of which were Hispanic-American.[33] The meeting programme lasted thirteen days, during which, in addition to the round table discussions, there were visits to historical sites related to the discovery and conquest of America and lectures by prominent members of the regime.[34]

[28] Archive of the Ministry of Foreign Affairs, R 2948, Exp. 72 '*Informe de la Sección Femenina sobre su paso por los diferentes países*', 1950.

[29] Royal Academy of History, *Archivo Asociación Nueva Andadura*, Blue Series, 167 B Doc. B.1.4. '*Acta de la reunión celebrada por representantes de los Círculos culturales de Argentina, Colombia, Chile, Méjico, Perú y Uruguay en Junta Central Coordinadora con su presidenta Pilar Primo de Rivera*', 6 June 1963.

[30] Delgado (1988): 167.

[31] Cañellas (2014): 84.

[32] '*Actividades del Primer Congreso Femenino Hispanoamericano*', *ABC*, 8 May 1951, p. 20.

[33] The number is available in the document of committees and conferences of the Hispanic-American Conference in the General Archive of the Administration, *Delegación Nacional Sección Femenina Departamento Coordinación* Caja 5808.

[34] General Archive of the Administration, *Delegación Nacional Sección Femenina Departamento Coordinación* Caja 5808, '*Acta de la junta del Primer Congreso Hispanoamericano Femenino*', 15 February 1951.

The ambition of the Female Section to become an action guide for Latin American women was manifested in two aspects of the organisation. The first was the scope of the agenda of the Congress, which sought to cover various aspects concerning women: women in religion, politics, education, war, and the Hispanic world; and the second was the proposal of a work methodology by thematic table to include reaching agreements for each of the topics discussed. These agreements would ultimately lead to action commitments by the attendees, who recognised themselves as representatives of their respective countries.

One of the most important challenges of the meeting was to develop a consensus definition of 'Hispanic woman'. In the vision of the Spanish National Delegate, it was impossible to deny women's increasing participation in the public space. For this reason, the Female Section assumed the responsibility of reflecting on and revising the new female role, bearing in mind the following:

> [T]he important thing now is to discriminate how all these activities should be channelled so as not to fall into dangerous deviations. But based on the fact that women are already, and it is not bad that they are, in the University, in the professions, in the bureaucracy, and in so many other things that were previously considered to be exclusively for men.[35]

In other words, Hispanist femininity should be based on contextual female conditions, without falling into models that, because they were static, had become anachronistic. This recognition of the multiplicity of female experiences echoes the Female Section's experiences in Spain. Therefore, the model of the Hispanic woman, beyond assigning women a role in society, comprised a proposal for a Hispanist interpretation of the importance of the feminine for society.

The 'archetype of the Hispanic woman' was agreed upon during the closing of the Congress, along with a Declaration of Principles and the conclusions of each thematic table. The archetype accounts for the basic characteristics of the Hispanic woman: she is in favour of marriage and motherhood, values purity, and the defence of the family is central, as well as the Hispanic-Catholic cultural values. Based on these convictions, the Declaration of Principles recognised the existence of an Ibero-American

[35] General Archive of the Administration, *Delegación Nacional Sección Femenina Departamento Coordinación*, Caja 5811, '*Discurso de inauguración Pilar Primo de Rivera*'.

way of being and thinking characterised by Christianity, respect for the human person, recognition of the family as the foundation of life in society and women as its pivot, and the religious action of women. Moreover, all of this focused on the need for women's education.[36]

For the female Falangists, a concrete political objective of the Congress was to contribute to the Hispanic Community by outlining a model of woman that would unite Spanish and Hispanic-American women. In this sense, we believe that the Female Section added a unique element to the creation of the Community by providing a gendered reading of Hispanidad. This issue allowed the integration of women as subjects of cultural Francoist diplomatic relations.

Falangist Strategies for the Creation of a Transnational Space

Along with the definition of 'Hispanic woman' and the attendees' commitment to defend the approved agreements in their countries of origin, the Female Section, supported by the Ministry of Foreign Affairs and the Institute of Hispanic Culture in Madrid, proposed several strategies to maintain the relationship that the 1951 Congress had commenced. The proposals that immediately arose were the creation of the Hispanic American and Filipino Women's Cultural Circles and of an Ibero-American and Filipino Women's cultural magazine, the foundation of Women's Residence Halls for Hispanic-American students, and the organisation of new congresses. In addition to these ideas, thanks to the financial aid provided by the Franco regime, a system of scholarships and exchanges was implemented. As a summary, the documentation shows us that more than twenty women's groups were formed in America from 1951 to 1977,[37] and the scholarship system reported nearly a thousand women improving their professional skills in Spain from 1952 to 1977.[38] The efforts made the Hispanic Community tangible, building it as a transnational exchange

[36] General Archive of the Administration, *Delegación Nacional Sección Femenina Departamento Coordinación* Caja 5808, '*Conclusiones Primer Congreso Hispanoamericano y de las Filipinas*'.

[37] General Archive of the Administration, *Delegación Nacional Sección Femenina Departamento Coordinación* Caja 5782-2.

[38] General Archive of the Administration, *Delegación Nacional Sección Femenina Departamento Coordinación* Caja 301, '*Relación de becas concedidas por esta Delegación desde 1948 a la fecha 1975*'.

and transfer space stimulated by cultural groups, trips, scholarships, and meetings.

The first initiative to operate was the Hispanic-American and Filipino Women's Cultural Circles (many of which were named '*Isabel la Católica*' in commemoration of the Congress), which were founded as early as December 1951. The idea of these groups was to protect and spread the model of the Hispanic woman, through this female social space. The patron saint of the Cultural Circles should be sought in the Medina Circles,[39] groups born during the post-war period, which sought the social and artistic dissemination of Spanish culture through recitals, conferences, and courses, inter alia. This cultural dissemination format was strategic for the public that the Circles sought to attract: young women and women from national elites.

The Circles expanded rapidly, and by the end of 1951, there were seven, which increased to 21 in 1957, and subsequently to 25 in the 1970s.[40] Many of these circles were born under the wing of the local Institutes of Hispanic Culture, and since 1958, their foundation was expressly requested. Others were supported in their operations by local diplomatic representations. Administration of the Circles was centralised in the peninsula through the Coordinating Board of the Hispanic-American Women's Cultural Circles, whose representation before the Institute of Hispanic Culture and the Board of Cultural Relations of the Ministry of Foreign Affairs was entrusted to Pilar Primo de Rivera.[41] In the organisation chart of the Female Section, the coordination was subordinated to the Foreign Affairs Office.

The internal organisation of the Circles replicated, in part, the structure of the Female Section. At the top was a board of directors and then Chiefs for each section of the Circle (folklore, foreign service, propaganda, etc.).[42] Despite this basic organisation, each Circle could add or remove sections considering their reality and aspirations.

[39] Aguilar (2012).

[40] General Archive of the Administration, *Delegación Nacional Sección Femenina Departamento Coordinación* Caja 5782-2.

[41] National Library of Spain, *Folleto Labor cumplida y relaciones con otros organismos de la Sección Femenina*, 1970, p. 4.

[42] General Archive of the Administration, *Delegación Nacional Sección Femenina Departamento Coordinación* Caja 5789, '*Reglamento Interno del Círculo Femenino Isabel la Católica*'.

In addition to the cultivation and promotion of the Hispanic culture, the Circles carried out local educational work. For example, we know that the Circle *Isabel la Católica* in Santiago (Chile) created the Chilean Folklore Association,[43] which cultivated the Spanish and national folklore, and dedicated efforts to its research.[44] Further, it developed an Art Course that focused on crafting with religious motifs and collaborated in the decoration of various churches throughout Chile. The teachers who taught the course were former scholarship holders of the Female Section.

Without a doubt, however, the most relevant task conducted by the Circles was the course of preparation and pre-selection of scholarship holders to study in Spain. The young women who wished to apply for a scholarship of the Female Section had to attend classes for a period of three to six months, after which they had to write a final report. The topics covered were Spanish art, religion, and history, the functioning of the Female Section, and national/local culture and history. The courses, like the conferences, were taught either by intellectual figures from the Hispanist sphere, close to the embassies, or by former scholarship holders of the Female Section. For example, from 1954 to 1960, some of the speakers at the Chilean Circle were Gisela Silva and Sara Philippi (both former scholarship holders), Manuel Gutiérrez Lea Plaza (architect and director of the Chilean Institute of Hispanic Culture), and a series of university professors. Spanish intellectuals passing through Chile were also invited to the Circle.

As part of this cultural dissemination, schools and teachers were also supported. At the end of 1951, the Circle of Santiago sponsored the creation of the Agricultural Family School that was organised following the Falangist example. In 1959, the Family School had 110 students and was celebrated by the Spanish Ambassador who, after participating in its anniversary, reported the following to the Ministry of Foreign Affairs:

> [The director of the school] referred in praiseworthy terms to the work being carried out in Spain to raise the moral and cultural level of rural women, pointing out that the Agricultural Family School was nothing other than the adaptation to Chile of the Spanish example. The interventions of the students in songs and dances were highly applauded and were based on

[43] Led by Raquel Barros Aldunate, whose vocation for folklore is related to the visit of the Spanish Choirs and Dances to Chile in 1949 and the subsequent scholarship to visit Spain to take the course of Music Instructor in the Female Section.

[44] Hispanic Library, '*Revista de la Agrupación Folklórica*', No. 2, December 1955.

many of the aspects of the work of Choirs and Dances of the Female Section, since two of the teachers of the aforementioned School are Spanish and trained in said Institution.[45]

The Circle also supported schools and colleges by donating brochures and maps of Spain, as well as giving lectures on Spanish issues. Further, it delivered copies of the magazine *Consigna* to pedagogical centres, which was published by the Female Section and aimed at students of Teaching.

In Argentina, the Circle of Córdoba created radio programmes that were broadcasted by *Radio Nacional Argentina* and formed the theatre group *Clavileño*. This Circle also took a stand on education. In 1956, it founded the Institute of Feminine Arts, a 'family training' school that enrolled thirty-seven students in its first year. The academic programme was in line with the subjects of the Female Section: social and family legislation, domestic organisation, psychology, gymnastics, decoration, cutting and sewing, cooking, religion, home medicine, women's recreation and culture, history of marriage, social service, music, folklore, and education, among others.[46]

The 1960s marked a change in the operation of the Circles. On the one hand, the context of the Cold War influenced Spanish policy towards Latin America, moving away from the Hispanic Community project in favour of a discourse close to anti-communism and technocracy.[47] On the other hand, the directors of the Circles sought to alleviate the little cultural interference of Hispanism, bringing it closer to the community. The members promised to increase their social work, set up Hispanist libraries, and sponsor schools with names related to Spanish culture.[48]

The second incentive for the transnational space that we will analyse was the participation of the Female Section in the system of scholarships and exchanges subsidised by the regime. The documentation indicates that the first Latin American scholarship holders arrived in Spain in 1947

[45] Archive of the Ministry of Foreign Affairs, R 9026, Exp. 5 '*Asunto: Escuela familiar agrícola Universidad Católica*', 2 September 1959.

[46] General Archive of the Administration, *Delegación Nacional Sección Femenina Departamento Coordinación* Caja 5789, '*Informe de actividades del Círculo Cultural Femenino Hispano Argentino de Córdoba 1955, 1956, 1957, 1958 y 1959*'.

[47] Del Arenal (1994): 58.

[48] General Archive of the Administration, *Delegación Nacional Sección Femenina Departamento Coordinación* Caja 5789, '*Congreso de 1963. Reunión con las dirigentas de los Círculos de Hispanoamérica. Sugerencias de los Círculos*', 3 June 1963.

and that the candidates were ordered and systematised from 1952. The objective of these scholarships was to enable young upper-class females, who were interested in Hispanism and the Falange, to become acquainted with the work of the Female Section, receive professional training in its schools, or pursue postgraduate university studies. They were expected to become propagandists for the Franco regime, disseminate Hispanism, and replicate ideas and models in their countries of origin. The class bias of these scholarships was also due to their economic conditions. Each scholarship holder had to pay for their personal and medical expenses, and more significantly, the ticket to and from Spain.

As previously mentioned, scholarship holders were recruited via the Cultural Circles. During the first years in this exchange system, the ideological profile of the young female outweighed her academic requirements.[49] After the evaluation of the respective course, the directors of the Circle would send a list of preselected candidates to Spain where the awards were decided. This ensured that the candidates had basic notions of Hispanism and passed the ideological standards of the members of the Circle. Other selection criteria included the age of the candidates (less than thirty years old) and their marital status, as they had to be single.

During the first years, the scholarship holders were encouraged to take the 'general scholarships', which comprised a five- or six-month stay in the Female Section and included the following: two months at the *Escuela Mayor de Mandos*, divided between the *Castillo de la Mota*, *Castillo de las Navas*, and *Granja Escuela de Aranjuez*; subsequently, a month in the National Delegation in Madrid visiting the *Escuelas Hogar*, *Cátedras Ambulantes*, and various training schools; one month in Barcelona studying the provincial organisation; and one month in the summer hostel.[50] This itinerary allowed the scholarship holders to learn about the organisation and doctrine. Moreover, it allowed them to become acquainted with the Falangist style and way of being through their participation in the ceremonial and the austere and disciplined life of the *Escuela de Mandos*, which was a boarding school.

[49] General Archive of the Administration, *Delegación Nacional Sección Femenina Departamento Coordinación*, Caja 5782, '*Carpeta Resumen de actividades del Servicio Exterior 1951–1973*'.

[50] Archive of the Ministry of Foreign Affairs, R 5238 Exp. 15 '*Condiciones de las becas otorgadas por la Sección Femenina de España*', 1952.

The candidates often requested concessions to the selection criteria that led to their reformulation, for example, in the maximum age of application. However, the most significant change was associated with the educational range to which the candidates requested access. In the late 1950s, the scholarship holders required professional preparation outside of the careers available at the Female Section. Consequently, the scholarship began to support doctoral courses, full degrees, and other specialisations in university centres, although always in areas accepted by the institution. The Falangists consented to Art, Education, and Social Sciences, which led to Doctorates in Fine Arts and Literature and Psychopedagogy and Social Work careers, among others. All of these 'female professions' were admitted by the institution due to their association with the standardised virtues in their female model.[51] These new conditions forced the subsidies to increase their temporary coverage to a full academic year with the possibility of annual renewal.

These scholarships had a significant impact. A total of thirty scholarships were being awarded annually in the late 1950s, most to Argentine women, followed by Chilean and Colombian. A decade later, the number of annual scholarships had risen to forty-four.[52] The last count of scholarships awarded by the institution that we have found in the institution's documentation dates from 1975 and accounts for 1065 scholarship holders.[53] However, despite the number of scholarship holders, there is little documentary information about the courses or careers taken, how they were evaluated by the institution, and what use they made of these studies upon their return to their countries of origin. In the case of Chilean women, for the first fifteen years of operation of the Circle, we know that the members received general scholarships and attended courses related to social assistance, rural studies, home education, youth instruction, pedagogy, literature, and nursing, among others.[54] We are also aware of cases

[51] Royal Academy of History, *Archivo Asociación Nueva Andadura*, '*Carreras para la mujer*', *Y revista para la mujer*, September 1941; 'La mujer universitaria', *Y revista para la mujer*, March 1942.

[52] General Archive of the Administration, *Delegación Nacional Sección Femenina Departamento Coordinación* Caja 5782-2 '*Resumen de labor exterior 1946–1970*'.

[53] General Archive of the Administration, *Delegación Nacional Sección Femenina Departamento Coordinación* Caja 301, '*Relación de becas concedidas por esta Delegación desde 1948 a la fecha 1975*'.

[54] General Archive of the Administration, *Delegación Nacional Sección Femenina Departamento Coordinación* Caja 5766, '*Chile 1951–1963*'.

of scholarship holders who formed a network of rural schools, trained social workers, joined groups advocating for religious education, and participated politically in defending nationalist, authoritarian, and conservative ideas.

Conclusions

The relationship between the Female Section and Hispanic America was also promoted by trips and visits of the Command that allowed Falangists and Hispanic-Americans to cross the Atlantic several times. Pilar Primo de Rivera visited America on tours that led her to strengthen ties with governments related to Franco and with the women's associations that supported female Falangism across the Atlantic in 1949, 1953, and 1968. The Command also travelled repeatedly, confirming the work of the Circles in loco. The Congresses of the Institutes of Hispanic Culture allowed the directors of the Circles to meet in America and Spain (1958 and 1963). Moreover, the Choirs and Dances made new tours to Latin America (1962 and 1968).[55]

However, once the idea of the Hispanic Community lost its usefulness for the government and disappeared to give space to other strategies of international projection, the support, particularly economic, which the Ministry of Foreign Affairs and the Institute of Hispanic Culture had provided to the Female Section to sustain these initiatives, declined. This resulted in difficulties in promoting new projects, such as organising a Second Hispanic-American Women's Congress or increasing the number of scholarships. Despite this, the transnational space that the Female Section had strongly promoted throughout the 1950s continued to operate, as did the initiatives that brought it to life. This relationship only ended when the institution ceased to exist in 1977.

In this chapter, we have learned about the Female Section that was engaged in Spanish international relations with Latin America, acting both as a vehicle for propaganda for the regime at different political moments, as well as an organisational and ideological example to follow. In this sense, the role played by the Female Section in Spanish cultural diplomacy has revealed it to be an actor in international relations. Further, it shows us that, during the creation of the Hispanic Community of Nations, the

[55] Tessada (2017b).

cultural Francoist diplomacy had a perspective that integrated women in a particular manner.

The participation of the Female Section in Spanish diplomacy had the institution's initiative as an important component, which contributed to shaping the relationship with America through innovative projects. For example, the first trip of the Choirs and Dances, the First Women's Congress, and the creation of the Cultural Circles emerged as projects from the offices of the Female Section. To this, we should add the contribution of the 'model of a Hispanic woman' to the creation of the Hispanic Community of Nations—and all this to retain and attract Hispanic-American women to its cultural action.

The model of femininity that the Female Section exported was not the same as that advocated on the peninsula. The Hispanist aura that surrounded the creation of the 'model of a Hispanic woman' gave it a strong Catholic imprint that overshadowed the influence of the Falangist style and way of being that was so present in Spain. Nevertheless, it emerged in the organisational structures, in the experiences of the scholarship holders, in the courses taught at the Circles, and in the calls for public participation of Hispanic-American women. Further, it undoubtedly emerged in the negotiation with the American context, in which young females demanded spaces for professionalisation and independence.

For the Female Section, this international projection also had political motivations. This is suggested by Pilar Primo de Rivera in her discourses, in which she refers to the political benefits of her international projection, both for the Dictatorship and for women. By this means, Pilar Primo de Rivera could achieve her Falangist goals on American territory. Finally, the Hispanic Community of Nations as a rhetorical fiction allowed the female Falangists to consolidate their work in Hispanic America. The long-term maintenance of the relationships for nearly thirty years is undoubtedly an important achievement for the institution.

References

Aguilar, Isabel (2012): "El programa cultural de la Sección Femenina: Vía de escape y mecanismo de control social de la mujer en la España franquista", Comunicación, X Congreso de la Asociación de Historia Contemporánea, Universidad de Granada.

Amador, María Pilar (2003): "La mujer es el mensaje: los Coros y Danzas de la Sección Femenina en Hispanoamérica", *Feminismo/s: revista del Centro de Estudios sobre la Mujer de la Universidad de Alicante*, N° 2: 101–120.
Barrachina, Marie Aline (1991): "Ideal de la mujer falangista. Ideal falangista de la mujer", Instituto de la Mujer, *Las Mujeres y la Guerra Civil Española*, Madrid, Instituto de la Mujer: 211–217.
Barrera, Begoña (2019): "Emociones para una identidad compartida. La Sección Femenina de FET-JONS entre la guerra y los años grises", *Historia y Política*, N° 42: 241–268.
Bengochea, Enrique (2012): "Las otras falangistas: guineanas y saharahuis en la Sección Femenina." Comunicación, XI Congreso de Historia Contemporánea, Granada.
Blanco-Camblor, María Luz (2005): "Similitudes y diferencias entre la 'Sección Femenina' en España y la 'Bund Deutscher Mädel' en la Alemania del Tercer Reich: Una aproximación", en Santo Tomás Pérez, Magdalena (Ed) *Vivir Siendo Mujer a Través de La Historia*, Valladolid: Secretariado de Publicaciones e Intercambio Editorial Universidad de Valladolid: 215–240.
Cañellas, Antonio (2014): "Las políticas del Instituto de Cultura Hispánica, 1947–1953". *Historia Actual Online*, N° 33: 77–91.
Casero-García, Estrella (2000): *La España que bailó con Franco: Coros y Danzas de la Sección Femenina*, Madrid, Editorial Nuevas Estructuras.
Cenarro, Angela (2017): "La Falange es un modo de ser (mujer): discursos e identidades de género en las publicaciones de la Sección Femenina (1938–1945)". *Historia y política: Ideas, procesos y movimientos sociales*, N° 37: 91–120.
Del Arenal, Celestino (1994): *Política exterior de España hacia Iberoamérica*, Madrid, Editorial Complutense.
Delgado Gómez-Escalonilla, Lorenzo (1988): *Diplomacia franquista y política cultural hacia Iberoamérica, 1939–1953*. Madrid, Editorial CSIC.
Delgado Gómez-Escalonilla, Lorenzo (2003): "La política latinoamericana de España en el siglo XX", *Ayer*, N° 49: 121–160.
Lacalle, Carlos (1957): "Hacia una Comunidad Hispánica de Naciones", *Mundo Hispánico* N° 85: 102–111.
Martín Artajo, Alberto (1953): "Spain in the present day world", *Pakistan Horizon* N° 6-1: 25–28.
Martín Artajo, Alberto (1956): "La supranacionalidad hispánica" Discurso pronunciado con motivo de la fiesta de la Hispanidad, 12 de octubre de 1954, en Martín Artajo, Alberto, *Hacia La Comunidad Hispánica de Naciones*, Madrid, Cultura Hispánica: 103–118.
Morales, Amalia y Vieitez, María (2014): "La Sección Femenina en la 'llamada de África': Saharauis y guineanas en el declive del colonialismo español". *Vegueta: Anuario de la Facultad de Geografía e Historia* N° 14: 117–133.

Morant I Ariño, Antonio. (2013): "Mujeres para una 'nueva Europa'. Las relaciones y visitas entre la Sección Femenina de Falange y las organizaciones femeninas nazis, 1936–1945. Tomo I". Tesis doctoral, Valencia, Universidad de Valencia.

Nerín, Gustau (2007): *La Sección Femenina de Falange en la Guinea Española, 1964–1969*, Barcelona, Ceiba Ediciones.

Primo de Rivera, Pilar (1983): *Recuerdos de una Vida*, España, Ediciones Dyrsa.

Rodríguez, Sofía (2005): "La Sección Femenina de FET-JONS: "Paños calientes" para una dictadura", *Arenal: Revista de historia de mujeres* N° 12-1: 35–60.

Sánchez, Rosario (1993): "Sección femenina, una institución en busca de investigador: análisis crítico de la bibliografía disponible", *Historia social*, N° 17: 141–154.

Stehrenberger, Cecile (2012): "Los Coros y Danzas de la Sección Femenina en Guínea Ecuatorial. Un caso de estudio del vínculo entre política de género y colonialismo". En Osborne, Raquel (Ed) *Mujeres bajo sospecha. Memoria y sexualidad 1930–1980.* Madrid, Editorial Fundamentos: 311–330.

Suarez, Luis (1993): *Crónica de la Sección Femenina y su tiempo*, Madrid, Asociación Nueva Andadura.

Tessada, Vanessa, (2017a): "La fundación de las Secciones Femeninas Exteriores en América Latina: retaguardia de mujeres durante la guerra civil", en Carrellán, Juan Luis (Ed), *La Guerra Civil Española: Estudios y reflexiones desde Chile*, Santiago: Centro de Estudios Bicentenario: 65–90.

Tessada, Vanessa (2017b): *Las estrategias de proyección internacional de la sección femenina española hacia Latinoamérica y su recepción en Chile: (1937–1977)*, Tesis doctoral, Valladolid, Universidad de Valladolid.

PART IV

Perspectives of Nationalization in Scientific Disciplines and the Arts

CHAPTER 10

On the Political Value of Science: The Three Lives of Spanish Mathematics in Early Francoism

José M. Pacheco

... the prosperity of a nation is directly related to the esteem that it has for itself and to the nurturing of mathematical sciences.
—José Doménech Estapá (1904)

This work was partially subsidised by Projects HAR2012-39260 and HAR2016-75559. The Catalan first names are presented in their Castilian version, which was used in those years in official documents, publications, and quotations.

J. M. Pacheco (✉)
Department of Mathematics, University of Las Palmas de Gran Canaria, Las Palmas, Spain
e-mail: josemiguel.pacheco@ulpgc.es

M. Janué i Miret, A. Presas i Puig (eds.), *Science, Culture and National Identity in Francoist Spain, 1939–1959*,
https://doi.org/10.1007/978-3-030-58646-1_10

Setting the Stage

As pointed out by José Doménech Estapá in his inaugural speech at the University of Barcelona in 1904–1905, the political value of science and its more applied variant, engineering, had been recognised since the mid-nineteenth century (Doménech 1904). After World War I, it became a criterion to measure the relevance—real or potential, along with geopolitics—of the various countries that would later participate in World War II. This is supported by the mathematical research output of Germany, France, Italy, the UK, and Spain in those years, as revealed in *Jahrbuch über die Fortschritte der Mathematik* (1868–1942).

Spain was not belligerent in the last global conflict, as it had not been in World War I, but this did not exempt it from repercussions, added to those of its own post–civil war. The war also devastated large areas of Europe, Asia, Africa, and Oceania, and a new world order was established (Judt 2008).

Practically ruined at the end of the nineteenth century after the loss of what was left of its once gigantic colonial empire, Spain had to wait several years before recognising—both in reality and legally—the importance of science, pure or applied, for the country. The work conducted in that period between 1915 and 1935 was so extensive that it came to be known as the 'Silver Age of Spanish Science', despite the political instabilities and endless wars of attrition in North Africa. Much has been written about this period (Otero 2014), both on its tragic end due to the civil war and the preservation of some of those ideals and achievements after the war by adapting to the political conduct of the General Franco regime, before giving way to internationalisation around 1960 as a habitual way of conducting science (Presas 1998).

Here, we will analyse what occurred in the small world of Spanish Mathematics after the civil war, and why. Despite the central role of this discipline in the world of science, it left only a faint trace in the Silver Age, and was even—perhaps deliberately—forgotten or erased. As an example of this forgetfulness, one can examine the commemorative book *Tiempos de Investigación* (Puig-Samper 2007) and compare it with the standing in other places based on the text of Helmut Hasse's inaugural lecture at the University of Wiesbaden for the 1951 course, whose title leaves no room for doubt: no less than *Mathematics as Science, Art, and Power* (Hasse 1953).

Spanish culture is prone to consider isolated personalities as icons to whom it attributes its successes or blames for its mistakes and failures, and this truism remains valid for all fields. To name a few, there are references to Ramón y Cajal (but fewer to his school), Ortega y Gasset and Blas Cabrera, Zubiri, and Menéndez Pidal as individuals, which depict them as almost solitary eminences. Such individuals were also to be found in Spanish Mathematics: in essence, it will suffice to consider Julio Rey Pastor (Rey from now on), a *factotum* in the Spanish and Latin American world of Mathematics between 1915 and 1935. Rey (1888–1962) has been the subject of many biographies with different scopes, all emphasising on his mathematical curiosity and entrepreneurial skills (Ríos 1979; see also Español 1990).

In our opinion, the biography of Rey that is closest to reality is his rather melancholic autobiographical reflections in the reply speech to Ricardo San Juan when he joined the Academy of Sciences in 1956 (S.A. 1956). Already in 1912, he was a full professor at the University of Oviedo with two scholarships from the Board of Advanced Studies (*Junta de Ampliación de Estudios*—JAE) in Germany. He was among the founders of the Spanish Mathematical Society (1911) and promoted the Mathematics Laboratory and Seminar of the JAE in 1918. He created mathematics journals, initiated long relationships with Spanish-American Mathematics, advised the doctoral theses of those who would form his group of disciples—and also a pressure group—for teaching positions, scholarships, and research grants. He was the author and mathematical texts, textbooks, a historian of science, and a successful lecturer.

Rey's prime, both mathematically (Rey 1932) and politically, ended with the civil war. He was in Argentina and refused to return to Spain both during and after the war despite offers from the Franco regime. He only returned—as a visitor—in 1948, and permanently after 1952. But both his scientific and personal influence had declined significantly, and he ultimately returned to Argentina, where he died in 1962.

The other leading figure of Spanish Mathematics, in its applied version, was Esteban Terradas (1883–1950), an individual as entrepreneurial, or even more so, as Rey, and certainly endowed with better political skills than the latter. An engineer by training and with a vast scientific culture, he shared the leadership with Rey for years, a collaboration that was harmonious both personally and scientifically. Back in Spain after the civil war, he was offered the directorship of the Aeronautical School from 1940, reviving a former Aeronautical Technical School founded in 1928 by the engineer, inventor, and mathematician, General Emilio Herrera (1879–1967), exiled after the civil war.

The structure of engineering studies in Spain required a solid preparation in basic sciences, particularly Mathematics (Fernández and Pacheco 2005). Admission exams to the technical schools were so difficult that candidates often chose to take two or three years of *Exact Sciences*, the old name of the Mathematics career path, to prepare them. This plan was only affordable to those with a good financial situation and, in return, many of those who followed this path found it feasible to become both engineers and mathematicians, favouring their teaching positions in technical schools and universities. The Mathematics career path could be followed since 1857 at Madrid, Zaragoza, and Barcelona, although the classical engineering schools, organised according to the French models, were already operating in Madrid: Mining (1777), Roads (1802), Architecture (1844), Industry (1845), and Agronomy (1855).

A Mathematical Community in Spain?

In her work *Orígenes internacionales de la política científica*, Mª Jesús Santesmases demonstrated the development trends of the scientific communities after World War II, forced by the influence of the USA in several scientific fields whose advances had accelerated due to the war effort in the first half of the 1940s. The ways and means of financing, the selection of study subjects, the channels for dissemination of results, and the implementation of control methods that survive to this day, understanding science as a political element on a much larger scale than before, particularly in view of its applications (Santesmases 2008).

The value of science for political purposes has changed throughout history: from the mid-nineteenth century, Physics, Chemistry, Biology, and their corresponding applications were recognised as having a highly strategic importance. The same was true for Mathematics as the core element of Astronomy and Navigation before beginning its path of abstraction away from its applied origins. For now, it will suffice to point out the widespread use of Mathematics by the USA during World War II to handle the volume of personnel, machinery, and logistics of war operations. Operational Research was thus created, later adopted in business planning. The more abstract aspects of Algebra were used from the start to decrypt messages, and something as familiar today as radar was implemented from studies of certain differential equations in the late 1930s, when war in Europe seemed inevitable (Cartwright and Littlewood 1945). The American methods in scientific activities were common by 1960.

In Spain's Mathematics, this evolution also took place, but a subdivision into three stages or 'lives' between 1939 and 1959 may be observed, correlated with the phases and strategies of the Spanish regime's incorporation into the international scene: (a) a first period, during World War II, characterised by the affinity with the Axis powers; (b) years of isolation according to Resolution 39 of the UN General Assembly (December 12, 1946) until sanctions were lifted by Resolution 386 of the same Assembly on November 4, 1950; (c) a third stage began around 1952, when Spanish-American relations were formalised through the Madrid Pacts, on Sept 23, 1953. Note that despite isolation, informal relations with the USA had existed since 1945.

The boundaries between the last two stages are rather fluid, but the Americanisation of the Spanish mathematical scene took slightly longer: in the late 1950s, it was not fully established due to the inertia of Rey's mathematical culture. Regarding mathematical topics, all three stages were marked by the contributions to Geometry of several kinds, classical Mathematical Analysis, ventures into Mathematical Physics, and Quantum Mechanics. In the third stage, Statistics emerged as an autonomous discipline, just as the first steps of Computer Science came into being.

Therefore, the 'three lives' of Mathematics correspond to the attitude of the Franco regime on matters of international integration and on how mathematicians took advantage of the opportunities of those political and strategic scenarios: it is not clear whether Mathematic were instrumentalised by the regime, which only allowed the occasional presence of some representatives in international meetings (Pacheco 2014).

The Three Lives of Spanish Mathematics

The 1940–1945 period—the 'first life'—was marked by autarchy, intense political repression, poor harvests, internal governmental intrigues, and structural reorganisation of the newly established regime (Beevor 2005; Preston 2007; Camprubí 2016; Pacheco 2017, 2018).

The remnants of scientific policies of the Second Republic (1931–1939) were recycled after the disappearance of the JAE and the creation of the Spanish National Research Council (*Consejo Superior de Investigaciones Científicas*—CSIC). In Mathematics, there was no scientific breakdown, only a reallocation of personnel to fill the vacancies produced by the civil war with people close to the regime. Technically, Mathematics remained in the hands of adherents to Rey. Older scholars, after the 'purging of responsibilities' and ensuring their adherence to the regime, continued working in the same subjects as before, generally far from considerations of direct applicability. Even so, retaliations were made against several mathematicians from that group, the most conspicuous being José Barinaga Mata, Olegario Fernández-Baños, and Tomás Rodríguez

Bachiller. The professorships of Barinaga (Cuesta 1966) and Bachiller were revoked, and Fernández-Baños lost his job as an advisor to the Bank of Spain.

The case of Bachiller illustrates something of the regime's nature: he had been discharged and could not hold high-ranking positions, but he was immediately assigned to urgent tasks—of great responsibility both academic and symbolically—such as advising young mathematicians from the Francoist zone, and to consolidate the 'Jorge Juan Institute of Mathematics' of the CSIC. However, he had to earn his living by privately tutoring students for admission to engineering courses and as a translator. Bachiller translated most of the textbooks of the 'Cambridge Mathematical Texts' collection.

Aeronautics was the reference engineering during the 'first life': in addition to technical challenges, it could provide some short-term economic benefit due to World War II by manufacturing parts or aircrafts for the competing countries. Terradas, in charge of the Aeronautical School from 1940, immediately appointed several of his former colleagues as professors of Mathematics. The school launched the *Revista de Aeronáutica*, whose first issues included contributions, particularly by Ricardo San Juan, on the theory of complex variables and the behaviour of wings and other aeronautical elements, an intrusion of Applied Mathematics during the 'first life'. In 1943—note the year—the school attempted to establish contacts with the US through Terradas but was dismissed by the Americans. Shortly thereafter, however, in March 1944, an agreement was signed with the USA allowing their aircraft to use the Cabo Juby airfield in the Spanish Sahara (Pérez-Marín 1970).

Through Rey, more modern Mathematics than those practised in the nineteenth century were introduced to Spain, and emphasis was also placed on improving the preparation of teachers at pre-university levels, promoting communication in journals, and suggesting the writing of texts for different academic levels, both for school and university use. From the early 1920s, Rey worked on both sides of the Atlantic, as a full professor in Madrid in the winter and Argentina during the summer. As theses adviser in Spain, Rey delegated José Mª Plans y Freyre—until he died in 1934—Álvarez Ude, and later Tomás Rodríguez Bachiller.

However, not everything was so simple. Bachiller himself, no longer so young, was forced to poorly write a thesis in 1934 to occupy a professorship from which to act as a director. Germán Ancochea entered the Rey group as a candidate for high school teaching, soon moved to Paris, concluded a thesis under Élie Cartan, and obtained a professorship in La Laguna (Tenerife), before going to Salamanca where he remained from 1936 away from the general stream of Rey's followers. In 1947, he moved to Madrid, without having fully reconciled with either the Franco regime or Rey.

An undated letter from 1963, from Antonio Rodríguez Sanjuán to Norberto Cuesta, comments on the death of Rey the previous year in the following terms (Rodríguez San Juan 1963):

> …what are we, then, you, Sunyer Balaguer, Rodríguez Salinas, Augé, Castro, Vidal Abascal, Gaeta, Abellanas and their successors, Ancochea wanting or not, but a school of disciples of Rey…? Flores de Lemus, who was almost never spoke with Don Julio due to the incompatible characters of two difficult men, wrote a letter to his brother in which he stated without blushing that everything we were we owed to Rey.

Rey's vision of Mathematics, of German and Italian origins, was preserved by the new incorporations, to the point that some were sent to Germany and Italy in the middle of the world war. Therefore, this 'first life' is characterised, we claim, by the scientific and thematic continuity of Mathematics in Spain after the civil war, a situation favoured by the scarce regime interest in promoting alternatives: perhaps, mathematicians had given up struggling against the government disinterest.

Mathematical research was published in Spanish university or academic journals with a low circulation, but particularly in the two former journals of the JAE inherited by the CSIC—the *Revista Matemática Hispano Americana* and the *Matemática Elemental*. There were a few papers in the Portuguese journals *Portugaliae Mathematica* and *Gazeta Matemática* and in Italian journals (Otero 2014; Outerelo 2009; Pacheco 2014, 2015).

Considering the foreseeable changes in the international perception of Spain after the imminent Allied victory in World War II, the regime alleviated before the end of the world conflict some sanctions applied after the civil war. Among others, the mathematicians Barinaga, Fernández-Baños, and Bachiller himself, who were already working in Madrid in the academic year 1943–1944, were finally reintegrated.

But for this inexorable passage of time, there were no general guidelines supporting this late reintegration, neither scientifically nor legally. The sanction of *absolute separation* from service was irrevocable, although *temporary separation* for four, five, or six years was more common, always accompanied by ***disqualification*** to ***hold public or trusted positions***. Sanctions were entirely complied with, although the regime used the services of sanctioned persons with little hesitation; for example, to commence the academic year 1940–1941, or the sanctioned Bachiller as acting director of the CSIC mathematical institute. For the regime, the sanctions had to be fully complied with, including, of course, the loss of wages. For example, Barinaga, who was sanctioned at the end of 1939, was reintegrated at the beginning of 1946. However, the restrictions resulting from the 1946 Resolution 39 again prevented scientific development, including its impact on everyday topics; this was evident in the scarcity of publications, printed on poor-quality paper, sometimes combining several issues of a journal in one, and the difficulties in diplomatic relations.

Professors of Mathematics and mathematicians present in Spain were not much affected: those with permanent positions retained them if they had been obtained after the civil war—or even before if they had been considered supporters of the regime after a thorough scrutiny. Most younger mathematicians had emigrated years before, immediately after the war or even before its end, to various organisations in South America inspired by Rey.

On the other hand, Mathematics, traditionally an inexpensive science, suffered from the scarce offer of new permanent positions at universities and high schools. Private academies dedicated to preparing access to higher education proliferated in academic cities where there were Science Faculties or Engineering Schools. Another resource was writing university or high school textbooks, as well as translations of books from German and French, for the vast South American market could not be forgotten. There were extended versions of R. Rothe's *Matemática superior para matemáticos, físicos e ingenieros*, whose German editions dated from the mid-1920s, while the *Análisis Matemático en Ingeniería* of von Kármán and Biot was translated from English and edited under the auspices of the Air Ministry (Von Kármán and Biot 1945).

It should be noted that while Spain was being punished at the UN for its fascist regime (*sic*), in neighbouring Portugal a similar regime unleashed a repressive wave in 1947–1948 that sent the best of Portuguese mathematicians into exile, and closed the Portuguese Mathematical Society for decades, although not its affiliated journals. This repression was a response to the new international situation: the enemy was now the USSR, and the existence of declared communists among Portuguese mathematicians was notorious. Several of the Portuguese exiles were also welcomed in Brazil and Argentina through Rey; the best-known case being António Aniceto Monteiro, who returned to Portugal after the fall of Salazarism.

There were efforts in Spanish mathematical circles to establish international connections. Luis Esteban Carrasco travelled in 1945 to the USA, attending courses in Chicago and Columbia. On his return, he obtained his doctorate with work developed there, advised by Sixto Ríos.

Amidst the blockade decreed by the United Nations, between 1946 and 1947 Bachiller—an alter ego of Rey in Spain after the civil war—was also in charge of maintaining contacts and travelled to the USA to visit several institutions and to establish scientific relations. Here is a letter from Bachiller to Karl Menger, who was in the USA, specifically Chicago, since 1938. The relationship between them stemmed from Antonio Flores Giménez's stay in Vienna, a JAE bursar before the civil war who published three articles in the *Ergebnisse* of Menger's Seminar before returning to Spain and then to Princeton in 1935 with another JAE scholarship. He returned to Republican Spain in 1937, therefore his later life was not easy.

The original letter is in French, and the ellipsis corresponds to family greetings (text courtesy of Bernhard Beham, Vienna. Source: *Autographenabteilung der Wienbibliothek (Korrespondenzen)*). Bachiller invites Menger to visit Spain and Portugal:

> Princeton, 11-VI-47
> Dear friend:
> I have just received your letter, and I am pleased with your excellent reaction to my proposal […]. So, I will start the preparations as soon as I return to Madrid, since the bureaucratic processes are always slow. Considering that the Summer University in Santander is already functioning and that next year, in August, we have a Mathematics meeting, it would be the right time for you; you would spend that month in a beautiful place with a good beach, and then you would go to Madrid, where you could stay until December […]. If you have no objections, I will also write to my friends in Coimbra, Lisbon, and Porto, so you can take advantage of your stay with us to organise other visits […]
> … … …
> Thanking you for your messages, receive, dear friend, my warmest regards.
> TR Bachiller

The importance of the letter derives from the fact that it revives old friendships forged prior to the civil war and fostered by the policies of the

JAE, highlighting how valuable those personal relationships could be for science in Spain. Nonetheless, Menger's planned visit did not occur.

A decisive event during this second life was, probably, the creation of the journal *Collectanea Mathematica* in Barcelona in 1948. It was founded by José Mª Orts, one of the first disciples of Rey, together with the young Enrique Linés, who had completed a doctorate after the civil war. The third founder was the architect Francisco Sanvisens, whose thesis was the first article in the inaugural issue of the journal.

The appearance of *Collectanea* decentralised the publication of mathematical papers that was until then concentrated in the *Revista Matemática Hispano Americana* (RMHA, series IV), published in Madrid since 1940 by the CSIC. *Collectanea* became the voice of the Barcelona Mathematics Seminar—authorised since 1941 by the regime and with an active Astronomy section—offering an alternative for the dissemination of Spanish mathematical research (Roca Rossell 2016). However, the RMHA and *Collectanea* turned out to be strikingly similar in content and even in typography, and Catalan authors would only move to the new journal much later. In its first years, it offered many contributions from foreigners, more than the RMHA. *Collectanea* is currently the oldest mathematical journal in Spain, published without interruption since its first issue.

Towards the 'Third Life': The Effects of the Pacts with the US

The transition from the second to the third life is not clear-cut but rather a process of adaptation and transformation promoted by the lifting of sanctions against Spain (UN Resolution 386, November 4, 1950) and the subsequent Spain-US Pacts of Madrid by the end of 1953.

The pacts basically included the provision by Spain of land for the installation of military bases 'for joint use' in Spanish territory. In exchange, the USA provided second-hand war material to Spain and implemented scholarships and funding schemes—including military studies—in that country (see contribution of Delgado Gómez-Escalonilla), and even shipments of dairy products for primary schools (Puig 2003).

The interaction with the USA led to a certain regime liberalisation with the appointment of Joaquín Ruiz-Giménez as Minister of Education (1951–1956), whose mandate ended abruptly due to a wave of student protests throughout the country. It was him who restored the possibility of obtaining doctorates at universities other than Madrid. The first one was Salamanca, in 1952, taking advantage of the celebration of the seventh centenary of the creation of the university in 1252, while others followed in a generalised manner in the subsequent year. Consequently, the academic world, and particularly Mathematics, was abandoning some centralist customs of the first post-war period. The first woman to obtain a doctorate in Mathematics was María Asunción Catalá Poch, who received it as late as 1970 from the University of Barcelona with her thesis on Astronomy.

The pacts opened the possibility of pursuing studies in the USA with Fulbright scholarships and some others, and small groups of students were immediately sent there, among whom there was no shortage of mathematicians, particularly recent graduates who went there to pursue their doctoral studies. However, we find also the already-well-established Ricardo San Juan with generous subsidies from the US Air Force for several works of a rather unapplied nature. Fernando Sunyer Balaguer, who can be considered a protégé of San Juan, also obtained financial support from the US Navy. In any case, internationalisation slowly commenced, and connections increased, yet it is not possible to speak of a Spanish school of Mathematics in those years, though afterwards Sixto Ríos foresaw some future in his 1963–1964 inaugural speech at the Academy of Sciences on 'The profitability of scientific research' (Ríos 1963).

The 'third life' of Spanish Mathematics is considered to have started from 1952. A representative journal, although not strictly mathematical, is *Theoria* (Series I: 1952–1955), founded by Rafael Sánchez-Mazas with an inaugural article by Rey: 'History, Science, Philosophy'. In scientific terms, this 'third life' was a waiting time before the widespread introduction of modes and methods imported from America, witnessing the slow academic extinction of the legacy of Rey and Terradas.

How to Become a Mathematician

After 1939, university studies in Mathematics in Spain followed the lines of a 1931 plan, in the style of Rey. With a classic basis, it was considered provisory but was used until 1942 through the civil war. It included contributions from Physics, Mechanics, and Astronomy and adapted well after the war to the scarcity and qualifications of the available teaching staff. During the 'first life', a new plan was prepared and launched in 1943, not very different from the previous one, although with different emphasis on some subjects and the inclusion of Religion, Sport, and Political indoctrination.

The 1943 plan allowed the few Faculties where Mathematics were studied (Madrid, Barcelona, Zaragoza, and shortly afterwards Santiago) to introduce some variations depending on the local teaching staff. The plans for 1931 and 1943 did not differ greatly from those in other European countries, at least in the first academic years. While the 'second life' did not change any of this, the third life did introduce a new Mathematics curriculum in 1952. It maintained the basic subjects (Mathematical Analysis, Geometry, Physics, and Astronomy), allowed the Faculties to offer a differentiated range of courses that should always include Analytical Mechanics, and introduced 'novelties' such as Abstract Algebra, Probabilities and Statistics, and Topology, subjects until then reserved for doctoral studies. The 1952 plan was changed in 1964, dividing the course path into several branches or specialties from the third year onwards.

The Two Journals: RMHA and *Collectanea Mathematica*

The following is a brief analysis of the two Spanish journals, the RMHA and *Collectanea* between 1945 and 1960 (*Collectanea* since 1948), based on both the texts and reviews of *Mathematical Reviews* and *Zentralblatt für Mathematik*. They supposedly had a scientific level comparable to other foreign journals. The RMHA disappeared around 1982 after a period of both budgetary and scientific instability but was the leading Spanish journal in Mathematics during the time covered by the study, along with *Collectanea*. As mentioned above, the journals were strikingly similar, differences only appear—in favour of the second—around the mid-1960s.

It has been pointed that Spanish Mathematics was in the hands of not only mathematicians but also engineers and architects. However, those with a technical background did not usually publish articles on Mathematics, and their written contributions can be found in technical and project reports, course materials, and various documents—what is now called 'grey literature'. On the other hand, whoever approaches these two journals will consider them to be pure Mathematics publications.

The RMHA published 45 Spanish authors, including two Hispanic-Americans despite the Spanish-American label, and 26 foreigners; the numbers are 27 and 24, respectively, for *Collectanea*. The primary language used in both journals is Spanish, although there are also works in German, Italian, French, and, to a lesser extent, English. Almost all articles have only one author: there are only five collaborative contributions, always with two authors. As was common at the time, abstracts did not appear at the beginning of articles.

There are few authors with works in the two journals (nine Spanish and four foreigners), nor are there many authors with numerous publications—only five authors have more than ten publications (the most prolific being Cuesta with 17 and Maravall with 16) and 35 with a single contribution. Both journals feature foreign authors, with one or two articles at most, except for the English theorist of Fluid Mechanics Milne-Thompson, who has five in the RMHA, and the geometrician and functional analyst Josef Weier, who has two in the RMHA and another five in *Collectanea*.

As the journals show, these Mathematics originated in Rey. He was above all a geometrician, and his aim was the academic replacement of the old theories from the mid-nineteenth century common in Spanish courses, by something more up to date (Santaló 1990). He almost succeeded, partly due to the advanced age of the corresponding professors: Miguel Vegas and Cecilio Jiménez, who represented the older Geometries, had been born in 1856 and 1858, respectively. Therefore, both classical Geometry and poorly understood versions of more modern trends are found in the journals. Real and Complex Analysis also had their part, including a curious fondness for studies on Integral Transformations announcing the late arrival of Functional Analysis. Mathematical Physics, Relativity, and Quantum Mechanics also received attention, as did Fluid Mechanics. Mathematical Statistics and the principles of Automatic Computing, the predecessor of Computer Science, were also introduced. The studies of Foundations and Theory of Science had some interest, and the advances in the History of Mathematics were primarily due to contacts—through Astronomy—with scholars of Arabic, Hebrew, and their cultural heritage in the Iberian Peninsula. Orús, Vernet, and Millás Vallicrosa could also be included among the Spanish mathematical historians, which would obviously raise the average level of Hispanic mathematical culture.

There are still questions that deserve special consideration. Regarding the number of foreign authors, especially German and Italian, it is not risky to suppose that most of them—former acquaintances from the JAE or escapees from the consequences of the world war—were invited to publish in Spanish journals, almost always with a single article. This was the case for Wilhelm Blaschke due to his relationship with Alberto Dou, and of Luigi Fantappié, who supervised the theses of Teixidor and Casulleras and whose close collaboration with fascism is well known. It should also be noted that several of the Germans had collaborated with the journal *Deutsche Mathematik* and were active in the Nazi party. *Deutsche Physik* and *Deutsche Mathematik* intended to publish 'Aryan science', untainted by Jews.

Another matter is the number of Spanish authors with only one paper. The most plausible explanation is that, for the purposes of university promotion, 'whoever has not published in one or both magazines is nobody'.

Presence in Foreign Journals

Only a few Spanish mathematicians published abroad during the years of this study, as confirmed by following their research output in *Mathematical Reviews* or *Zentralblatt*. Through personal contacts, letters, and occasional visits to conferences, Spaniards can be found in non-Spanish journals. There are two cases of scholars whose research was almost entirely published abroad: Germán Ancochea and Federico Gaeta. The first is the most obvious absence, and although he had written in the RMHA before 1945, thanks to Blaschke he published two articles in the journal of the Hamburg Mathematics Seminar. His research output includes 30 other works in several foreign journals. Gaeta, much younger, after having published in the RMHA, opted for Italian journals, as he had been a disciple of Francesco Severi. Two analysts, Ricardo San Juan and Sunyer Balaguer, both with more publications abroad than in Spain, were known and recognised outside the country. Spanish mathematicians appear in German, Italian, Portuguese, Swedish, US, Brazilian, Argentinian, and French journals, some of them well-respected ones.

However, the presence in foreign publications was scarce and sporadic, except for the examples mentioned above. This was to change from 1960 onwards with a younger generation of mathematicians who had benefitted from a long training abroad.

Awards and Recognitions to Mathematicians

The political context of the mathematical disciplines would not be complete without information about the awards and distinctions granted to its professionals.

Six positions at the Academy of Sciences, of which Rey was a member since 1923 and Terradas since 1933, were eclectically allocated among those close to Rey (3) and other independents. Four associate members were also appointed, including the exile Luis Santaló, a gesture made taking advantage of the Ministry of Ruiz-Giménez. Vernet has also been included to highlight the fact that he was belatedly elected—in 1974—as an associate member of the Academy of Sciences.

The *Gran Cruz de la Orden Civil de Alfonso X el Sabio*, a distinction created in 1939 by the Franco regime, was awarded to five people, all of them quite old: Torroja, 1947; Terradas, 1949, one year before he died; the priest and astronomer Aller (born in 1878) had to wait until 1955; and the remaining three (San Juan, Rey himself, and Millás Vallicrosa) were awarded in 1959. Millás is included for his relationship with Orús, Vernet, and Astronomy.

The 'Francisco Franco' National Science Prize was awarded to two mathematicians: San Juan (1949 and 1954; the only case for these prizes) and Sunyer Balaguer (in 1956), who did not collect it.

Juan March was a businessman who played an important, even decisive, role in financing Franco's military uprising. In 1955, a foundation named after him was created, with a scholarship and funding programme for scientists. In three years (1957, 1958, and 1959), ten mathematicians were funded by the Foundation to develop specific works or projects, justified with a final report. Here is Rey himself in 1959—he was awarded 500,000 pesetas (the report has not been found)—and his collaborator Pedro Pi Calleja, returned to Spain in 1956 and who was going through financial difficulties. Pi was already old for the time, and it was difficult for him to settle in his native Barcelona. The rest were younger, born in the 1920s, including María Wonenburger (1927), who received a scholarship in 1958. There is also an stipendium to San Juan in 1962—when Rey died—to write an essay on him (San Juan 1962).

Scarce Female Presence

Only seven women, five Spanish and two foreigners, appear in the previous sections. Five Spanish women is not a large group, relevant is not the number but rather the difficulty of following and interpreting their existence and scientific careers: scientifically some disappear very early, while still young. Born between 1924 and 1927, all of them completed their higher education after the civil war, a time that was not particularly favourable for women pursuing university careers in Spain, much less for starting research. Some were later appointed to teaching positions

The first is Ignacia Gómez Aguilar, author of an article in the RMHA on Topology, a new subject in Spain at the time, who ended up as a teacher at the *Escuela Normal* in Segovia. María de la Cinta Badillo Barallat, who devoted herself to questions of multipurpose logic, with three articles on *Automatic Calculation and Cybernetics*, did not continue with her research and reappeared in the mid-1970s applying to high school teaching positions. María Wonenburger Planells, who published twice in the RMHA, was the first female mathematician to receive a Juan March scholarship (1958). After being awarded a Fulbright scholarship, she developed her entire career in the USA until 1983, when she returned to Spain for family reasons and retired from Mathematics. The astronomer Mª Asunción Catalá, who was active since the early 1950s, was the first female Doctor of Philosophy (DPhil) in Mathematics in Barcelona, and had a long and distinguished career, publishing regularly in collaboration with Vernet and Orús. Griselda Pascual Xufré, who published in *Gaceta Matemática*, was a teacher in high schools and at the University of Barcelona, being highly respected and influential in the Methodology and Didactics of Mathematics.

The foreigners: Silvia Martis-Biddau (Italian, Cagliari) was a person of a certain age (1896) with many articles in Italian journals, who wrote about a problem of Fantappié in *Collectanea*. Jeannine Viard (French, Paris) was a collaborator of the Science theorist Jean-Louis Destouches and an editor of the Philosophy of Science course given by him in 1943 in Paris, in addition to several individual contributions to the *Comptes Rendus* between 1945 and 1947. Her 98-page article in *Collectanea* is probably her doctoral thesis, although this has not been confirmed. She then disappeared from the records of the *Mathematical Reviews* and the *Zentralblatt*.

Conclusions

During early Francoism, Spain was marked by the conflict between economic hardships and the forced acceptance of a regime whose main characteristic was the ideological undefinition derived from its own origin: a heterogeneous mix of military, Catholics of several factions, Falangists, landowners, monarchists, and opportunists. The result was a pragmatic conglomerate to maintain power, and apparently with remarkable success. In these circumstances, prospects for scientific activity were not favourable, and even less so for Mathematics, a speculative science whose development and attempts at updating were interrupted by the civil war.

The 'first life' until 1945: With the remnants of Rey's group, it was possible to maintain a certain scientific continuity, submit doctoral theses, and even send several of the new doctors abroad—in practice reduced to Germany, France, Italy, and Portugal, a common practice before the civil war, as well as reviving the journals of the former Mathematics Laboratory Seminar, now controlled by the CSIC. At this stage, Applied Mathematics played an important role, and was almost exclusively in the hands of engineers, who were committed to the creation of an aeronautical industry.

The 'second life' included the period from the end of the world war until 1952, and was marked by the isolation imposed on Spain in 1946 by the victors of the world war through the UN, which strongly reduced foreign access for most Spanish. Despite this, at the end of 1946, Ancochea and Bachiller travelled to the USA on a mission to internationalise Spanish Mathematics. The latter remained there until well into 1947. In the meantime, in Spain, the new doctorates after the civil war were consolidated in their positions, and the journal *Collectanea Mathematica* of the Barcelona Mathematics Seminar was created.

The 'third life' reflects the influence of the pacts with the USA. Since 1945, it was known that the future enemy of the USA would be the USSR, but in 1946 the United Nations approved a blockade of anti-communist Spain. However, for some issues, there were bypasses to international isolation. The semi-secret political manoeuvres previous to the 1953 pacts became visible with the lifting in 1950 of the sanctions decreed in 1946.

Among mathematicians, there was a younger generation, born in the 1920s and trained entirely after the civil war, with access to scholarships and visits abroad, now identified with the USA. In the wake of this increased accessibility to foreign countries, in 1955 the extensively mentioned Ancochea even attended a conference in Moscow. The newer generation was responsible for a significant improvement from 1960 onwards, and later for the establishment of new Mathematics Faculties.

Due to tradition and idiosyncrasy of mathematicians, Mathematics has been until recently an individual matter. Collaborations were not usually reflected in the authorship of the articles, but rather consisted of letter exchanges or conversations in meetings. In Central Europe, even through world wars and Nazism, visits were commonplace, but not in Spain.

The analysis of the journals clearly demonstrates the individualism of the Spanish mathematicians of that time, justifying the official neglect as a group. In the absence of sufficient critical mass, obtaining the scarce funding was difficult and its granting depended on who the applicant was: San Juan received all possible awards and funding, whatever their origin, despite the fund shortage his colleagues lived in. He even received a grant in 1962 to write a biography of Rey when the latter died.

Authoritarian regimes have a good ally in individualism: they can choose who to favour and, simultaneously, maintain a very comfortable division in certain weakly organised groups.

Female presence in the mathematical scene of those years is very scarce: Only five women in the world of Mathematics during early Francoism, but this was the standard in those years, not only in Spain. The only two female disciples of Rey—in the late 1920s—Carmen Martínez and María Capdevila, did not have brilliant careers either. Martínez, who held the first Spanish doctorate in Mathematics (1928), ended at a high school in Seville—there is a street named after her—and Capdevila had a similar fate in Barcelona.

References

Anonymous (1956): "Hijos ilustres de Toledo: Ricardo San Juan Llosá", *Provincia de Toledo*, 44: 64–75.

Beevor, Anthony (2005): *La guerra civil española*. Crítica, Madrid.

Camprubí, Lino (2016): *Los ingenieros de Franco*. Crítica, Madrid.

Cartwright, Mary; Littlewood, John (1945): "On Non-linear Differential Equations of the Second Order, I", *Journal of the London Mathematical Society*, 20(3): 180–193.

Cuesta, Norberto (1966): "Don José Barinaga Mata, in Memoriam", *Gaceta Matemática*, 18: 63–86.

Doménech, José (1904): *Concepto pedagógico de la Ciencia Matemática*. Tipografía "La Académica", Barcelona.

Español, Luis (ed.) (1990): *Estudios sobre Rey*. Instituto de Estudios Riojanos, Logroño.

Fernández, Isabel; Pacheco, José (2005): "On the Role of Engineering in Mathematical Development", *European Journal of Engineering Education*, 30(1): 81–90.

Hasse, Helmut (1953): *Mathematik als Wissenschaft, Kunst und Macht*. Verlag für angewandte Wissenschaft, Wiesbaden. (Inaugural lesson at Wiesbaden, 1951).

Judt, Tony (2008): *Postguerra: Una historia de Europa desde 1945*. Taurus, Madrid.

Otero, Luis (ed.) (2014): *La Universidad Nacionalcatólica, la reacción antimoderna*. Universidad Carlos III de Madrid.

Outerelo, Enrique (2009): *Evolución histórica de la Licenciatura en Ciencias matemáticas (Exactas) en la Universidad Central*. Facultad de Ciencias Matemáticas de la Universidad Complutense de Madrid. Internet: webs.ucm.es/BUCM/mat.doc13002.pdf.

Pacheco, José (2014): "Mobility and Migration of Spanish Mathematicians During the Years Around the Spanish Civil War and WW II", *Science in Context*, 27(1): 109–141.

Pacheco, José (2015): "The Life and Mathematics of Norberto Cuesta (1907–1989)", *Folia Canariensis Academiae Scientiarum*, XXVI: 85–126.

Pacheco, José (2017): "La ciencia olvidada: Matemáticas y matemáticos en las construcciones nacionales durante las dictaduras de España y Portugal", *International Symposium Ciencia y cultura como recursos simbólicos del nacionalismo en la España del siglo XX*. Barcelona, December 2017.

Pacheco, José (2018): "On the Attitude of Iberian Dictatorships Towards Mathematics and Mathematicians", in Saraiva, Luis (ed.): *Mathematical Sciences and 20th Century Dictatorships*, 55–65. Sociedade Portuguesa de Matemática, Lisboa.

Pérez-Marín, Antonio (1970): "Don Ricardo San Juan en relación con la Ciencia Aeronáutica". *Revista de la Real Academia de Ciencias de Madrid*, 337–341.

Presas, Albert (1998): "Nota histórica: una conferencia de José María Albareda ante las autoridades académicas alemanas". *Arbor*, 631–632, (July–August).

Preston, Paul (2007): *El holocausto español: Odio y exterminio en la guerra civil y después*. Debate, Barcelona.

Puig, Núria (2003): "La ayuda económica norteamericana y los empresarios españoles", *Cuadernos de Historia Contemporánea*, 25: 109–129.

Puig-Samper, Miguel (ed.) (2007): *Tiempos de Investigación. JAE-CSIC, cien años de ciencia en España*. CSIC, Madrid.

Rey, Julio (1932): "Teoría de los algoritmos lineales de convergencia y sumación", *Publicaciones (series B) de la Facultad de Ciencias Exactas, Físicas y Naturales de la Universidad de Buenos Aires*, 12: 51–222.

Rodríguez Sanjuán, Antonio (1963): Letter to Norberto Cuesta from Madrid (n.d.) Archive "Cuesta Dutari", Biblioteca General de la Universidad de Salamanca. Box 116, folder 5, item n°468.

Ríos, Sixto (1963): *Rentabilidad de la Investigación Científica*. Inaugural address (1963–1964), Academy of Sciences, Madrid.

Roca Rossell, Antoni (2016): *La recerca en ciències exactes i enginyeria a l'IEC: El cas del Centre de Estudis Matemàtics*. Institut d'Estudis Catalans, Barcelona.

Ríos, Sixto (1979): *Julio Rey, Matemático*. Instituto de España, Madrid.

San Juan, Ricardo (1962): "Julio Rey, su vida y su obra vistas por un discípulo", *Revista Matemática Hispanoamericana*, XXII(2): 60–93.

Santaló, Lluís (1990): "La obra de Rey en Geometría y Topología", *Revista de la Unión Matemática Argentina*, 35: 3–12.

Santesmases, Mª Jesús (2008): "Orígenes internacionales de la política científica", in: Romero de Pablos, Ana; Santesmases, Mª Jesús (eds.), *Cien años de política científica en España*: 293–322. Fundación Bbva, Bilbao.

Von Kármán, Theodor; Biot, Marcel (1945): *Métodos de Análisis Matemático en Ingeniería*. Instituto Nacional de Técnica Aeronáutica, Madrid.

CHAPTER 11

The Influence of French Fundamentalist Nationalism on the Ideology of the Generation of 1948

Sara Prades Plaza

The Importance of Defining Spain in the Franco Dictatorship

From the beginning of Francoism, specifying the nature of the Spanish nation became a major concern of intellectuals who supported the regime that had won the Spanish Civil War (1936–1939) and that now needed legitimisation. For this reason, examining the disputes that occurred around defining the nation allows us to understand the conception of Spain by the various political cultures that coexisted in the authoritarian ideological commitments of Francoism.

Some time ago, Saz, Ferrary, and Juliá pointed out that Spanish fascists, national Catholics and monarchists had different ways of understanding history, culture, and territory, although they did not agree on the name

S. Prades Plaza (✉)
Jaume I University, Castelló de la Plana, Spain
e-mail: sprades@uji.es

M. Janué i Miret, A. Presas i Puig (eds.), *Science, Culture and National Identity in Francoist Spain, 1939–1959*,
https://doi.org/10.1007/978-3-030-58646-1_11

they gave to these competing political-cultural projects. These studies questioned the monolithic and uniform image of the Francoism, demonstrating the existence of a diversity of cultural, political, and economic values that shaped the characteristics of the regime and that persistently attempted to establish themselves as national ideals with which to lead the state.[1]

Therefore, the study of the different cultural references of Francoism is an important subject for understanding the nature of the regime, since the struggle to gain monopoly of the official culture comprised the main political conflict between sectors of the authoritarian ideology.[2] Some of these rivalries within the dictatorship arose around the national question, since the regime involved nationalist ideologies and a nationalist dictatorship.[3] That being the case, the study of these conflicts is of paramount importance for understanding the relevance of Francoism in the construction of the contemporary Spanish nation.

For all of these reasons, it is necessary to analyse the historiographic and nationalist discourses of a series of historians or philosophers of history who were among the victors of the Civil War. Our attention falls on the Generation of 1948, who claimed that year to be the 50th anniversary of the disaster of 1898, which marked the end of the Spanish Empire. It was also the 100th anniversary of the Revolutions of 1848 in Europe and the 300th anniversary of the signing of the Peace Treaties of Westphalia. For the intellectuals of the Generation of 1948, the latter event was the starting point of a long European decline, since it represented the end of the Christian *Universitas*, a position defended by Spain. This *Universitas* comprised the Pope's participation in European politics since he considered that all states should submit to his power. This prompted one of its members, Pérez Embid, to write the article '1648, 1848, 1898, 1948', in which he proposed closing the parenthesis that had been opened since the abandonment of the post-Renaissance Christian project.[4]

The academic works that emerged in this context were related to Spain's political and social environment, and contributed to its redefinition during Francoism in the period after World War II, when the regime sought to move away from its association with the defeated Axis powers.

[1] Saz (2003): 366*ff*.; Ferrary (1993): 115*ff*.; Juliá (2004): 358*ff*.

[2] Ferrary (1993): 314.

[3] Saz (2004): 262.

[4] Pérez (1949).

They presented themselves as a political-cultural group when the Francoism accelerated its defascisation, since, despite all the makeover operations carried out by Franco, the Western democracies continued to identify the dictatorship with fascism. Its accelerated cleansing of the fascist and totalitarian macula did not prevent international ostracism, in response to which Francoism assumed a 'policy of waiting' and Numantine resistance. Also contributing to this were the fading of the Falange in the circles of power and the promotion of nationalist Catholic fundamentalist sectors.

The defeat of fascism would restore to the intellectual debate issues such as Spanish normality regarding the European path, the origin of Spanish decline, or the national Catholic identity. Once again, the definition of Spain became a matter of discussion, and it was essential for the survival of the regime that it identify itself as essentially anti-communist and Catholic. The Generation of 1948 dedicated itself to this task, a group of men with their own cultural project influenced by their religious and ideological convictions.[5]

As with other groups of intellectuals during Francoism, the Generation of 1948 recognised the hegemony of a leader, usually a university professor, who had the authority to appoint his assistants, hire them, and provide them with access to publications or public official positions.[6] Their undisputed leader was Rafael Calvo Serer, who set out the objectives of the group in his article 'Una nueva generación española', which served as a presentation of the collection *Biblioteca de Pensamiento Actual* of the Rialp publishing house and as a founding manifesto that preluded what would become the future editorial line of the journal *Arbor*, in which the Generation's intellectuals published.[7]

They had three main demands. First, they demanded change in the Spanish way of thinking, as well as abandonment of the problematic vision of Spain; to this end, they wrote a national history that aspired to become the hegemonic narrative of the Spanish past. They were not particularly devoted to Prehistory or Ancient History, and the few works dedicated to the Middle Ages were concerned with highlighting the plurality of the state in the Iberian Peninsula. Above all, their modernist historiographic discourse focused on praising the reign of the Catholic Monarchs for having successfully combined the governments of each of the peninsular

[5] Díaz (2008); Prades Plaza (2012).

[6] Peiró (2016) has called these professors 'little dictators'.

[7] Calvo (1947).

kingdoms, according to their specific idiosyncrasies, with a unitary policy abroad. Likewise, they praised the political position of the Habsburg dynasty that opted to defend Catholicism against the Protestant Reformation, which paved the way for a Spanish path to modernity that marked the national destiny until the Contemporary Age.

They sought continuity with eighteenth-century Spain, condemning the subsequent liberal era and accentuating the desire for social reform present in the monarchy of enlightened despotism. In general, they condemned the nineteenth century for being excessively centralist and revolutionary, two characteristics considered to be foreign, but they did not avoid studying this century for that reason. Lastly, they criticised the totalitarianism and liberal democracy of the twentieth century, describing these political systems as illegitimate, and against which the rebellion of the people was justified.

Second, subordinated to the main objective, they sought to strengthen relations with the European intelligentsia, with whom they wanted to promote the exchange of collaborations and researchers. Lastly, they advocated the revitalisation of the scientific potential of the Spanish regions, which was the only way, in their opinion, to achieve a strong science in the same manner as the rest of the European countries.

Their main centre of action was the Department of Modern Cultures of the CSIC (*Consejo Superior de Investigaciones Científicas*—Spanish National Research Council), from where they promoted scientific exchange through scholarships, courses, conferences by Spanish teachers outside the country and by foreigners in Spain, or collaboration in international conferences. This was one of the few institutions with which to overcome international isolation, which was, as noted above, one of the main problems that concerned the political elite in the 1940s.

From the CSIC, they took advantage of Jaume Vicens Vives's motivation to organise the study of Modern History in different research centres across the country, dividing the different contexts to be analysed between various centres, such as the University of La Rábida and the Schools of Modern History in Valladolid, Madrid, Barcelona, Valencia, Bilbao, and Seville.[8] In this way, they intended to centralise the entire scientific production of the country in the CSIC, integrating institutions that had emerged from non-state initiatives. This desire to monopolise all of the

[8] Letter from Calvo to Vicens of 31.12.1949, Archivo General de la Universidad de Navarra, Rafael Calvo Serer: 1/32/618.

scientific research produced in Spain explains why the relations between the CSIC and university were sometimes strained.

The truth is that the Generation of 1948 was considered a different group than those in its national environment and it proposed a cultural policy to be imposed by state institutions. It was presented as an alternative to the group led by Laín Entralgo and Tovar in the *Revista de Estudios Políticos*, which represented a Falangist sector that reflected the legacy of the Generation of '98 and praised the integration of the Spanish ideology of winners and losers. As instruments to achieve their objectives, they used their own works, the research carried out by this working group, and the organisation of new conferences and summer courses.

They returned to the ideas of Maeztu on the reverential meaning of money and to his attempt to apply Max Weber's theories to Catholicism, thus assuming the economic assumptions of modernity without accepting its politicians.[9] They attempted to develop a discourse that would make capitalist development compatible with the Spanish tradition, with theses similar to those of the reactionary European nationalists.[10] In this sense, they were in line with the ideology of national Catholicism, which had evolved during the nineteenth and twentieth centuries to support this nationalist, Catholic, monarchist, regionalist, Europeanist, and economically modernising political-cultural project during the Francoism.

Background: The Legacy of *Action Française* in *Acción Española*

National Catholicism is a political culture that emerged in the nineteenth century to update the European counter-revolutionary thought, with similar discourses coexisting in Spain, France, Italy, and Portugal.[11] It was born at the end of the nineteenth century, with the decline of some colonial empires. This encouraged the emergence of numerous nationalist discourses and the awareness by intellectuals of their social function, during a period of ideological changes against the principles of positivism and the Enlightenment. In this sense, European reactionary nationalism did not question capitalism and economic modernisation, but rather despised political liberalism, and provided, after the liberal revolutions of the

[9] Botti (2005): 196.
[10] Saz (2007): 141.
[11] Botti (2005): 206*ff.*

nineteenth century, an updating and vindication of the governing institutions of the Old Regime, the Church, and the Monarchy, so that they would lead a society articulated around corporations and regions.[12]

In this context, left and right extremist ideologies emerged, among which integral nationalism is particularly interesting; this was a faction of French conservatism that was the most influential example for Spanish national Catholicism. Its radical ideological discourse prioritised exaltation of the homeland. Led by Charles Maurras and Maurice Barrès, the most characteristic elements of its discourse were exaltation of the nation and its regions, as well as criticism of liberalism and the secularisation of society.

To spread their message, they organised themselves around *Action Française* (*AF*), which was established in 1899 as a league that, unlike political parties, aimed at achieving its goals not only in politics, but also at a religious and ideological level. Its initial objectives were intended to be achieved through direct action, mobilising its militants in street parades and demonstrations of force, without acting in the parliament, since they defined themselves as anti-parliamentary.[13]

Around 1905, *AF* formed their own political party, and developed a more careful political-cultural project than that defended by the encyclopaedia, thanks to the means at their disposal: writers, a journal, an institute at the Sorbonne for the education and training of future elites, the *Cercle Foustel de Coulanges* of university students, the publishing house *La Nouvelle Librairie Nationale*, and very active propaganda organisations, such as the *Camelots du Roi*, *Étudiants d'Action Française*, *Dames Royalistes*, and the *Association des Jeunes Filles Royalistes*.[14]

Their programme seduced many voters opposed to the French Republic, such as noble landowners, Parisian Catholics, the bourgeoisie, monarchists, and anti-Semites. However, they never became a mass organisation, since *AF*, whose vice president was Maurras, was less competent in carrying out a political strategy than in creating states of opinion. This was particularly true during its most vigorous period, prior to World War I, when ideas such as decentralisation, rebirth of the regions, restoration of the monarchy, and defence of intermediate bodies and corporatism were revived in France.

[12] Saz (2003): 59*ff.*
[13] Berstein (1992): 62.
[14] Weber (1985).

Later, in the interwar period, although the dominant discourse of the French right was Maurrassian in terms of its coherence and prestige, they did not procure as much support as previously. During the 1930s, authors close to *AF* published in *Je suis partout*. These included Robert Brasillach, Bertrand de Jouvenel, Henri Massis, Thierry Maulnier, Hilaire Belloc, and especially Maurras, but also Benito Mussolini. Among other things, Mussolini's collaboration in *Je suis partout* has led many authors to reflect on the relations that *AF* established with fascism, having anti-Semitism, racism, and xenophobia as the main points in common,[15] but also nationalism and anti-liberal and anti-Marxist radicalism.[16] Likewise, both ideological movements exalted the social community to the detriment of Kant's individual freedom or Hegel's subjective freedom.[17]

Some authors, such as Zeev Sternhell, even claimed that *AF* became fascistic during the interwar period, and, although this process did not imply that it adopted all the attributes of a structured fascist movement, it participated in what was at the time the essence of fascism. In other words, *AF* had the ambition to destroy the heritage of the Enlightenment, individualism, liberal utilitarianism, and democratic egalitarianism. Furthermore, according to Sternhell, *AF* saw in Mussolini a leader who could put Maurrassian doctrines into practice, as, for Maurras, fascism represented a radical part of the anti-Enlightenment European discourse in which integral nationalism was inserted. Sternhell also stressed that there were supporters of *AF* who openly saluted the victory of Nazism and that Maurras was tolerant towards those attitudes.[18]

However, Sternhell's views have been rejected by most of the historiography, and although some, such as Milza, recognise certain convergences between fascism and *AF*, they refute Maurras's approach to fascism.[19] Paxton, Griffin, and Eugen Weber, among others, have shown that, when Mussolini came to power, *AF* was strongly opposed to the importance it afforded to the state.[20] His territorial claims, his alignment with Nazi Germany, and his defence of centralism separated fascism from integral nationalism. Furthermore, just before World War II, Maurras strongly declared that France's worst enemies were Hitler and the Republic.

[15] The first work to study the relationships between these political cultures was Nolte (1963).
[16] Saz (2016): 150.
[17] Sutton (1994): 274–275.
[18] Sternhell (2000).
[19] Milza and Decleva (1993).
[20] Paxton (1972); Griffin (1991); Weber (1985): 157.

Moreover, *AF*, unlike the fascists, never tried to become a mass party, since it did not believe in the power of the masses.[21] These and other issues led Saz to affirm that, like so many other conservative political cultures, they became more or less fascistic during the Interwar period, but this fascisation did not make them lose their idiosyncrasy, since they stemmed from different political cultures.[22]

The papal condemnation of *AF* in 1926, accused of instrumentalising religion for political purposes, was an obstacle to the dissemination of Maurrassian thought in Spain, even though it had been influencing the country since its first writings. During the Second Spanish Republic, *El Debate*, a conservative Catholic spokesperson, launched a campaign condemning *AF* and Maurras, calling them opportunists for using Catholicism as an instrument. Unamuno expressed himself similarly in *La agonía del cristianismo* and Father Arboleya in an article published in the journal *Renovación Social* on 15 February 1927.

However, a political-cultural group was soon to emerge that did not seem to strongly care about the Vatican condemnation of *AF*, unlike what would happen among the ranks of the CEDA, a party founded in 1933 and for whom *El Debate* would become a spokesperson. This different assessment gave rise to a discussion during the early 1930s between *El Debate* and *Acción Española* (*AE*), a Catholic fundamentalist journal and advocate of an anti-republican uprising, which has been understood as a clear distancing of positions between the two branches of Spanish national Catholicism.[23]

With the establishment of the Second Spanish Republic in 1931, many monarchists went into exile in France, where they maintained a strong contact with the ideas of *AF*. It was then that *AE* was formed, which looked at Maurrassian integral nationalism as an example to follow in its fight against the Republic. Although they tried not to identify themselves with any of the anti-republican parties, most of their members held leadership positions in the monarchical party *Renovación Española*, such as Goicoechea, Calvo Sotelo, Maeztu, and Sainz Rodríguez. Their reflection was based on the historical and political discourses of the Spanish traditionalists of the nineteenth century, introducing the contributions of various currents of European thought influenced by the political and cultural

[21] Dard (2013): 205.
[22] Saz (2016): 154.
[23] Dewaele (2003).

context of the end of the century. For all these reasons, *AE* was the school of thought that made the greatest contributions to the political culture of Spanish national Catholicism in the 1930s, as it served as a channel for transmitting the main European counter-revolutionary currents.

Although, at times, they did not want to acknowledge it, the influence of *AF* was felt in the organisational model of *AE*, which they gave so unoriginal a name as the Spanish translation of the French association.[24] The Count of Santibáñez del Río, Vegas Latapié, and Maeztu had planned to publish a journal of the same name, which materialised during 1931 and contributed to the revitalisation of the European counter-revolutionary ideology in Spain. Its directors were the Marquis of Quintanar and Maeztu, and from this project emerged the Cultura Española association, directed by Pemán, with the aim of organising courses, conferences, and book publications.

As the *Institut d'Action Française* had done in its time, they organised a series of weekly conferences from February 1932. In the following year, they founded the publishing house *Cultura Española*, in the same line as the *Nouvelle Librairie Nationale*, which translated the main works of the French anti-republican doctrinaires, including the following: *La revolución francesa* and *Lo que podría hacer la Monarquía* of Gaxotte; *Las leyes de la política francesa* of Benoist; *Hacia un orden social cristiano* of La Tour du Pin; *El fin del Imperio español en América* of Marius André; *La tercera república francesa* of Bainville; *La letra y el espíritu (La función social de los oficiales)* of Marshal Lyautey; *La quiebra de un Régimen. Ensayo sobre el Gobierno de mañana* of the Count of Paris; and *Encuesta sobre la Monarquía* of Maurras.

AE was focused on the organisation of *AF* to constitute its political-cultural platforms, but this does not mean that they accepted all the Maurrassian doctrine, since this was not even defended by all the members of the French group. Vegas Latapié[25] and Joan Estelrich were its most

[24] Pemán pointed out the inconvenience to some of *AE*'s founders that the name of his journal was very similar to *AF*, since the latter was having problems with the Vatican for its positivist tone. Pemán (1970): 70–71.

[25] Vegas, who was a French native speaker, was the most Maurrassian member of *AE* and, as he describes in his memoirs, he dedicated two hours a day to reading *AF*. He also affirmed to be amazed by the conquest of the *Quartier Latin* by the *Camelots du Roi* under the charge '*Vive le Roi!*', avoiding the sale of the republican press. Vegas (1983): 46–47.

devoted followers, but other members of *AE*, such as Maeztu,[26] Sainz Rodríguez, and Pemán, only assumed some aspects of the views of the French theoretician. Even so, its understanding of the French nation, Catholicism, and the monarchy was, among the European ideologies, the most similar to the national Catholic ideology in Spain. For this close connection between the two discourses, Maurras's ideology deserves special attention for understanding Spanish national Catholicism.

Although *AE* published its last anthological issue in 1937, the publishing house *Cultura Española* continued publishing in the 1940s, including *En vísperas de la tragedia* of Maeztu and *El pensamiento político de Calvo Sotelo* of Vegas Latapié. However, it was soon relegated by publishers of a similar ideology, such as Rialp and especially its *Biblioteca de Pensamiento Actual*, both of which were cultural platforms dominated by their ideological heirs during Francoism: The Generation of 1948.

Influences of the Historical Maurrassian Discourse on the Generation of 1948

As seen above, *AF* had a clear influence on *AE*, and, through it, on its ideological successor: the Generation of 1948. However, it should also be noted that some Maurrassian ideas reached members of the group directly, without intermediaries. In this sense, the aspect that interests us most is his unique rewriting of the History of France, which exalted its classical values.

Maurras saw in the past the idealised image of what he was striving to build in the present, the exact reflection of the nation's needs for eternity. For that reason, he attributed great importance to the nineteenth-century historian Fustel de Coulanges, who privileged the political history of institutions over the social, economic, or mental history. Fustel saw the monarchy as an institution with the motivation to unify the nation, and considered that the trilogy of classicism, realism, and Catholicism was a unique opportunity from which France had benefited. Like him, Maurras defended the usefulness of studying history, since, in his opinion, it repeated itself.

[26] In the personal archive of Maurras, there are two letters written by Maeztu in 1934, when he published *Defensa de la Hispanidad*, so an epistolary exchange of a certain frequency is inferred. Centre Historique des Archives Nationales (CHAN), Fonds Charles Maurras, 576 AP, 63, Dossier Espagne.

Like the French theoretician, the Generation of 1948 saw certain aspects of the past as examples to be followed in the present, in their attempt to rewrite the History of Spain.[27] However, unlike most of the Generation of 1948, for whom the Modern Age was the climax of European history, Maurras believed that classical Antiquity was the stage to be imitated, since it meant classification and order within chaos. Ultimately, it was the moment when man had installed his sovereignty over nature, and, therefore, it meant the peak of Humanity.[28]

In particular, the defence of Greek Antiquity was central to his ideology, since he regarded this civilisation to be the founder of the arts, languages, and progress of the northern Mediterranean, but at the same time he considered Greece to be an exotic land that heralded the East. As can be observed, this defence of Antiquity was accompanied by the demand for a certain European unity.

For most *AF* ideologues, Europeanism would have started with the Roman Empire, so that Latinness could reconcile Europeanism with nationalism. Maurras believed that, since Romanised countries shared the Catholic religion, this could be the foundation of an eventual alliance of Latin countries. For him, Catholicism was the foundation of a third force that would defeat both materialisms, communism and liberalism, which he considered equally harmful.

In the effort to rebuild Catholic Europe, Maurras also had another influence in the historical discourse of the Generation of 1948. For this group, the revival of the Christian *Universitas* would restore Spain to its deserved prominence in world politics. Spain would then attain its true value in relation to Europe, and the same would happen to the old continent, which would have to return to the Spanish model to correct its evident decline in World War II. Catholic Spain would have to play a leading role in the reconstruction of Christian Europe, and Paniker even stated that 'Europe cannot advance any further as it has reached the final end of the historical dissolution and now Spain presents itself with the solution. If Europe wants to survive, it will have to return to the Spanish solution'.[29]

Following this line of reasoning in defence of Catholicism, Maurras even proposed revising the educational contents to restore interest in the

[27] Pérez (1949).

[28] Some reflections on Maurras's claim to classicism can be found in Nguyen (1991): 757 and Sutton (1994): 37.

[29] Paniker (1951): 112.

privileges of the Roman Catholic Church. In the opinion of the French theoretician, there was a fundamental difference between Catholicism and Protestantism, given that the first was immersed in classical Antiquity, which is why, thanks to the Catholic Church, France had inherited a taste for rationality and order. On the contrary, Protestantism had been the root of all of France's political problems, having caused the sixteenth-century civil wars, the Thirty Years' War, the seventeenth-century English Civil War and Revolution, and, lastly, the French Revolution itself.[30]

It should be noted that, for Maurras, the Reformation and Revolution came from Germany, which was considered the homeland of barbarism and Lutheranism, while for *AE* and the Generation of 1948, but not for the Germanophobe Menéndez Pelayo, they came from France, which was considered the homeland of freedom. As an example of this line of reasoning, it is worth mentioning Suárez Verdeguer's study 'Génesis del liberalismo político español', in which he reflected on the consequences and meaning of the Peace Treaties of Westphalia, which, in his opinion, had represented the military triumph of the Reformation and the political predominance of France in Europe, considering both events as part of the same political phenomenon and placing them at the centre of the target of what would have to be fought.[31]

Returning to Maurras's claim of Catholicism, it can be said that for him the church was a decisive force in the fight against the secular Republic, since the legitimation of power could only come from God. In this sense, his motto '*politique d'abord*' meant that he considered politics to be essential, above morality, since it tended towards national exaltation. Other aspects of society were subordinated to politics, including religion, which was considered a useful element to legitimise political power. It was this instrumentalisation of Catholicism by integral nationalism that led Pope Pius XI to condemn them, suspending priests who openly supported *AF* and coming out against religious burial and marriage for the rebellious laypeople of the same association.[32]

Despite the influence of the Maurrassian ideology on the Generation of 1948, they contained some significant differences. Contrary to the Maurrassian ideology, the Generation of 1948 placed Catholicism at the centre, proposing a Christian vision of history and advocating the

[30] Sutton (1994): 63.

[31] Suárez (1947): 349.

[32] Rémond (1982): 114–132.

superiority of Catholic morality over politics. Furthermore, Calvo Serer, who led the group, stated that Catholicism was the only element of national unity, without which it would be impossible for the different members of the nation to coexist.[33] On the contrary, for Maurras, the nation was the most solid and fruitful bond in the contemporary world after the destruction of the seventeenth-century Catholic order. He argued that the nation had been defined by Renan and Taine as something natural and superior to the will of men, so much so that they could not question it. For him, the nation was an organic, hierarchical, and decentralised community that required the establishment of the traditional monarchy and the values of Catholicism and Classicism.

As may be seen, integral nationalism placed the monarchy at the service of the nation, and not France at the service of the king, a fundamental rule of the ancient monarchical tradition. According to Maurras, all the necessary measures aimed at national strengthening converged in the establishment of the monarchy, an institution that looked extremely useful to the French theoretician. In *Encuesta sobre la Monarquía*, he proposed using force to quickly reinstate the king as a legitimate procedure, considering the ineffectiveness of legal actions to 'cure France of the democratic poison'. In his view of the monarchy, the personal interest of the ruler and the public interest, far from being opposed, necessarily coincided. Lastly, he understood that the monarchy had to be representative, but not parliamentary, with the monarch holding all the power and the nation being represented in corporate parliaments.

Maurras's influences can be found in López-Amo, one of the authors of the Generation of 1948 who wrote about the monarchy. His definition of the monarchical institution is clearly influenced by Maurrassian ideas, given that, for both authors, the French Revolution had been a useless process that led to an illegitimate political system: democracy. Unlike the monarchy, democracy, which lacked legitimacy, could not solve the problems of the people, and, for that reason, it had led to World War II.[34] In general, the Generation of 1948 considered Spain a monarchical State in which the elites' role was to guide the population and in which Franco and his party could provisionally lead society, indoctrinated by the church, until economic development was achieved.

[33] Calvo (1953): 301.
[34] López-Amo (1952): 259.

To continue analysing the Maurrassian nation, it should be noted that this concept lies at the antipodes of liberalism, since it considers that executive power can only be strengthened through decentralisation, understood as the 'nationalisation of the nation'. In this sense, Girardet noted that Maurras's nationalism was regionalist, since he understood that respect for regional languages reinforced and consolidated the unity of France, as demonstrated by its historically plural reality.[35] His provincial idea of region was not opposed to the nation, but he considered that the region must be preserved in order to face the dangers that threatened the disintegration of the homeland. He also believed that decentralisation and federalism were not possible in the Republic, another reason to oppose this form of government.

His provincialism attracted many *félibres* and regionalists to *AF*, who were progressively marginalised by the royalists. When Maurras arrived in Paris, he became aware of his provincialism and wrote with Amouretti his first political text, *La Déclaration des félibres fédéralistes*, published by Mistral on the front page of the Parisian newspaper *L'Aioli* on 22 February 1892. This manifesto demanded the restoration of communal liberties, the replacement of the districts by the historical provinces and the strengthening of relations with other federalist groups in France. Moreover, the *félibres* longed for the union of Latin Europeans thanks to their common Mediterranean, Hellenistic, and Roman character. This idea was embraced by Víctor Balaguer, a friend of Mistral, as well as by *La Veu de Catalunya*, which reproduced the *félibre* manifesto on 13 March 1892, commenting on it in favourable terms, and *L'Avens*, which summarised it. Due to its significance, it can be considered the main manifesto of the entire history of *Félibrism* since it marked the starting point of a federalist effort in the Mediterranean area.

This is why Maurrassian thought was also introduced in Spain through the Catalan regionalism of the *Renaixença*. Their regionalist demands brought André, Barrès, and Maurras closer to Catalanism, introducing their ideas through culture and art, rather than through politics. Subsequently, there were discourses close to integral nationalism in the rest of Spain among the critics of the Disaster of '98, of which one that stands out is that of Maeztu, who stated that the national problem was due to the scarce economic development of some of its regions.

[35] Girardet (1983).

This idea of a pluralistic regional Spain was resumed by the Generation of 1948, opposing Castilianism, which was criticised as being centralist, revolutionary, and foreign.[36] Consequently, in the publications and cultural platforms controlled by them, this national idea appeared in articles on History and historiography and in studies on the main thinkers of Spanish and European traditionalism. This was a characteristic feature of the Generation of 1948 throughout its existence, establishing itself as a reference for conservative regionalism.

The reviews of works that reflected on the meaning of the Spanish nation and studies on intellectuals or prominent figures of the past who had defended the Spanish regional plurality were the spaces selected for presenting the idea of the regionally plural Spain of the Generation of 1948. They argued that centralism had to be overcome to find the authentic essence of Spanishness. However, the demand for regional complexity should not be confused with the defence of other nationalism than the Spanish one. In fact, the Generation of 1948 defined itself as anti-nationalist, since it considered nationalism a strange impulse to the Spanish tradition that responded to the individual ambitions of each state.

For these historians, nationalism had led Europe into World War II, which resulted in a triple crisis: military, political, and spiritual. They considered nationalism to be a consequence of the Lutheran Reformation, and that its final result had been the Nazis' domination in Europe. Therefore, to save the European identity, Catholicism had to be revalued, and the foundations of the Christian *Universitas*, which would have been in danger, had to be reconstructed.[37] It should be noted, however, that the Generation of 1948 advocated the defence of the Christian *Universitas*, and opposed the selfish satisfaction of the ambitions of each nation for contributing to the annihilation of the union of Christians. This, however, does not imply that they were not nationalists, since the denial of one's own nationalist condition is one of the most characteristic features of Spanish nationalism in the twentieth century.[38]

Returning to the historical discourse of Maurras, it's significant that he considered the seventeenth century to have been the greatest period in

[36] Later, some of the Generation of 1948 evolved into more radical positions, and Calvo in even accepted the Catalans' right to their political independence if they so desired. Martí and Ramoneda (1976).

[37] Calvo Serer (1953): 301. This thesis was also indebted to Christopher Dawson's contributions on the break, since the Reformation, of the European unitary order.

[38] Saz (2003): 409.

French history, as this was the time of Descartes, La Fontaine, and Bossuet, who had contributed to its national greatness. This *grandeur* did not reside in the personal merit of its kings, but in the typically French ideas embodied in the literature, works of art, and political and social institutions of that century. He considered that the revolutionary ideas had come to overshadow this greatness of seventeenth-century France, which had emerged with Voltaire, on his return from England, and Montesquieu, who had also resided in the perfidious Albion, where they would have been imbued with 'constitutional anglomania' and the precepts of English liberalism.[39] Even so, Rousseau was the main responsible for the French Revolution, since he had introduced the three 'Swiss ideas: equality, fraternity, and freedom' that gradually came to infest the entire country. Later, for mortification of Maurras, the Romantics continued the work of 'political anarchy' that Rousseau had started.[40]

Despite all this, Maurras considered that curative action was possible, as demonstrated by the persistence of personalities in the second half of the nineteenth century, such as Auguste Comte, Le Play, Renan, Taine, Fustel, Mistral, and Anatole France. Therefore, Maurras systematised the scattered currents of French conservatism and traditionalism that had emerged during the nineteenth century in opposition to the social and political significance of the French Revolution. At the beginning of the following century, he proposed a new synthesis of counter-revolutionary criticism following De Maîstre, Bonald, and Taine.[41]

However, for the Generation of 1948, the beginning of the problems of the homeland had previously been established, following the line opened by Donoso Cortés who believed that the Peace Treaties of Westphalia of 1648 were the culmination of a historical process that would have eliminated the old political order of medieval Christianity and replaced it with an order regulated by the ambitions of each modern State.[42] Westphalia, according to this interpretation, consecrated the triumph of the Lutheran Reformation, which would have its ultimate success in the French Revolution and the end of the Old Regime. Therefore, although Maurras and the Generation of 1948 differed on the date on which they considered the decline to have started, they agreed on the

[39] Sutton (1994): 67.
[40] Sutton: 72.
[41] Rémond (1982): 168–180.
[42] Donoso (1851).

diagnosis that the Revolution was the foundation of that decline of the nations of Europe.

Maurras's claim can be seen in the pages of the journals close to the Generation of 1948, *Arbor* and *Ateneo*, on the occasion of his death on 15 November 1952. The head of the French Section of the Department of Modern Cultures of the CSIC, Jean Roger, wrote two extensive articles on *AF*.[43] In both articles, he apologised for Maurras and his achievements, denouncing his persecution at the hands of Christian Democrats and Communists. In Roger's view, Maurrassianism remained the only real alternative to Marxism and democracy.

Shortly after this, Santiago Galindo wrote a prologue in the *Editora Nacional*, which directed between 1953 and 1962, to *Mis conversaciones con Maurras y su vuelta a la Iglesia*, the work of Cormier, who had accompanied Maurras in his final days.[44] As the title of the Spanish version indicates, Galindo remarked that, in his final moments, Maurras had once again believed in God.

Later, the work of Massis *Maurras et notre temps*, which appeared under the title *La vida intelectual en Francia en tiempos de Maurras*, was translated in the *Biblioteca de Pensamiento Actual*.[45] Moreover, Hericourt, a disciple and collaborator of *AF*, intervened in January 1953 at the *Ateneo de Madrid*, another platform of the Generation of 1948, with a lecture that was later published in the collection *O crece o muere*.[46]

However, Calvo Serer, who had not hesitated to proclaim *AE*'s legacy for the ideology of his group by claiming to be its heirs during the Francoism, did sometimes deny that his ideas were influenced by Maurras. It could be that he did not want to consider himself influenced by someone whom the Pope had condemned and who openly declared his nationalist views. In any case, although the leader of this political-cultural group denied Maurras's influence and, on the contrary, inflated the true relations with the extinct *AE*, the influences of both groups on the ideology of the Generation of 1948 are evident; thus, it can be concluded that this was a political-cultural group in line with the European fundamentalist nationalism of the twentieth century.

[43] Roger (1952a); Roger (1952b): 8*ff*.
[44] Cormier (1955).
[45] Massis (1956).
[46] Hericourt (1953).

Conclusions

As previously noted, the concept of nation of the Generation of 1948 was strongly based on the ideas of Maurras, who sought to rewrite national history by seeing in the past the idealised image of what he wanted for the present. Similarly, the Generation of 1948 accepted the Maurrassian conviction that all measures aimed at strengthening the nation would require the establishment of a monarchy, an institution that should have a non-parliamentary representative, as Maurras had pointed out. This idea had also been embraced by *AE*, which considered monarchy the only legitimate form of government, through which the nation projected itself into the past and into the future, preserving its essential identity despite apparent changes.

Both Maurras and *AE* believed that the decline of the nation had hit rock bottom with the Revolution and with the end of the Old Regime. This national decline was also considered a consequence of the existence of an 'anti-nation', which parasitised the true national essence, so that this decline would not be overcome if the enemy were not first eliminated. This idea, together with the Generation's claim for restoration of the monarchy due to the progress it had promoted throughout history, was a reason for its discussion with less monarchical sectors of the Francoism.

Similarly, this monarchical alternative was completed by a regionalist perspective that criticised Castilian centralism for being revolutionary and foreign. In the same way that Maurras had done, the Generation of 1948 considered that the nation was composed of a group of regions with its own personality that provided all its wealth to Spain. In their opinion, one of the problems faced by Spain was the inadequate conception of the place of the regions in the organisation of the state.

Moreover, they were all counter-revolutionary since they considered that the combination of the Spanish tradition and fundamentalist Catholicism should be the approach taken to leading the cultural life of the country. Not only that, they considered this alternative to be Europe's only option to save its identity and rebuild the foundations of the Christian *Universitas*, which the Generation of 1948 believed to have been in danger since World War II.

In short, their historical reflections were based on a conception of nation that updated the thinking of Menéndez Pelayo, Maurras, and *AE*, claiming its centrality in the definition of the contemporary concept of Spain. Therefore, the Generation of 1948 combined the Spanish reality of that time with its historical tradition, so that the conclusions they reached legitimised those, and no other, ideological principles of Francoism.

In this legitimisation, they were not in agreement with the fascists, who sought to create a totalitarian state based on the authority of the party and its leader, the *Caudillo*, in which politics prevailed over the economy through popular participation. Furthermore, the Generation of 1948 also opposed the possibilist Catholics of *ACNP*, who opted for a more open-minded and complacent Catholic doctrine compared to those defeated in the civil war.

These discrepancies and the growing political commitment of the Generation of 1948 exceeded the limits allowed by the regime, and they were forced to lower their acrimony from 1953. However, they could mitigate the influence of the Ministry of National Education and demonstrate the impact that an opening, such as the one that Ruiz-Giménez was attempting to carry out, could have in a regime like the Francoist.

Moderating their positions was sufficient for this political-cultural project to preserve the essence of its agenda at a time when the legitimacy of the regime was no longer so much in its origin as in its exercise. This project had to be conducted by other means in a cultural panorama in need of new blood, given that a new period of Francoism was beginning in which the emphasis would shift from the doctrinal and political to the economic. It was the triumph of a sector that still hoped to continue what had been achieved on 18 July 1936 through a traditional Catholic monarchy that pondered the weight of the regions in the Spanish configuration and the need for integration in Europe through economic modernisation.

References

Berstein, Serge (1992): "La Ligue," in SIRINELLI, Jean-François (ed.): *Histoire des droites en France, T.2. Cultures*, Paris, Gallimard.

Botti, Alfonso (2005): "Algo más sobre nacionalcatolicismo," in De La Cueva Merino, J. y López Villaverde, J. A. (coords.): *Clericalismo y asociacionismo católico en España, de la Restauración a la transición: un siglo entre el palio y el consiliario*, Cuenca, UCLM: 195–212.

Calvo Serer, Rafael (1947): "Una nueva generación española", *Arbor*, 24.

Calvo Serer, Rafael (1953): "La Iglesia en la vida pública española desde 1936," *Arbor*, 91–92: 289 y ss.

Cormier, Aristide (1955): *Mis conversaciones con Maurras y su vuelta a la Iglesia*, Madrid, Editora Nacional.

Dard, Olivier (2013): *Charles Maurras*, Paris, Armand Colin.
Donoso Cortés, Juan (1851): *Ensayo sobre el catolicismo, el liberalismo y el socialismo, considerados en sus principios fundamentales*, Madrid, La Publicidad.
Dewaele, Hélène (2003): *Les relations entre droites autoritaires Françaises et espagnoles de 1931 à 1940*, Paris, EHESS.
Díaz, Onésimo (2008): *Rafael Calvo Serer y el grupo "Arbor"*, Valencia, PUV.
Ferrary, Álvaro (1993): *El franquismo: Minorías políticas y conflictos ideológicos 1936–1956*, Pamplona, Eunsa.
Girardet, Raoul (1983): *Le nationalisme français. Antologie. 1870–1914*, Paris, Points-Seuil.
Griffin, Roger (1991): *The nature of fascism*, London, Routledge.
Hericourt, Pierre (1953): *Charles Maurras, escritor político*, Madrid, Ateneo.
Juliá, Santos (2004): *Historias de las dos Españas*, Madrid, Taurus.
López-Amo, Ángel (1952): *El poder político y la libertad (La Monarquía de la Reforma Social)*, Madrid, Rialp.
Martí, José y Ramoneda, Josep (1976): *Calvo Serer: el exilio y el reino*, Barcelona, Laia.
Massis, Henri (1956): *La vida intelectual en Francia en tiempo de Maurras*, Madrid, Rialp.
Milza, Pierre y Decleva, Enrico (1993): *Italia e Francia, i nazionalismi a confronto*, Milan, Franco Angeli.
Nguyen, Victor (1991): *Aux origines de l'Action Française. Intelligence politique à l'aube du XX siècle*, Paris, Fayard.
Nolte, Ernst (1963), *Der Faschismus in seiner Epoche*, München, Piper.
Paniker, Raimundo (1951): "Una cautela a los historiadores españoles," *Arbor*, 69–70: 112–113.
Paxton, Robert (1972): *Vichy France: Old Guard and New Order, 1940–1944*, New York, Barrie and Jenkins.
Peiró, Ignacio (2016): "Catedráticos franquistas, franquistas catedráticos. Los *pequeños dictadores* de la Historia", in Caspístegui, Francisco Javier y Peiró, Ignacio (eds.): *Jesús Longares Alonso: el maestro que sabía escuchar*, EUNSA: 251–291.
Pemán, José M^{a} (1970): *Mis almuerzos con gente importante*, Barcelona, Dopesa.
Pérez Embid, Florentino (1949): "1648, 1848, 1898, 1948," *Arriba*, 10.6.1949.
Prades Plaza, Sara (2012): *España y su historia: la generación de 1948*, Castelló, UJI.
Rémond, René (1982): *Les droites en France*, Paris, Aubier.
Roger, Jean (1952a): "El affaire de la Acción Francesa," *Arbor*, 56.
Roger, Jean (1952b): "Charles Maurras," *Ateneo*, 22.
Saz, Ismael (2003): *España contra España*, Madrid, Marcial Pons.
Saz, Ismael (2004): *Fascismo y franquismo*, Valencia, PUV.
Saz, Ismael (2007): "Mucho más que crisis políticas: el agotamiento de dos proyectos enfrentados," *Ayer*, 68: 137–163.

Saz, Ismael (2016): «Entre la reacción y el fascismo: las derechas europeas en la primera mitad del siglo XX», in Fuentes, Maximiliano et al. (coord.): *Itinerarios reformistas, perspectivas revolucionarias*, Zaragoza, IFC: 143–160.

Sternhell, Zeev (2000): *Ni droite ni gauche. L'idéologie fasciste en France*, Bruselas, Complex.

Suárez Verdeguer, Francisco (1947): "Génesis del liberalismo español," *Arbor*, 21: 349–395.

Sutton, Michael (1994): *Charles Maurras et les catholiques français. 1890–1914. Nationalisme et positivisme*, Paris, Beauchesne.

Vegas Latapié, Eugenio (1983): *Memorias políticas. El suicidio de la Monarquía y la Segunda República*, Barcelona, Planeta.

Weber, Eugen (1985): *L'Action Française*, Paris, Fayard.

CHAPTER 12

Of Queens, Soldiers, Nuns, and Bullfighters: Nationalist Narratives in the Fiction Films of the Franco Regime (1939–1963)

Gabriela Viadero Carral

Introduction

Although research on Spanish national identity and nationalism has been conducted for decades, the literature on Franco's nationalist cultures[1] is not particularly extensive, and even less so, although in increasing numbers, is the focus on nationalisation processes during the Franco regime.[2]

For this reason, we consider it relevant to study the messages conveyed in the Spanish cinematography produced during early Francoism. On the one hand, cinema is a means of generating and disseminating knowledge; on the other hand, it was controlled and guided by the nationalist Franco

[1] On nationalism and Francoism and the two anti-liberal nationalist cultures that coexisted during the dictatorship (fascism and national Catholicism): Saz (2008): 153-174.

[2] An approach to the state of this question can be found in Moreno Almendral (2014): 7–36.

G. Viadero Carral (✉)
Universidad Complutense de Madrid, Madrid, Spain

M. Janué i Miret, A. Presas i Puig (eds.), *Science, Culture and National Identity in Francoist Spain, 1939–1959*,
https://doi.org/10.1007/978-3-030-58646-1_12

regime,[3] and unsurprisingly, therefore, it played an important role as a nationalising tool.

Cinema, particularly fiction, appeals directly to the emotions. Our identification with certain characters and the apparent plausibility of the narrative can influence the mental frameworks with which we represent reality. This characteristic, combined with its broad reach, made cinema one of the most powerful communication tools of the twentieth century, and as Hobsbawm notes, 'unlike the press, which in most parts of the world only interested a small elite, cinema was, almost from the beginning, an international mass medium'.[4]

The Franco regime was a military dictatorship of national Catholic and Falangist inspiration that was established in Spain in 1939; one of its legitimising arguments was the defence of the Spanish nation, after a war caused by the uprising of part of the army against a democratically elected government. Therefore, and knowing that the Franco regime influenced the production of fiction films through censorship and economic incentives, it should come as no surprise, as we will explain below, to find a nationalist interpretation in some of the cinematography of the time.

Our research will focus precisely on this nationalistic interpretation, never losing sight of the fact that the idea of Spain that was presented in these films served, on many occasions, to fulfil one of the main objectives of the Franco regime: to remain in power as the guarantor of the nation's unity and glory.

Cinematography: Guidance and Censorship

As in other areas of culture, the Franco regime adopted an authoritarian position regarding cinematography. Although it considered that film production could never be an exclusive activity of the state,[5] it nevertheless controlled such activity through subsidisation, censorship, and repression, particularly until 1963. The regime could control film production through a series of laws and institutions, laying the first stone in 1938 with the creation of the National Department of Cinematography (*Departamento Nacional de la Cinematografía*—DNC), whose objective was to immedi-

[3] We understand that nationalism is, as Gellner wrote, a 'political principle that states that political and national unity must be consistent'. Gellner (1983): 1.

[4] Hobsbawm (1994): 198.

[5] Diez Puertas (2002): 275.

ately produce propaganda in accordance with the needs of the new state. The report on the constitution of the department states:

1. Film production can never be an exclusive activity of the state.
2. Conversely, the state must promote private initiatives for the development and growth of the national cinematography.
3. The state shall supervise and guide the cinema so that it is worthy of the spiritual values of our motherland.
4. The state, in any case, retains the production of news and propaganda documentaries.[6]

This stimulation of private initiatives, together with the supervision and guidance mentioned above, created a protectionist system in which the production companies depended more on state institutions than on commercial results.

The guidance and protection carried out by means of filming permits and economic incentives that could be either direct, such as credits and prizes granted by the National Entertainment Union, or indirect, such as the screen quota. As for supervision and guidance, censorship took over.

The system was organised by combining the production of Spanish films with the import of foreign films, the real business, the one that provided guaranteed profits.[7] The production companies that sought to obtain these foreign films so that they could be seen in national territory had to make one or more films in Spain, at a cost of not less than 750,000 pesetas. For each film produced in Spain, a classification commission determined how many permits—between two and five—would be granted to the production company to import foreign films, based on the quality, value, and provenance of these foreign films.

Moreover, for each imported film, the production company would have to pay a royalty to the National Entertainment Union.[8] The amount varied depending on its Classification Board category, with royalties of

[6] *Ibid*: 275.

[7] Caparrós Lera says that 'many films were made only to obtain import licenses, which would ensure safe profits. Then, investors would leave the world of cinema, frowned upon, and invest in land or apartments on the Costa del Sol'. Caparrós Lera (1983): 31.

[8] The National Entertainment Union (*Sindicato Nacional del Espectáculo*—SNE) first appeared in the law of 23 June 1941 of the National Headquarters of the Movement, by which the unions were classified by branches of production, and the SNE was number 23. The union was responsible, among other things, for granting or refusing credit to producers, as well as for awarding prizes to the best Spanish films.

75,000, 50,000, or 25,000 pesetas for the first, second, and third categories, respectively. This money was then used to finance the films (up to 40%) and prizes—two of 400,000 pesetas and four of 250,000 pesetas. With this approach, the state ensured the production of Spanish films in which it was interested and controlling those that came from abroad.

Another approach to promoting the production of films of interest to the state was through the granting of 'national interest' by the National Delegation of Press and Propaganda. By Order of 15 June 1944 from the deputy secretary of Popular Education, films of 'national interest' would be films:

> produced in Spain, whose artistic and technical works are essentially Spanish. It is also essential for the issuance of said title that the film contains unequivocal evidence of exaltation of racial values or teachings of our moral and political principles.[9]

These films had a series of privileges, such as a preference 'for all contracting purposes in cinemas located in national territory'.[10] On 13 May 1946, also by Ministerial Order, the Ministry of the Army created the 'military interest' as an award for films that promoted this type of value.

In 1952, an important change occurred, which aimed at improving the quality of the films. By Order of July 16, national production was separated from the import of foreign films and financially subsidised. Shortly before this, on March 21, the Board of Classification and Censorship had been created, in charge of censoring national and foreign films, as well as classifying them according to their technical and artistic qualities. Similarly, the export of low-quality national films was prohibited, so that only those that reached a minimum quality could be exported.

The classifications 1A, 1B, 2A, 2B, and 3 determined the amount of non-refundable protection: 50%, 40%, 35%, 30%, and 25% for national interest films, 1A, 1B, 2A, and 2B films, respectively; and no protection for films with a classification of 3. However, few changes occurred because, as Agusto M. Torres explains, the principles that governed the Board remained unchanged, and so the film industry continued to produce films that were neither good nor commercial, but which pleased the Board. What could have been a significant advance, therefore, came to nothing.[11]

[9] Sindicato Nacional Del Espectáculo (ed.) (1945): 1141–1142.

[10] *Ibid*: 1141.

[11] M. Torres (1973): 16. Quoted in Pozo Arenas (1984): 86.

The truly important change came about in 1963, with the appointment of García Escudero as the director general of Cinematography. From that date, and coinciding with the campaign of the controlled opening up of the regime, the Francoist institutions authorised the production of a new cinema, which, as can be inferred from the research of Arangüez Rubio, although not of particular interest to the regime, provided a new image and made a positive impression in international circles.[12] Even so, the protection scheme continued to control film production—although more objective formulas were approved, related to the films' box office performance—and censorship continued to determine what could and could not be seen.

If we analyse the films produced on Spanish soil between 1939 and 1963, we will see that many of them explore the idea of Spain: either in a discourse that glorifies the national past, and that often indirectly depicts the regime as a guarantor of all things Spanishness, or as a reflection of a certain national identity, as in films starring singers[13] and bullfighters, which idealised a rural and conservative society.

To facilitate the analysis of this entire cinematographic set, we have classified each film according to the theme that it addresses: Reconquest and Imperial Spain (with 1492 as the axis),[14] the War of Independence (1808–1814), and the Civil War (1936–1939), the Spain of the colonies, Catholicism, and the Spain of bulls and flamenco.[15]

Reconquest and Imperial Spain

The first theme is evident in the following films: *Amaya* (1952), *El valle de las espadas* (1963), *Inés de Castro* (1944), *Locura de amor* (1948), *La leona de Castilla* (1950), and *Jeromín* (1953). *Amaya*, by Luis Marquina, is a historical drama set in the eighth century and centred on the

[12] Arangüez Rubio (2006): 77–92.

[13] Flamenco singer.

[14] We use the word Reconquest to refer to the period of history of the Iberian Peninsula from the eighth to the fifteenth centuries, which began with the Umayyad conquest of Hispania and ended with the fall of the Nasrid Kingdom of Granada due to the Christian kingdoms' expansion. Imperial Spain comprised a group of Spanish territories ruled by Hispanic dynasties from the sixteenth to the nineteenth century. Spanish nationalist historiography has placed special emphasis on the Catholic Kings and the discovery of America as the foundation of that imperial Spain.

[15] Viadero Carral (2016).

disagreements between the Goths and Basques, who were in permanent conflict. *El valle de las espadas*, by Javier Seto, is a Spanish-American co-production with an internal chronology set in the tenth century in peninsular territory, a time when Spain was divided into kingdoms and invaded by Muslims. *Inés de Castro*, by Manuel Augusto García Viñolas and José Leitão de Barros, is a Spanish-Portuguese co-production about the marriage of the Infant of Portugal to the Castilian Constanza, daughter of Don Juan Manuel, Prince of Villena, to end the disputes between Portugal and Castile in the fourteenth century, and also about the love story of this Infant with the Galician Inés de Casto. *Locura de amor*, by Juan de Orduña, takes place in the fifteenth century and narrates the turbulent relationship between Juana, Queen of Castile, and her husband Felipe. *La leona de Castilla*, also by Juan de Orduña, concerns the sixteenth-century Revolt of the *Comuneros*, as well as the figure of María de Pacheco, widow of Padilla, one of the *Comuneros* executed by the Imperial troops.[16] Finally, *Jeromín*, by Luis Lucía, also takes place in the sixteenth century and focuses on the figure of Don Juan de Austria.[17]

Historical events shown in *Locura de amor* (1948) and *La leona de Castilla* (1950) provide a perspective that is very close to the context of the dictatorship. Both films depict internal conflicts, the rebellion of some Spaniards against other Spaniards, and the inevitability of armed intervention if the motherland is in danger. The parallelism with the uprising of the national side and subsequent civil war, facts understood by Francoism as in the name of national defence, is more than evident and demonstrates the regime's need for legitimacy. In both films, the warriors/military are the national heroes, exactly as in Franco's perspective of the civil war.

The film *Amaya* (1952) starts from a time of splendour and unity, in which the system of government is the Spanish Catholic monarchy, which will be interrupted by the Jews and the Muslim invasion. To return to the golden age, Goths, Basques, and Asturians, chosen by God and under the sign of the Cross, will have to start the Reconquest. The *El reino de las espadas* (1963) repeats this loss of Spain, a consequence of the Muslim

[16] The Revolt of the *Comuneros* was a rebellion led by the Castilian social elites in the sixteenth century due to the discontent and suspicion caused by the arrival of Carlos V of Germany at the Cortes of Valladolid and that made them fear a significant loss of power.

[17] Don Juan of Austria was an illegitimate son of Emperor Carlos V, who was recognised as a member of the royal family after the death of his father. Don Juan de Austria was one of the main people responsible for the victory of the *Liga Santa* in Lepanto, defeating the Ottoman Empire.

occupation and the disputes between the Castilians, Leonese, and Navarrese. To defeat the common enemy, the infidel, the different kingdoms must unite to achieve redemption. Led by Santiago and San Millán de la Cogolla, they attain a victory that marks the beginning of the end for the Muslims. Finally, *Locura de amor* (1948) is centred on the national fall,-- after the death of the Spanish Queen par excellence, Isabel (Isabella I of Castile)--, caused by the Flemish invaders, and facilitated by the lack of unity among Spaniards. Juana, her daughter, as a symbol of national virtues, and despite her failure in life, was to leave a legacy that Emperor Charles would subsequently turn into an empire.

Another common theme is the idea of Spain as a chosen nation, as depicted in *Amaya* (1952), *El valle de las espadas* (1963), and *Jeromín* (1953). In these films, Divine Providence defines Spain as a nation, by being chosen to defend the Catholic religion. This defense would be the basis of the unity among the different Spanish people. Different warriors/military figures (Iñigo, Teodomiro, Pelayo, Fernán González, Carlos I, Felipe II, and Don Juan de Austria) will fight for her. The national Catholic discourse reaches its climax in *Jeromín* (1953), by the exaltation of a Hispanic Cesar whose militarism in favour of the cross turns the nation into an empire.

Finally, on several occasions, Spain appears to be identified, and, therefore, reduced to Castile and its imaginary, a land of bare moors, a forging landscape of a strong, honourable, and austere character, and of virtuous women, such as Constanza of *Inés de Castro* or María de Pacheco of *La leona de Castilla*. The identification of Spain with Castile was already evident in the nationalist discourse of the generation of '98,[18] from which Spanish fascism would later be inspired.

Defensive Wars

The War of Independence against Napoleon (1808–1814) and the Civil War (1936–1939) were both conflicts defined as wars of defence and interpreted in national terms in the films of the period. The former relates

[18]The Generation of '98 is the name that has traditionally been used to refer to a group of intellectuals and writers, marked by what became known as the Disaster of '98 and the political and social crisis that it triggered. The Disaster refers to the military defeat of Spain in the Spanish-American war and the consequent loss of Cuba, Puerto Rico, and the Philippines in 1898.

the struggle of a civilised and peaceful nation (Spain) to defend its freedom against a barbaric invader (France). Catholicism is depicted as the source of Spanish civilisation, while the barbarism of the French can be explained by its atheism, the lack of a ruling god.

In the films that address the religious question in greater depth, the Spanish nation fights for God, who leads them to victory. We find here, again, the idea of a chosen nation. On the other hand, the church as an institution takes the side of the patriots.[19]

The Spanish rebellion is represented as a popular uprising. The people, unarmed, face the best army in the world, with scarcely any chance of winning. Against all odds, however, the Spanish expel the army from their territory.

The myth of successive ages also appears, woven from conservative nationalism: paradise (a united and Christian Spain) is followed by fall (an invasion by atheistic foreigners who promote the disunity between patriots and Frenchified) and redemption (a belligerent nation, expulsion of the French, and a return to unity). The War of Independence would represent a new moment of refoundation, as the heroic Spaniards would save the nation, just as their ancestors did in the Reconquest.

As for the Spanish identity, it is presented as a sum of localisms, generally associated with areas with a strong historical identity imprint, such as Castile, Aragon, Andalusia, Madrid, and the Basque Country, whose unity is generated in opposition to the French otherness. Therefore, regional differences do not limit Spanish identity, although there is a clear preference for a certain way of understanding Andalusia, strongly influenced by nineteenth-century Romanticism and represented by the bullfighter, the gypsy singer, the bandit (bandolero), and so on.

This Spanish community identifies itself with the *Rojigualda*,[20] which becomes its incarnation. In this context, Eliade's comment about archaic thought is very interesting: for him, a certain object can become sacred, that is, saturated with being, through participation in a certain symbol.[21] In this way, a two-coloured cloth became a sacred object, the symbol of a nation.

[19] Being the patriots those Spaniards against the French invasion. Some Spanish people were Frenchified (Francophile), and defended French ideas from the Enlightenment and a French rule in Spain.

[20] The name given to the flag of Spain, which comprises three horizontal stripes (red, yellow, and red), with the yellow stripe twice as wide as each of the red ones.

[21] Eliade (2015): 16.

To promote the idea of a united and Catholic Spain against an invading and atheistic France, implicitly criticising the Enlightened ideas that France defended and questioned the Francoist ideology and power of the church, it was necessary to read and interpret the facts according to these criteria. Owing to the importance of demonstrating that this was a war of the Spanish against the French, the cinematography ignored the prominent role played by England and Wellington in the War of Independence, as well as the presence of Polish, Italians, and Germans in the English army. For the same reason, the great battles, such as Ciudad Rodrigo, Vitoria, or Arapiles, which were won thanks to the English, receive little attention, while the battle of Bailén, led by the Spanish commander Castaños, does appear prominently.

Another armed conflict, this time closer to the time of the films' production, was the Civil War (1936–1939). Immediately after this war, cinema not only maintained its focus on the rebel side's propaganda but also sought to mythologise the event, replete with heroes and deeds. Thus, the conflict was portrayed as a refoundation of the national order, which was once more threatened, this time by communists and republicans.

In these 1940s' films, the protagonist and hero, who risks his life for the national community, is a soldier. In the 1950s, however, the figure that is exalted is the martyred priest.

This shift in focus, from the military to the religious, can be associated with the growing presence of Catholic groups in government (1946–1957). In addition, Franco proposed a concordat to the Vatican in 1951, which was signed in 1953. On the other hand, a military alliance was established with the United States, reaffirmed in the Pact of Madrid of 1953, with the aim of moving away from the fascist rhetoric of the first post-war years and basing its legitimacy on the fight against communism, in the context of the Cold War. In fact, the fiction films of the 1950s show the Spanish soldier in World War II but not related to belligerence. This soldier does not attack but is attacked, and the plot focuses on his stay in Soviet concentration camps from which he attempts to escape.

An entire collection of anti-communist films were produced in the 1950s. In these films, Spain is depicted as a country of peace, built around family, the Catholic religion, and the homeland, as opposed to Soviet Russia and its satellites, which are characterised by atheism, the hierarchy and supremacy of the party, and an absence of feelings—a place where it is not possible to live without violence. There is, therefore, a brutal clash between the imaginary of the 1940s, in which the soldier dies killing in a war, and the peaceful and Catholic Spain of the 1950s' films.

Colonial Spain

Another topic of the cinematography of the time is Spain and its colonies; specifically, the discovery, conquest, and colonisation of America, the occupation of Africa, and the subsequent loss of America and the Philippines Spanish colonies.

These films, particularly those related to the occupation of America, highlight the figure of Isabel *la Católica (Isabella I of Castile)*,[22] who is identified with the motherland and whose reign marks the national golden age. She determines what the *Colombino* project should be[23]: a Catholic venture through which the Spanish people share blood, religion, and culture. The queen, as a role model—built on the principles of what is defined as Castilian, that is, austerity and spirituality—serves to support the whole mystique of poverty, which will come to define Spanishness.

It is interesting to note that the different films depict the Other—that community that would come to oppose the values of the exalted group—as oblivious to the idealism and austerity that define Isabel, and only interested in obtaining an economic return from the colonies, in accordance with their materialism.

On the other hand, the American Indians, who never rebel against the Spanish, are not depicted as Other. Conversely, they are grateful for all that this 'superior' culture can offer them. In this case, the Other is personified by the French and Jews, as well as by the northern European powers, whose interest in the colonies is purely economic.

In this sense, there is a strong influence of the generation of '98[24] and the casticism turn[25] that took place at the beginning of the century regard-

[22] Isabel *la Católica Isabella I of Castile* was queen of Castile from 1474 to 1504 and queen of Aragon after her marriage to Fernando de Aragón from 1479. She is known as Isabel *la Católica*, the title that Pope Alexander VI attributed to her through the papal bull *Si convenit* in 1496.

[23] *Colombino* refers to Christopher Columbus, the navigator who was seeking funding for a trip to the Indies by an alternative route and ended up finding it in the monarchs of Castile. This trip led to the discovery of America in 1492.

[24] The Generation of '98 is the name traditionally used to refer to a group of intellectuals and writers, marked by what became known as the '98 Disaster and the political and social crisis that it triggered. The Disaster refers to the military defeat of Spain in the Spanish-American war and the consequent loss of Cuba, Puerto Rico, and the Philippines in 1898. The members of this generation constantly asked themselves about the being of Spain, its identity, its past, and its future.

[25] Casticism, in this context, refers to the traditional Spanish identity. The casticism turn refers to the exaltation of the traditional Spanish identity versus the European ideas of progress that took place at the begining of the 20th century.

ing Europeanism. Álvarez Junco explains Spain's realisation that it had high-quality values to offer, such as mysticism or Don Quixote, but that these differed from European values. This encouraged the latent anti-modernism in the intellectual elites of the beginning of the century, as materialised in the outburst 'Let them invent!'[26]

At heart, this meant making scientific, economic, and political backwardness a sign of identity. This identity fitted like a glove to the context of the period that concerns us, since, from 1939 to mid-1955, the country, under a regime of autarchy, underwent an economic crisis, to which was added its international marginalisation. Isabel (Isabella I of Castile), the national queen par excellence, served to legitimise the reality of such poverty under this early Francoism, as a consequence of the idealism and austerity inherent in the Spanish spirit.

Despite the importance of the discovery and colonisation of America in the discourses that exalted the Spanish nation as a great empire, few films covered this subject. However, Spanish colonialism in Africa, a much lesser subject, did feature prominently in cinematography.

It seems simple to relate these facts to Francisco Franco, whose performance in the war in Morocco launched his career in the army and whom the propaganda defined as the successor of the Catholic Kings, the Spanish kings par excellence, who would have already marked Africa as an objective of the Spanish imperial mission. Franco and his regime would recover that African empire that dated back to the RRCC (Catholic Monarchs of Spain (Isabella I of Castile and King Ferdinand II of Aragon)) and that was presented to us in several films and documentaries.

The internal chronology of the films that we have grouped under the African theme focuses on the first decades of the twentieth century, and the films develop their plot in the context of the African colonies, mainly Morocco. The main character is the soldier, who, in this case, is defined as a special caste within the army, who moves in a quasi-mythical world, far from the metropolis and, therefore, from civilisation. His death in combat represents the glorification of the hero, creating a kind of national guardian angel. In this type of film, the flag is again depicted as a sacred object, the defence of which can lead to death.

The nation appears as a belligerent, attacking force, led by the Africanist military, while the feminine is depicted in these films as an obstacle to the

[26] Álvarez Junco (1997): 55.

soldiers' duty, as a conservative force. On the other hand, politicians and the bourgeoisie are defined as materialists, as opposed to the military, who are idealists. Let us remember that materialism was a quality that defined the Other, the foreigner, while idealism was inherent to the Spanish.

This otherness is also represented by Moroccans who rebelled against the Spanish authority. However, since the intention is to exalt the Spanish-Moroccan brotherhood, this enemy is virtually absent; it has no face. Furthermore, many Moroccans fight on the Spanish side, albeit in inferior conditions. On the other hand, although, at first, the characters seem interested in learning about Morocco's culture and traditions, the films show us that their only real objective is domination. In this type of film, the idea of Spain as a civilising nation emerges once more, from which it follows that Moroccans are barbarians who need to be protected, and therefore, controlled.

We find here a contradiction in the Spanish identity discourse, as, although the aim indeed is to promote the Spanish-Moroccan brotherhood, playing with the Islamic past of the peninsula, it is undeniable that the infidel was considered a historical enemy.

Finally, in a few films, the plot unfolds in the Spanish colony of Guinea and their internal chronology corresponds to the twentieth century. In these, the idea of a civilising Spain with a national Catholic tradition emerges once more, as already observed in relation to America, and through which religion and culture are shared. Again, the otherness is characterised by materialism, in this case in the English and French who traffic black slaves. The Spanish, although fighting to eradicate slavery and preaching the equality of the races, advise against miscegenation.

These films also show an evident disregard for black people, who are depicted as inferior. In contrast to civilised Spain, Guinea is described as an inhospitable, barbaric, and violent region. As mentioned above, there is an opposition of concepts (Spain–civilisation vs Otherness–barbarism).

To conclude, we will refer to a series of films whose plot takes place in the context of the loss of the Philippines and the last American colonies. In these films, again with the soldier as the main character and defined as the national hero, the flag also has a strong presence as a symbol to honour.

Some present the archetype of the resistance to the siege that we also find in the narratives related to the Civil War or War of Independence films, and that repeat the scheme of other similar events, such as Sagunto or

Numancia,[27] thus reinforcing the idea of cyclical time and national refoundation. On the other hand, and despite the loss, 'the military defeat is rewritten as a moral victory'.[28] The military are hailed as heroes on their return to Spain, leaving the nation in good standing.

While otherness is rarely depicted in these films, in any case, it is represented by the independentists, who are treated differently depending on their skin colour: much greater respect is shown to the whites than to the Indians, who are accused of using the revolution for social ascension. The depiction of the Tagalogs is very similar to that of the blacks since they are also considered inferior beings. To this is added their 'ingratitude' for not appreciating the opportunity that Spain is offering them to learn from the Spanish.

Catholic Spain

There is an entire collection of films focused on Catholicism that we analyse in terms of its main characters: saints, missionaries, priests, and nuns. The first group comprises men and women chosen by God who possess all the Christian virtues: faith, hope, charity, prudence, strength, justice, and temperance. They are role models, whose conduct is, in turn, inspired by Jesus Christ. They can do miracles, and their greatness is recognised in life, as they die surrounded by honours.

These men and women are ascetics far removed from any materialism, which, as mentioned above, is part of the Spanish traditional identity. In this sense, it is noteworthy that most of the saints who appear in these films are of Spanish nationality, thus establishing the country as a nation of saints. This is national Catholicism in all its splendour.[29]

[27] The Siege of Saguntum was a military confrontation that took place in 219 BC between the Carthaginians, led by Hamilcar Barca, and the Saguntines. This was a long siege that lasted for approximately eight months, due, among other things, to the tenacity of the Saguntines. The Siege of Numancia was a military confrontation between the Romans and Numantines that ended after a 15-month siege in 133 BC. The Numantines preferred suicide to surrender and burned the city so that it would not fall into Roman hands. The few survivors were sold as slaves. The Spanish nationalist historiography has used these two military confrontations as evidence of the bravery of the Spanish and their love for independence from the invader.

[28] Colmeiro (2000): 301.

[29] On national Catholicism as a current of thought that attempted to combine religion and nation, see Álvarez Junco (2001): 433–496, and the reference work on the subject Botti (2008).

While the military world only has space for men, women play an important role in the field of religion and many women appear in Spanish cinematography, in one guise or another. *Catalina de Inglaterra* (1951) is a great example of this, since the main character, Catherine of Aragon, represents the Castilian values of austerity and purity in a materialistic and lustful English court, which will eventually depart from the Catholic Church.

Another outstanding figure of the Christian imaginary that is shared with Spanish nationalism (primarily in its national Catholic aspect) is that of the missionary. The missionaries are responsible for taking the 'true' religion to other countries, risking their lives to do so. Once more, there is the idea of God's chosen people to disseminate his doctrine, as well as the absence of material interest, which aligns with the Spanish identity founded on idealism mentioned above.

More than heroes, the missionaries are depicted as martyrs, since they allow themselves to be killed, just as Jesus Christ did. They share the same Christian virtues as the saints, but outside their territory, surrounded by disease, indigenous violence, contempt for pagans, and self-remoteness. Many will die evangelising, defending the word of Christ, son of the 'only and true God'.

The last figures of this imaginary are the Spanish priests and nuns, who are depicted as central elements of the Spanish society of the time. These characters are responsible for maintaining the values of the group, for the success of the common life; specifically, to defend social justice, to fight against the powerful, to take care of the poor and underprivileged, to attend to women whose honour has been disgraced, and so on.

It is interesting to note that, in this type of film, God is the origin of life, with reason and science placed in a very secondary position, and sometimes even discredited.

Romantic–Folkloric Spain

Said says that Westerners had developed a discourse in which the Orient was identified with the exotic, the irrational, and the dangerous.[30] In turn, the romantic travellers of the nineteenth century had created an image of Spain that we can describe as romantic–folkloric, using the criteria of that orientalist discourse to which the region of Andalusia had contributed due to its Arab past, the gypsy race, and the singing and dancing, an image

[30] Said (1978).

endorsed by the Spanish artists and intellectuals of the early twentieth century and that was reflected in all kinds of creations, including paintings, dramas, novels, music, and, with the arrival of the film industry, cinema.

For this period, and although this romantic–folkloric film genre was not a Francoist creation, it served perfectly to reflect with beauty a conservative and religious society. Furthermore, considering the harsh post-war reality, it was good entertainment, while justifying the country's economic backwardness since it corresponded to the Spanish identity, represented by Isabel and her mystique of poverty, and as also reflected in the colonising venture. The Andalusian way of dancing and singing showed that it was not necessary to achieve material success to be happy, and that the Spanish carried within them a certain joy.

However, we should not overvalue the ideological component, since the production of this type of film followed different logics, the most prominent of which was the economic logic.[31]

The main character in these films is usually the personification of the national essence—a woman of exotic beauty, marked ethnic traits (black hair, dark eyes), with great jest and self-confidence. The sun attains a definite importance as a source of joy, since it evokes love for life, which translates into singing and dancing.

While the song does appear as an expression of happiness, it also appears as a vehicle of fatality, as in the case of *La Lola se va a los puertos* (1947), a film starring a female singer[32] condemned to enchant men with her voice but incapable of experiencing love. This romantic universe woven around Spain was governed by a tragic destiny.

The stories are contradictory. One is more romantic, with the characters risking their lives by letting themselves be carried away by their irrational side and not integrating into the system; another is more folkloric, based on a superficial image of manners and advocating the cause of conservatism.

On another note, many of the films' main characters travel abroad to succeed in singing, thus exercising a patriotic-cultural mission. Their success is based on their Spanishness, which is understood as their talent for singing and dancing. In this sense, the main character acquires heroic overtones, since a hero is someone who gives his life for something greater

[31] Gallardo (2010): 28.

[32] Flamenco singer.

than himself, who risks himself for the community[33]; in this case, these diplomats at the forefront of the cultural conquest suffer from their remoteness from the homeland and the loneliness of someone lost in a new territory. In contrast to the American dream that is based on the material, the Spanish dream is based on the cultural.

Catholicism appears as an integral part of this Spain, marked by a set of rites in the form of processions, pilgrimages, and festivals, such as the *Rocío* or the *Cruz de Mayo*, a popular religiosity that contributes to a picturesque image of the country.

Apart from Andalusia, other identities also have a certain presence, such as Aragon and Madrid. The first focuses on a series of rites, customs, and festivals of a popular and folkloric nature. Madrid, for its part, is defined as the stage of the Goyesque, of the *majas* and the *manolos*.[34]—an image imbued with music, dance, and popular entertainment, such as bullfighting—or the more castizo: the Madrid of the *chulapas y chulapos*[35] that had been immortalised in the operettas.

Another heroic figure of the Spanish imaginary is the bullfighter, a man with a mythical aura who faces death by vocation and not for economic success. In fact, many of the films on this subject depict the bullfight as a highly symbolic ritual, a tragic space in which death and life merge.

The world of bullfighting is dominated by masculine values, such as bravery, courage, and aggressiveness, in which women play a secondary role: either as a conservative force, opposing the men who are risking their lives, or encouraging them in their mission to overcome fear, but never as a heroic main character.

As a collective ritual, the audience plays a fundamental role, as it gives continuity to the *fiesta*, ratifying its existence. To this end, it penalises the bullfighter for his fear and leads him to death if necessary. Every Spanish person is a potential bullfighter, and, by jumping into the arena on a Sunday afternoon, the bullfighter shows his courage and becomes a hero. Bullfighting is viewed from a national perspective, representing the Spanish identity, a popular spectacle, an expression of the living forces of the community, which always transcends the individual.

[33] Campbell (2015): 167.

[34] *19th century working-class Madrileños. They represent the real Spanish identity as opposite to the new trends that came from France, adopted by the upper-classes.*

[35] working- or lower-middle-class group of people from Madrid who were famous for their elaborate style of dress and cheeky attitude.

Finally, religion and the Catholic institution appear linked to the world of bullfighting, to the extent that Saint George is defined as the first bullfighter—a fascinating relationship if we consider the pagan origin of this type of spectacle.

Final Considerations

As mentioned above, in 1938, the *bando nacional* laid the first stone for the foundation of a system to guide cinematography, with the creation of the National Department of Cinematography. However, in 1941, and following Fanés, 'we could say that the situation was identical to that of the Republic, were it not for three characteristic facts of the time. The accentuation of censorship; the almost ruined state of the film studios [...] and the disappearance, on the way to exile, of a significant contingent of film professionals [...]'.[36]

The fact is that film censorship was not a specific practice of the Franco regime. It was implemented much earlier, already in the early twentieth century, during the reign of Alfonso XIII, and remained in the Republic, despite Article 34 of the Constitution that recognised freedom of expression and explicitly prohibited any regulations that limited that freedom. In fact, as Paz y Montero points out, 'it was the usual way of proceeding in those years in all countries [in relation to cinematography], with differences that related only to the way that censorship was carried out and the direct role of the state administration in it'.[37]

Censorship increased during the Franco regime, but, as explained above, it also created a system that directed and guided film production, particularly until 1963. Before 1941, the legislation had paid no attention to cinema other than to tax it—although it was recognised that protection was necessary for economic and political reasons.[38]Therefore, we can state that the close link between cinema and political structures that occurred during the Franco regime is a unique case in the history of Spanish cinema.

As explained above, Spain is the main character in a great number of films produced during these years, the defence and unity of which was in the hands of the military, religious entities, and kings. Spain also appears in the form of regional identities, mainly Andalusian, elevated to the category of national.

[36] Fanés (1982): 75.

[37] Paz Rebollo and Montero Díaz (2010): 370.

[38] Pozo Arenas: 38.

This should not surprise us if we consider that the different families (Carlists, Falangists, Juanists, military, and Catholic institution) that grouped against the Republic had as common ground the defence of Spain, to the point that they called themselves the *bando nacional* and qualified their enemies as anti-Spain. According to these groups, the main issue was the unity of Spain, which they saw threatened by the Republicans, by the autonomy statutes of the Basque Country and Catalonia, and the need to fight against what they interpreted as the communist drift of the Republic, with the victory of the *Frente Popular*[39] in the 1936 elections.

However, cinema did not give voice to those groups that could endanger the Franco regime, such as the Carlists or Juanists: the Spanish nation was in the hands of Franco and his regime, and kings were obsolete. The Falangists, the military, and the Catholic Church did have their share of power but, not all of them were depicted in film. The characters who starred in films with nationalist content were, above all, medieval and Imperial Spanish kings, or military and religious personages.

As for kings, it was Isabel *la Católica* (Isabella I of Castile) who had the greatest presence. Her reign marked the golden age to be achieved, and her character traits, such as austerity, supported a whole mystique of poverty which defined the Spanish. From Isabel's time, there would be several cycles of fall and redemption. One of these moments comes with the War of Independence, when Spanish values were threatened by a barbaric and violent community. The civil war marks another such moment of refoundation when the unity of Spain was once more in danger. Not surprisingly, Republicans and Communists, like the French of the War of Independence, are depicted as atheists and as violent, barbaric, and uncivilised, while the nationals have their God. In the 1950s, there is a change of discourse in relation to the Civil War, interpreted as the pioneering conflict that disrupted the communist plans.

Finally, it is interesting to contrast the anti-modern discourse that defines the Spanish identity and that is so often reflected in the cinematography of the time with the demands, by certain sectors, for modernisation and investment in new technologies. In fact, from 1963, the conflict between tradition and modernity was reflected in a great number of films. Reconfiguring an entire national identity built on tradition was not going to be an easy task.

[39] The *Frente Popular* was a Spanish electoral coalition created in January 1936 by the main left-wing parties.

References

Álvarez Junco, José (1997): "El nacionalismo español como mito movilizador: cuatro guerras", en CRUZ, Rafael; PÉREZ LEDESMA, Manuel (eds.): *Cultura y movilización en la España contemporánea*, Madrid, Alianza: 35-68.

Álvarez Junco, José (2001): *Mater Dolorosa*, Madrid, Taurus.

Arangüez Rubio, Carlos, "La política cinematográfica española en los sesenta: la propaganda del régimen a través del nuevo cine español (1962-1967)" en *Sociedad y utopía: Revista de ciencias sociales*, n° 27, 2006, pp. 77-92.

Botti, Alfonso (2008): *Cielo y dinero: el nacionalcatolicismo en España, 1881-1975*, Madrid, Alianza Editorial.

Campbell, Joseph (2015): *El poder del mito*, Madrid, Capitán Swing.

Caparros Lera, José María (1983): *El cine español bajo el régimen de Franco (1936-1975)*, Barcelona, Edicions de la Universitat de Barcelona.

Colmeiro, José F. (2000): "Nostalgia colonial y la construcción del nuevo orden en *Los últimos de Filipinas*", en SEVILLA, Florencio y ALVAR, Carlos (eds.), *Actas del XIII Congreso de la Asociación Internacional de Hispanistas* (Madrid, 6-11 de julio de 1998), vol. 4, Madrid, Asociación Internacional de Hispanistas-Castalia-Fundación Duques de Soria: 294-302.

Diez Puertas, Emeterio (2002): *El montaje del franquismo*, Barcelona, Laertes.

Eliade, Mircea (2015): *El mito del eterno retorno*, Madrid, Alianza.

Fanés, Félix (1982): *Cifesa. La antorcha de los éxitos*, Valencia, Institución Alfonso el Magnánimo.

Gallardo, Emilio José (2010): *Gitana tenías que ser*, Sevilla, Centro de estudios andaluces.

Gellner, Ernest (1983): *Nation and nationalism*, Oxford, Basil Blackwell.

Hobsbawn, Eric J. (1994): *Ages of Extremes*, London, Michael Joseph.

Moreno Almendral, Raúl (2014): "Franquismo y nacionalismo español. Una aproximación a sus aspectos fundamentales", *Hispania Nova*, 12, pp. 7-36.

Paz Rebollo, María Antonia and Montero Díaz, Julio (2010): "Las películas censuradas durante la Segunda República: valores y temores de la sociedad republicana española (1931-1936)", *Estudios sobre el mensaje periodístico*, 16, pp. 369-393.

Pozo Arenas, Santiago (1984): *La industria del cine en España: legislación y aspectos económicos 1896-1970*, Barcelona, Universitat Barcelona.

Said, Edward (1978): *Orientalism*, New York, Pantheon Books.

Saz, Ismael (2008): "Las culturas de los nacionalismos franquistas", *Ayer*, 71, pp. 153-174.

Sindicato Nacional Del Espectáculo (ed.) (1945): *Anuario del espectáculo 1944-1945*, Madrid.

Torres, M. Augusto (1973): *Cine español, años 60*, Barcelona, Anagrama.

Viadero Carral, Gabriela (2016): *El cine al servicio de la nación (1939-1975)*, Madrid, Marcial Pons.

CHAPTER 13

The Nationalisation of the Avant-garde during Francoism

Jorge Luís Marzo

Introduction

It was not until relatively recently (the 1990s) that the relationship between Franco's nationalism and the artistic avant-garde that emerged in Spain after the Civil War (1936–1939) began to be included in the repertoire of interests of Spanish historiography.[1] Some works of art, particularly the post-informalist, and a significant part of the artistic and literary critique, focused on this relationship for decades, but it had little place in the history of art, unless to celebrate or demonstrate the Spanishness (or otherwise) of a particular artist.[2] As this relationship was intrinsic to the theoretical and casuistic core of the Spanish history of art, the academy's

[1] A brief review of this update process can be found in Albarrán (2019), and Marzo and Mayayo (2015).

[2] See Díaz Sánchez and Llorente (2004).

J. L. Marzo (✉)
BAU College of Design, Barcelona, Spain
e-mail: jorge.marzo@bau.cat

M. Janué i Miret, A. Presas i Puig (eds.), *Science, Culture and National Identity in Francoist Spain, 1939–1959*,
https://doi.org/10.1007/978-3-030-58646-1_13

approach to it soon became engrained, as it implied considering the academy itself as an object of study. It was, therefore, necessary to change all the instruments to understand the institutional role in the process of communion between the avant-garde and nationalism. To start with, it was necessary to approach the political competition of the avant-garde in a different manner.

The avant-garde in Spain was politically relevant, thanks to the role assigned to art and culture from the mid-1940s. The case of surrealism illustrates both the tensions to which pre-war styles were subjected in the first two decades of the Franco dictatorship, with those styles originally being linked to ideological manifestos, and the effects that those tensions had on the way that the avant-garde was defined. After the Civil War, many of the young artists returned to the surrealist and abstract heritage, avoiding at all times provoking political effects and short-circuiting all memories of the side adopted during the war by the surrealist movement, which gave majority support to the legitimate government of the Republic. Many groups in Almería, the Canary Islands, Valladolid, Barcelona, Madrid, Santander, and Valencia followed this direction to a greater or lesser extent. The cases of Santander and Barcelona are particularly revealing of those twists in the style and substance that it carries, and the role demanded of art to transcend the miseries of political life. Personalities such as Rafael Santos-Torroella played a relevant role in the seduction that these avant-garde practices exerted in the cultural political wastelands of the dictatorship. In the morning, he would insist to his friend and poet Luys Santa Marina, then leader of the Falange in Barcelona, that there was nothing to fear from post-surrealism after it had been purged from Breton, and that he turned a blind eye to certain exhibitions and magazines. In the afternoon, he attended literary *soirées* with members of the bourgeoisie of the Eixample (a neighbourhood of the Catalan capital), and invited them to bankroll the emergence of a new artistic scene that no longer looked outwards but inwards. It should be remembered that Santos-Torroella had been sentenced to death in 1939 for his role as political commissary at the front, which he escaped thanks to the efforts of the Falangist and philologist historian Antonio Tovar in Valladolid.

In 1948, the School of Altamira was founded in Santander, dedicated to abstraction, which strongly captured the attention of Eugenio d'Ors, who was, at that time, the main official factotum of the arts. Art, according to the School, had to free itself from the shackles of worldly events and undergo a process of 'essentialisation', the same that could be found in the simple but profound signs featured in the Cantabrian cave paintings. In 1949, d'Ors created the *Salón de los Once*, a showcase for post-surrealist abstraction in Madrid. In 1953, the First Congress of Abstract Art of the Menéndez Pelayo International University of Santander was finally held, an event that certified the pairing process between a part of the avant-garde and a sector of the Franco regime, thanks to the full public sponsorship of Manuel Fraga, secretary general of the Institute of Hispanic Culture.

That process illustrates the success of the 'incompetence' policy of intellectual Francoism with respect to the avant-garde and to art and culture in general: not only did it drain the works of ideological life but sought its legitimacy in the promotion of its immersion in the wake of the national tradition, and there was no better body for this than the Institute of Hispanic Culture, created by Franco in 1945 to shelter the Falangist intellectuals, who had left the government and whose militarism had to be discouraged after the defeat of fascism in Europe. Its first director, Joaquín Ruiz-Giménez, who was attached to the Vatican sector of the regime, advocated for the end of the harangue and the adoption of a 'spiritual' diplomacy that could find allies in Catholic countries, particularly in Latin America, and that would contribute to the creation of a story that would underline the civilising aspect of the Hispanic tradition. The informalist abstract avant-garde was immediately interpreted academically as essentialist and transcendent, humanist and existentialist, adjectives that conveyed a supposedly national act, practically ahistorical. The Spanish artistic nation was, therefore, blended within the framework of the cultural policies of the Franco regime, through a phenomenological narrative by which any expression of the truly national genius could be described in relation to the effects of consensus that the great figures of its past had conceived on a global level, even beyond the fact that the artists had political intentions, as was evidently the case of Goya, Picasso, or Miró. Moreover, to articulate those effects, who better than the academy to separate the wheat from the chaff, to decide who was 'competent' and who was not. If the analysis of Spanish art was, above all, a hermeneutic project, the role of the academy became relevant by adopting a fully competent function as an arbitrator of the national. Hence, the need, as noted above, for the university to radically change the instruments it used to explore the history of art, based both on analysis of the causes and mechanisms whereby the institutions and their followers assumed such a political function and on the exploration of a whole set of artistic ideas and practices whose objectives were not to comply with the nationalist order of cultural production but to provide spaces for debate and criticism of social production. The present text is certainly situated within the wake of these historiographic transformations.

Francoist Policies and Artistic Memory: Phases

Some studies have pointed out that the efforts to nationalise a large part of Spanish artistic production cannot be reduced to the Civil War and the subsequent dictatorship (1939–1977).[3] In fact, the nationalist policies of cultural memory were gradually shaped through the sanction and canonisation of certain cultural formats, first in the hands of the academies of Fine Arts themselves, with their origins in the Enlightenment, and, later, in those of the regenerationist and noucentist movement and the Generation of '98, which contributed to the homologation of the 'Spanish problem' to the intense nationalist debates that occurred worldwide in the late nineteenth and early twentieth centuries, and in fact comprised a true war of narratives. These debates discussed the role of culture as a guarantee of consistency and survival in the new global brand market, still imperialist, but increasingly linked to Soft Power, and the persuasion that can be exerted based on culture as opposed to the traditional repressive mechanisms[4]. The progressive influence of stories based on 'obvious fate' in certain countries in the late nineteenth century, such as American-style Pan-Americanism, Pan-Germanism, Francophony, or the Commonwealth, promoted the emergence in Spain (and in sectors of the upper-class Latin American bourgeoisie, conservative and Catholic) of the intellectual and diplomatic phenomenon of ***Hispanidad***, whose fundamental doctrine was defined by the defence of a historical and cultural space that presented signs of identity rooted in language, miscegenation, and religion, which were proposed in open confrontation with the 'Nordic' doctrines considered utilitarian and mercantilist. This issue became one of the main concerns of the Generation of 1898, a group of writers and artists who portrayed and analysed national demoralisation after the loss of the last overseas colonies under the pressure of the United States. Ángel Ganivet and Josep Pla stated that 'our historical role obliges us to transform our

[3] Tusell (1999); Cabrera (1999); Marzo (2010)
[4] Nye (2004)

material action into a spiritual one'.[5] José María Pemán and Miguel de Unamuno were sure, even from different perspectives, that language and culture must confront any technocratic empire to come.[6] Unamuno's famous 'Let them invent!', which must be considered within the context of the controversy with Ortega y Gasset between 1906 and 1912 regarding the Europeanisation of Spain and the role of positivism in this process, would become the hallmark of an entire era: 'It is useless to go around it, our contribution is above all a literary contribution [...] and since there exists and must exist a differentiation of the spiritual and physical work, both in communities and in individuals, we have this task [...]'.[7] Unamuno, therefore, would claim that art was the infallible mirror of the Spanish: 'If we want to penetrate the Spanish soul, we should be attentive to its painters, for the Spaniard sees better than he thinks.'[8]

In this context, certain events fomented the narrative process of what should be defined as Spanish culture, and the role of this culture in the projection of the country's history: at the National Exhibitions of Fine Arts (since 1853), the Universal Exhibitions of 1888 and 1929 (Barcelona), or the Ibero-American Exhibition in Seville (1929), the ideology of the expressive forms that portrayed the national idiosyncrasy were slowly taking shape both in Spanish (with the Baroque as its main theme) and Catalan (in this case, embracing the Romanesque) or Basque nationalism. The underlying idea in this long process was that, in the face of the failure of politics (understood as the backwardness of capitalist democracy and social science in relation to Europe), Spain recognised itself in its culture. Cultural expressions substitute the political, which cannot be expressed in any other manner. Cultural expression, thus, acquires a strong public and national competence.

However, it was undoubtedly the emergence in Spain of a radical nationalism in the late 1920s that gave rise to a profoundly essentialist reformulation of culture that would have full public competence with the victory of Franco's troops in 1939. We can distinguish three clear stages in that process:

(1) In the first stage, already in the years before the Civil War and later during the war, the Spanish Falange, a party of fascist ideology founded in

[5] Pérez Montfort (1992): 24
[6] *Ibid*: 17
[7] Unamuno (1911): 329
[8] Tusell (1999): 98

1933 by José Antonio Primo de Rivera, defined culture in ideological terms as a civilising fight against all that was considered 'anti-Spanish' and 'alienating', and that had the effect of dividing the Spanish intellectual and cultural production in a Manichean manner, between those whose purpose was to continue the old imperialist and Catholic traditions supported by a centralised state and those others considered weak or corrupt, either for a decadent liberalism or for being considered to serve revolutionary objectives of social and political transformation, and supported by supranational interests. The 'intellectuals' became priority targets of the attacks by the fascist forces: 'Let the intellectuals die!' was loudly shouted by the founder of the Legion José Millán Astray to Miguel de Unamuno at the auditorium of the University of Salamanca in 1936. In the opinion of Romualdo de Toledo, deputy for the Traditionalist Communion, the country had risen up in arms before 'the lack of essential education and of doctrinal and moral training, the alienating mimicry, the Russophilia and effeminacy, the dehumanisation of literature and art, the fetishism of the metaphor, and the verbalism without content'.[9] In this way, what were considered 'appropriate cultural forms' were identified to establish the national as opposed to those cultural expressions that denied it. The monarchist politician Eugenio Vegas Latapie clarified with great precision what forced the differential 'factor' of the intellectuals: 'It was one thing to senselessly shoot people, but quite another to purge the education that forms the conscience of a nation, and whose control is vital.'[10] The Marquis of Quintanar was also clear: 'There are not two sides here that can parley; on one side is the Army and the Spanish people, on the other is a collection of traitorous intellectuals and professional assassins. They must be exterminated without mercy.'[11] In his speech on 1 October 1936, General Franco justified his own cultural war in the same manner: 'A current of mistaken intelligentsia that, despising everything that truly national thought meant, had a preference for everything bizarre that was generated in other countries.'[12]

In this context of repression, both physical and semantic, Unamuno's interpretation of the cultural character of the Spanish nation was notoriously updated. The 'national' artist was to represent the interests of the

[9] Claret (2006): 24
[10] Claret (2006): 26
[11] *Ibid*: 348
[12] Reproduced in the Spanish newspaper *ABC*, Seville edition, on 2 October 1936.

homeland; he was to symbolise the inextricable union between the people and the leaders, whose epitome was the glorious painting of the Golden Age, when the 'intimate communion' between genius and monarchy was achieved, and which could be so vividly perceived in the Prado Museum. Hence, the discredit with which the nineteenth and twentieth centuries were seen, when the link between artists and power was severed, a consequence of an autonomous and anti-transcendent conception of artistic practice. The artist, in particular that practitioner of modern styles, became immediately suspicious of national heresy (ignoring that many of the artists who supported Francoism had been militants in the pre-war avant-garde), since he had replaced the transmission of spirituality with social transformation and iconographic iconoclasm. Therefore, the Falangist mural painter José Aguiar questioned in 1940 what should be interpreted when speaking of 'social' art: 'Social art? Yes; this meant political art, pedagogy with a luscious graphic. Conversely, art is social simply for nurturing the collective, that is, the national, of its purest emotion, of its mystique.'[13] The artist had to give his individuality to a higher purpose, which was the new state and its project to re-catholicise society. The modern novelty was, therefore, closed, or at least suspended, pending official confirmation of its functionality. It is in this context that we should understand the words of the Falangist philosopher and politician Adolfo Muñoz Alonso when he stated that 'novelty, however traditional we may suppose it to be, is anti-Catholic'[14].

(2) The second stage takes place from the end of the war until 1945. First, it should be noted that the repression and exile of a great number of creators and cultural technicians (approximately 70,000 people linked to the arts, teaching, and cultural industries had to go into exile[15]) intensified the desire to make a clean sweep of cultural notions from the immediate republican past. The cultural desert that arose with the repression and exile of so many creators and intellectuals, together with the first post-war academic, iconographic, and architectural reconstruction programmes, provided an apparent basis for the implementation of unified policies by the Falange, which would allow them to create a new aesthetic model that was already completely nationalised.

[13] Llorente (1995): 60

[14] Echeverría Plazaola and Mennekes (2011): 75–77

[15] Marzo and Mayayo (2015): 24–65

The cultural areas of the first Franco government in Burgos, created during the war, had already been taken over mostly by Falangists, including Ramón Serrano Suñer, Dionisio Ridruejo, Juan Antonio Giménez Arnau, Antonio Tovar, Gonzalo Torrente Ballester, Martín de Riquer, Luis Rosales, Manuel García Viñolas, Leopoldo Panero, Felipe Vivanco, Eugenio d'Ors, and Agustín de Foxá. The Plastic Arts Section of the National Propaganda Service was appointed to aesthetically guide the appearance of the new state. The rather vague objective of all of these figures was to constitute a fascist state art that would celebrate the dynamics of the Christian crusade; the condemnation of academicism as a flawed and spiritless form, but also of an important part of the avant-garde as castrator of the national and imposer of the foreign; and the joyous choral militancy of a movement at the forefront of the rebirth of the national and imperial memory. To centrally manage these objectives, the Council of the *Hispanidad* was established in 1941, and refounded at the end of 1945 as the Institute of Hispanic Culture (*Instituto de Cultura Hispánica*—ICH), which we will discuss below.

The ephemeral Falangist aesthetic project was, therefore, scaffolded on various fronts. In addition to the strictly propagandist media, such as the cinema, radio, and press, it is worth mentioning, first, the architecture and monuments, within the framework of reconstruction programmes, that pursued the idea of a 'national style', as in the lines adopted by the General Directorate for Devastated Regions and Restorations that was created in 1939. In the early 1940s, several National Assemblies of Architecture and conferences of the Federation of Town Planning and Housing were held, promoted by publications such as *Reconstrucción* or the *Revista Nacional de Arquitectura*, in order to establish unitary criteria for action, attempting, without success, to mimic certain formulations of aesthetic unity from Germany and Italy.[16] In fact, the Spanish aesthetic traditionalism defended by this early Francoism advocated for the Herrerian model derived from the Monastery of El Escorial—a sober, monastic, and imperial style—as a precursor of the new fascist orders based on the implementation of a grandiloquent state rationalism. The Spanish fascist culture was presented, therefore, with its own traditions, without wanting to seek inspiration from foreign models. The project of *El Valle de los Caídos*, promoted by Franco himself in 1940, was born from those arguments. Muralism also became a way of monumentally expressing the new national order, thanks

[16] López Díaz (2003)

to the impulse given to this type of format by magazines, such as *Jerarquía* and *El Español*, which endorsed the multiple commissions requested from José María Sert, José Aguiar, Daniel Vázquez Díaz, Francisco Baños, and Ramón Stolz Viciano.

The second aspect of the cultural policies of this stage was the exhibitions, particularly of sacred art, which served to focus the propaganda debate on the reconstruction of the religious iconography devastated during the war and to support the religious discourse of the state and Church's role in the management of culture. The exhibitions of religious art held in 1939 in Valencia and Vitoria, or the one in Madrid in 1944, were aimed at reaffirming the profession of faith of the new state, searching for anti-communist support in international forums, and preventing possible disagreements in the reconstruction of the destroyed temples and religious objects, so that this would be channelled into the 'truly national' styles. This issue strongly reveals the development of the narrative regarding the constitution of a supposed national style. In January 1943, the art critic Tomás Borrás requested that sculptors refuse to reconstruct the religious carvings based on 'standard effigies in an anodyne, deliquescent, and mirrored series', and not to lose sight of 'the pathetic force and the successful decision of chanting that they had with Berruguete or with Gregorio Hernández' (artists of the fifteenth and sixteenth centuries, respectively).[17] The imagery tradition of the early imperial years represented the basis on which to rebuild, thus avoiding modern idiosyncrasies, which, either for their spiritual insignificance or for their formal banality, short-circuited the message of the state.

We termed this first Falangist project of fusing culture into a state style as 'ephemeral'. In fact, since 1945, everything had been left to chance. The German and Italian defeat put an end to the Spanish fascist chimera. Gabriel Ureña pointed out the causes of that failure. First, the defeat of Hitler and Mussolini led to a retraction of the most clearly fascist impulses in Spain, thanks to a diplomatic strategy aimed at projecting neutrality, in which the Falangist ideologies had no place. Second, the impoverished post-war economy did not contribute to the artistic excesses proposed by some state bodies, usually expressed in magnificent urban and architectural works. Third, and significantly, the deep traditionalist nature of Spanish fascism built certain historical works and ensembles that already

[17] BORRÁS, Tomás (1943): 'Conjectures on Plastic Arts', Spanish newspaper *El Español*, No. 10, 16 January 1943, p. 8. Reproduced in Díaz Sánchez and Llorente (2004): 195

existed in the Spanish heritage.[18] Mª Isabel Cabrera has already demonstrated that the historicist nature of Spanish fascism is much stronger than that of the rest of the European totalitarian regimes.[19] There is no better expression of the cultural legacy of tradition than the words of the minister of national education, José Ibáñez Martín, in 1941: 'To give back to Spain the sense of its responsibility to the world, all Spanish had to be reminded of the imperial value of our classical culture and the universalist sense of Spanish Catholic thought.'[20]

How was this 'responsibility to the world' that Spanish culture assumed from 1945 channelled? Which steps were taken to merge post-war cultural diplomacy, the emergence of avant-garde movements, its official acceptance, and the role of tradition that was so precious for the dictatorship? It is now time to unfold the third stage of the process of the nationalisation of culture, the stage that has to do with the post-war avant-garde that we intend to describe here.

Tradition and De-ideologisation

How was it possible for the most important avant-garde artists and critics of the Spanish post-war period, most of them opposed to the dictatorship, to openly collaborate with the regime for the implementation of its cultural policies? This question, which was openly disregarded by Spanish historiography until relatively recently, creates such an accumulation of paradoxes, disappointments, and twists that it can certainly be considered the 'black hole' of the history of art that was written in Spain. We will not address here the important role that the definition of culture by public authorities and the world of art had on that issue during the Transition and the 1980s; they constructed a de-ideologicalised narrative of the cultural expressions produced under the dictatorship to generate a new value of consensus in the heat of a fragile democracy.[21] Instead, we will focus on the process that enabled an unprecedented merging of the avant-garde and (Francoist) nationalism, especially throughout the 1950s, thanks to a peculiar understanding of aesthetic memory and transcendence.

[18] Ureña (1982): 23
[19] Cabrera (1999): 199
[20] Álvarez Casado (2002): 49
[21] Marzo (2010b)

On 18 July 1945, after the defeat of the European fascisms, Franco changed the leadership of the government, replacing the most exalted Falangists with the 'liberal' Falangists. In December, the Council of the *Hispanidad* was refounded, converted into the ICH under the Ministry of Foreign Affairs, and provided with a very generous budget. The ICH was born with the intention of 'maintaining the spiritual ties between all individuals who composed the cultural community of the *Hispanidad* [...] through an instrument of cultural dialogue',[22] although in private it stated that 'it is not about creating culture, but of using the existing one as a support abroad to mobilise aid and alliances'.[23] The then minister of foreign affairs, Alberto Martín Artajo, described it as follows: 'The role of the Institute is to use all the weapons of diplomacy and its intelligence in the defence of the isolation decreed by the United Nations [...] through the hosts of thought and culture'; he also declared that the institution must vindicate 'the reason of Spain against the unreason of half the world.'[24] In fact, the function of the ICH was none other than to prepare a cultural state narrative in which the 'naturalness' of the traditional narrative of the marriage between state and culture was a guarantee of success, exploring its applicability in an international cultural environment exposed to a battle of post-war ideological narratives, kindly offering the artistic Hispanic inheritance as a symbol of durability and good neighbourhood. The ICH was a school of cultural propaganda in which some of the most relevant cultural managers of the regime were trained, such as the following: Manuel Fraga Iribarne, secretary of the Institute in 1951; Luis González Robles, art commissioner, affiliated to the General Directorate of Cultural Relations; or Carlos Robles Piquer, director of the University Section.

Why is the role of the ICH relevant in the history of communion between the avant-garde and Spanish nationalism that is being addressed here? Because it was the main instrument used by the regime to convert the fascist enthusiasm, hastily abandoned after the end of World War II, into a national-existentialist discourse, which would erase the strong 'exceptionalist' features of the regime, projecting a cultural normality that could somehow be homologated abroad. The ICH produced a narrative of authenticity, anti-communist, culturalist, and anti-materialist, that would sustain its value as a cultural, spiritual, and ecumenical power, with

[22] Law of 31 December 1945, Official State Gazette, 2 January 1946
[23] Delgado Gómez-Escalonilla (1992): 461
[24] *Ibid*: 458–459

its sights set on two diplomatic fronts: the first, the European countries with the greatest Catholic roots (Italy, France, and Belgium) and Latin America; the second, the United States, from whose geostrategic and anti-communist interests in the emerging Cold War Spain hoped to obtain returns.

One of the main arguments of the ICH was the capacity of the state to redirect the same traditions on which Spanish Falangism was based, but now towards the interior of the individual, towards existentialism, without appeals to structural and collective changes. The dictatorship began a process of social depoliticisation based on precisely the same Falangist arguments that it now intended to abandon: that styles pass over Spain without affecting it; that the style of Spain is itself; that Spain, despite the modernisation that it is pursuing, continues to be naturally and naively counter-modern, understanding this adjective as the nation's desire not to abandon a certain humanist anti-utilitarianism with religious roots. With no risk of exaggeration, we could describe it as a process of 'trans-fascisation' by maintaining the basic concepts of the traditional nationalist discourse but located in a naturalistic and culturalist narrative that appears normal.

The promotion of that difference had much to do with the new opposition against the political aspects of the modern in art during the Cold War: how to merge nationalism, freedom (or its absence), and style in the expressionist path of abstraction as the supreme Western means against the fascist and Stalinist totalitarianisms that promoted spiritual automatism, or against the commercial product seen by this apparent humanism as a mere leash of banality? Here, it should be borne in mind the importance of the international debate on the constitution of national styles based on ideological affiliations, a dispute led by the United States but in which a good part of the European countries participated.[25]

While in France, the informalist movement which emerged in the early 1950s advocated a radical surrealist and expressionist shift that focused on the volcanic manifestation of existentialist subjectivity to cope with the horror of the recent war tragedy, in Germany an abstract expressionism was cultivated to not only claim the full creative freedom annihilated during Nazism but also offer the individual path as the only way to overcome collective guilt, without forgetting the post-surrealist experiences led by groups of artists in Amsterdam, Copenhagen, and Brussels. In the United States, on the other hand, the debate was conducted under a notable

[25] See Guilbaut (2007)

ideological imprint, in which the government played a prominent role. American Abstract Expressionism was presented as clear evidence of the desire to do and of the communicative and unmasking desire of the radical and liberating individualism that post-war America intended to represent against the stylistic and political guidelines present in the communist bloc. The works of critics such as Clement Greenberg and Harold Rosenberg and artists such as Jackson Pollock, Mark Rothko, Franz Kline, Clyfford Still, Robert Motherwell, Willem de Kooning, and Mark Tobey were presented as a genuine expression of the historical momentum of the American empire through the 'moral embodiment' of a new global Enlightenment project, based on 'social individualism', symbolic of the marriage between democracy and capitalism. Thus, democracy and art were rescued from the ruins of Europe by the United States, which offered them to the world as the supreme form of free expression. Alfred d'Harnoncourt, director of the Museum of Modern Art in New York (MOMA), a central institution in the construction of that narrative, defended the American *civilitas* in 1948 in the following terms: 'a society incredibly enriched thanks to the full development of the individual for the sake of society as a whole. I believe that the ideal name of that society is democracy, and I also believe that modern art, in its infinite variety and in its incessant exploration, is its most outstanding symbol'.[26]

The US government soon embraced that conceptual scaffolding. Through international itinerant exhibition programmes, conferences, and publications, American Abstract Expressionism unfolded as a cultural Marshall Plan during the 1950s and 1960s.[27] The Spanish authorities, eager to abandon the autarchy of the 1940s and prone to the use of cultural diplomacy as an instrument of international alliances, immediately understood the advantages of embracing the American programmes. Therefore, for example, on 27 September 1955, the Promotion of the Decorative Arts (*Fomento de Artes Decorativas*—FAD) in Barcelona hosted a lecture by d'Harnoncourt, the aforementioned director of MOMA, who, along with other members of the museum, such as Alfred Barr (founding director), Porter A. McCray (director of the itinerant exhibi-

[26] D'HARNONCOURT, René (1948): 'Challenge and Promise: Modern Art and Modern Society', lecture read at the Annual Meeting of The American Federation of Arts. Quoted in Guilbaut (2007): 340

[27] For the cultural programmes and strategies of the US government and its main art institutions during the Cold War, see Guilbaut (2007); STONOR SAUNDERS, Frances (2013): 'The CIA and the Cultural Cold War', Barcelona, Debate.

tions department), and Frank O'Hara (chief curator), became notorious figures in the relations that were established between Franco's government and the American artistic circles. The argument of d'Harnoncourt's lecture in Barcelona could not be more appropriate: the construction of tradition and artistic identity in the context of a universal program: "We should not be afraid of what we are told that universal tendencies annul the personality."[28] D'Harnoncourt raised the issue of the *Kunstwollen*, whereby the desire of the artist (their 'blessed individuality') must lead the interpretation of national identities, an idea that would quickly find its place in the numerous texts promoted by the ICH. It was precisely in 1955 that one of the star exhibitions of the American cultural programme, *Modern Art in the United States*, made its appearance in Barcelona. The event *A selection of the collections of the Museum of Modern Art in New York*, within the framework of the III Hispanic-American Art Biennial produced by the ICH, confirmed the intense relations of cultural diplomacy between the governments of Spain and the United States. It began with the opening of the Institute of North American Studies in Barcelona in 1951, and culminated with the opening, in 1960 in New York, of the exhibitions *Before Picasso, After Miró* (Guggenheim Museum), and *New Spanish Painting and Sculpture* (MOMA).

We started this text with surrealism, and the process that led to its de-ideologisation in the hands of the new managers and critics who were the protagonists of the Spanish art circles in the late 1940s and early 1950s. The influence of the international debate on the de-ideologisation of creative practices did indeed have a significant influence in that process. If the new international styles were promoted from the perspective of depoliticisation as an instrument of a cultural policy typical of liberal, individualistic, capitalist, and 'benign' societies (as opposed to the communist agenda), surrealism and abstraction promoted by the ICH and its followers could be argued on essentially the same terms, as they, in fact, were. However, the coincidence of approaches was also manifested in another aspect: the need to subvert the banality of a mechanical and commercial modernity to reveal the undeniable power of the individual, presented in the context of telluric vocation and destiny. Moreover, there was nothing like the term 'destiny' to support the essence of the national and its cultural memory.

[28] Conference reproduced in *La Vanguardia Española*, Barcelona, 28 September 1955, p. 17; and in the Bulletin of the Promotion of the Decorative Arts, IV quarter, Barcelona, 1955, p. 96.

This doctrine should not be understood as an underground current that we can only perceive as time has passed, but as part of the official manifestations of the regime. In the opening speech of the First Hispano-American Art Biennial, held in Madrid in 1951, the Minister of Education, Joaquín Ruiz Giménez, declared, in Franco's presence, that "radical individual vocation", "spiritual transcendence", "universality", and "historical sense" bring together the various elements of the equation that allowed the regime and the avant-garde to merge.[29] In this, the role of the ICH was central and Felipe Vivanco was one of the first to start working on this merger. In 1940, he wrote: 'Rejected the isms and discarded the art of the nineteenth century, the reference for our century will undoubtedly be the painting of the seventeenth century, whose greatness lies in the fact that the artist maintains an attitude of "service" to the themes proposed by the spirit'[30]. Luis González Robles, the official commissioner of the regime, begins to identify the work of Tàpies with the essential characteristics of 'Spanishness': 'an ethical attitude towards life and a mystical vision of the world, the aridity and austerity of the lands of Spain and the realism, the textures of the earth, the dark colours and the faint hues of the Spanish artistic tradition'.[31] Simultaneously, the informalist critics unfold a whole range of assumptions by which the new style rises thanks to its counter-modernity, its existential anti-utilitarianism in the wake of the Spanish tradition. For example, Alexandre Cirici already valued the appearance in 1948 of the Catalan post-surrealist group *Dau al Set* for its 'conservative and anti-modernity values', since 'they fight against the systematic destruction of the spirit that occurred in western Europe, during centuries of Renaissance pedantry, scientific progressivism, Caesarism, and utilitarianism'.[32] Here reappears the old *motto* of Ortega, Unamuno, and Chueca Goitia on the ability of the Spanish baroque culture to transgress the rationalist order of the pagan Renaissance, recovering the lost potentialities of a medieval mysticism that can combine unity, transcendence, and verticality.[33] In 1957, Joan Teixidor, who was also an art critic, pointed out

[29] The lecture of Joaquín Ruiz Giménez, entitled 'Art and Politics; Relations between Art and State', was reproduced in *El correo literario*, year II, No. 34–35, Madrid, 1 November 1951, p. 1.

[30] Díaz Sánchez (1998): 56

[31] Borja-Villel (1988): 203

[32] CIRICI PELLICER, Alexandre (1951): 'Dau al Set', *Vint-i-dos* [*Ariel*], February 1951, p. 60; quoted in Selles (2007): 75

[33] See Ortega y Gasset (1911), Chueca Goitia (1947)

that informalism merged with the baroque 'when it tears and contorts the serene surfaces of the Renaissance, to introduce this crack of anguish and doubt that the "other art" is now trying to reproduce again as a warning to men'.[34] The abstract avant-garde then recovered the traces of the Golden Age tradition to demonstrate the validity of that myth when it comes to questioning pre-war modernity. In 1959, the painter Rafael Canogar attributed to abstract painting the concrete function of 'finding again the true essences of the Spanish painting of all times'[35]; two years earlier, the artist Antonio Saura had also argued that informalism was 'a true asceticism of colour and expression, of telluric violence, under forms of cosmic and baroque synthesis', which fully coincided with the publications of the ICH, for whom informalism was 'the Spanish tradition of austerity, tenebrism, mysticism, and expressiveness of the seventeenth century'.[36]

The Baroque as a Conclusion

There are numerous references to the baroque during this period. We cannot present here even a minor part of those references that appeared in magazines, newspapers, catalogues, and academic books. To summarise, and as a sort of culmination of that whole journey, it is worth recalling the content of number 57 of the magazine *Papeles de Son Armadans*, published in 1960 and dedicated exclusively to the work of Antoni Tàpies.[37] The magazine was a literary and graphic project promoted since 1956 by the writer and later Nobel Prize winner Camilo José Cela in Palma de Mallorca, subsidised by the family of the francoist banker Juan March, and which immediately became one of the main intellectual art references at the time. The issue on Tàpies, in which important national and international figures participated, was the most orderly effort to situate the Catalan artist as the heir of a long tradition of Spanish arts. Cela, the editor, was responsible for connecting the universe of Tàpies with that of Menéndez Pidal, Azorín, and Picasso. The critic Alexandre Cirici situated Tàpies within the great Spanish baroque tradition but also within the

[34] Teixidor (1957): 44

[35] Canogar (1959): 70

[36] Saura (1957): 52

[37] *Papeles de Son Armadans*, monographic about Antoni Tàpies, Madrid-Palma de Mallorca, No. LVII, 1960

Catalan Romanesque-gothic tradition, in romanticism and in existentialism, following the ahistorical premises of the baroque established by Eugenio d'Ors.[38] Vicente Aguilera Cerni, Udo Kultermann, Friedrich Bayl, and Juan Antonio Gaya Nuño also situated Tàpies in the wake of the seventeenth-century art. Gaya Nuño even argued that the artist's baroque ancestry was such that it was not even possible to give another name to the style in which he painted: 'Three hundred years ago, pessimism used skulls and putrefactions. Now, he uses parietal surfaces, also in putrefaction. So, there is not much reason to mention any other art in relation to Tàpies' painting.' Kultermann, on the other hand, perceived his work as the necessary brake on superficiality and anecdotism.

This narrative of transcendence, vocation, and tradition acquired a progressive international homologation. James J. Sweeney, director of the Guggenheim Museum, stated in 1960 that 'the Spanish pictorial prodigy consists of the fact that when the abstract is becoming stagnant or exhausted within the founding or pioneering countries due to a lack of new talents, Spain produces them profusely thanks to the tradition that accompanies them'.[39] Simultaneously, Frank O'Hara of MOMA declared that the quintessence of Spanish informalist painting lay in its direct connection with the artistic tradition of the baroque.[40] The British critic Paul Grinke, while analysing Tàpies' exhibition at the ICA in London in 1965, alluded to the artist's concern with realist putrefaction, something 'essentially Spanish' that was associated with the tradition of the seventeenth century. Similar texts can be found everywhere, and we will not dedicate more attention to them now. Nor is it difficult to trace numerous opinions in European and American circles that were suspicious of these historicist references by abstract artists, and wondered to what extent all those works were not the result of some form of institutional propaganda or tourism marketing.[41] Thus, we will not elaborate further.

However, why the baroque? In previous works, we have already analysed the tremendous influence of the baroque narrative in the cultural construction of Spanishness and *Hispanidad*.[42] The baroque, not so much as a style, but as a visual and linguistic politics of memory—what we have

[38] See d'Ors (2002)
[39] Tusell (2002): 50
[40] O'hara (1960)
[41] Marzo (2007)
[42] Marzo (2010a)

called the baroque effect in other works[43]—will become the main argument to support a counter-modern discourse and thus link it to the essences of the nation. By considering the baroque as a meta-style, that is, a project that opposes humanism to the very positivist sequence of the history of art, Francoism would succeed in bringing various artistic programmes together. Therefore, the baroque tradition, thanks to slight tactical differences of appreciation, would serve both to build imperialist fanfares, such as the Basilica of the Holy Cross of the Valley of the Fallen, and to combine the existentialist narratives that would give rise to Informalism as early as the 1950s.

The nationalisation of the informalist avant-garde can only be understood, therefore, in the historical commitment that the regime and the artists assumed to provide themselves—each faction with rather different objectives—a common ground in which to structure culture and tradition as a solution to the problems of representativeness and representability of what was absent in the country. Tradition was exhibited by artists as a competent source of legitimacy against the lack of freedom, and as a form of survival in an ecosystem that was opposed to modernity and linked to essences; culture, in the hands of the regime, was proposed as a substitute for the political, and as a ground on which to foster a homologous appearance of well-being. There, the national agreement materialised, giving rise to a fertile collaboration that would last almost 15 years, until the new circumstances of mass culture in the 1960s and a greater awareness of the dissident role of creative practices suspended it, opening a path of disagreement by which the national art competence gave rise to greater social competence.

References

Albarrán, Juan (2019): *Disputas sobre lo contemporáneo. Arte español entre el antifranquismo y la postmodernidad*, Madrid, Exit.

Álvarez Casado, Ana Isabel (2002): *Bibliografía artística del franquismo: publicaciones periódicas, 1936-1948*, tesis doctoral, Madrid, Universidad Complutense.

Borja-Villel, Manuel (1988): "Los cambios de gusto. Tàpies y la crítica". *Tàpies. Els anys 80*. Barcelona, Ajuntament de Barcelona.

Cabrera, María Isabel (1999): *Tradición y vanguardia en el pensamiento artístico español (1939-1959)*, Granada, Universidad de Granada.

[43] Marzo and Badia (2010)

Canogar, Rafael (1959): "Tener los pies en la tierra", *Papeles de Son Armadans*, no. 37, pp. 66-70.

Chueca Goitia, Fernando (1947): *Invariantes castizos de la arquitectura española*, Madrid, Dossat.

Claret Miranda, Jaume (2006): *El atroz desmoche. La destrucción de la universidad española por el franquismo, 1936-1945*, Barcelona, Crítica.

D'ors, Eugenio (2002): *Lo barroco*, Madrid, Tecnos&Alianza.

Delgado Gómez-Escalonilla, Lorenzo (1992): *Imperio de papel. Acción cultural y política exterior durante el primer franquismo*, Madrid, CSIC.

Díaz Sánchez, Julián (1998): *La oficialización de la vanguardia artística en la posguerra española: el Informalismo en la crítica de arte y los grandes relatos*, Ciudad Real, Universidad de Castilla-La Mancha.

Díaz Sánchez, Julián, & Llorente, Ángel (2004): *La crítica de arte en España (1939-1976)*, Madrid, Istmo.

Echeverría Plazaola, Jon, & Mennekes, Friedhelm (2011): *Intrusos en la Casa. Arte Moderno, espacio sagrado*, Pamplona, Fundación Jorge Oteiza.

Guilbaut, Serge, ed. (2007): *Bajo la bomba. El jazz de la guerra de imágenes trans-atlántica, 1946-1956*. Barcelona, MACBA.

Llorente, Ángel (1995): *Arte e ideología en el franquismo (1936-1951)*, Madrid, Visor.

López Díaz, Jesús (2003): "Vivienda social y Falange: ideario y construcciones en la década de los 40", *Scripta Nova. Revista electrónica de Geografía y Ciencias Sociales*, Universidad de Barcelona, vol. VII, n° 146. URL: http://www.ub.edu/geocrit/sn/sn-146(024).htm

Marzo, Jorge Luis (2007): *Arte Moderno y Franquismo. Los orígenes conservadores de la vanguardia y de la política artística en España*, Girona, Fundació Espais.

Marzo, Jorge Luis (2010a): *La memoria administrada. El barroco y lo hispano*, Madrid, Katz.

Marzo, Jorge Luis (2010b): *¿Puedo hablarle con libertad, excelencia? Arte y poder en España desde 1950*, Murcia, CENDEAC.

Marzo, Jorge Luis, & Badia, Teresa (2010): *El d_efecto barroco. Políticas de la imagen hispana*, Barcelona, CCCB.

Marzo, Jorge Luis, & Mayayo, Patricia (2015): *Arte en España (1939-2015). Ideas, prácticas, políticas*, Madrid, Cátedra.

Nye, Joseph (2004): *Soft Power: The Means to Success in World Politics*, New York, Public Affairs.

O'hara, Frank (1960): *New Spanish Painting and Sculpture*, New York, The Museum of Modern Art.

Ortega y Gasset, José (1911): "Arte de este mundo y el otro", *El Imparcial*, Madrid.

Pérez Montfort, Ricardo (1992): *Hispanismo y Falange*, Ciudad de México, Fondo de Cultura Económica.

Saura, Antonio (1957): "Espacio y gesto", *Papeles de Son Armadans*, n° 37, Palma de Mallorca, pp. 50-54.

Selles, Narcís (2007): *Alexandre Cirici Pellicer*, Barcelona, Afers.

Teixidor, Joan (1957): "Arte otro en Barcelona", *Destino*, 16 de febrero de 1957, p. 44.

Tusell, Genoveva (2002): "La proyección exterior del arte abstracto español en tiempos del grupo El Paso", *En el tiempo de El Paso*, Madrid, Centro Cultural de la Villa, pp. 87-117.

Tusell, Javier (1999): *Arte, historia y política en España (1890-1939)*, Madrid, Biblioteca Nueva.

Unamuno, Miguel de (1911): "Sobre la tumba de Costa", *Nuestro Tiempo*, Madrid, n° 147, pp. 325-336.

Ureña, Gabriel (1982): *Las vanguardias artísticas en la posguerra española. 1940-1959*, Madrid, Istmo.

PART V

Internationalization of Science and Culture in the Franco Regime

CHAPTER 14

French Hispanism and Spanish Cultural Diplomacy during the Franco Regime

Antonio Niño

In this chapter, we will discuss the difficulties encountered by the Francoist authorities to rebuild a collaboration with French Hispanism, a fundamental university movement since the early twentieth century which aimed to recover the prestige of Spanish culture in Europe.

The foundations of the treaty with the French Hispanism were set from the outset on an implicit, mutually beneficial pact, whereby each party found support in the other for its own purposes: the leaders of Hispanism could count on the enthusiastic collaboration of the Spanish intellectual elites in their crusade to extend Spanish language teaching in France, firmly establishing their discipline in the university world. The Spanish authorities and academics were very interested in fighting the negative stereotypes about Spain persisting in French public opinion and, by extension, in Europe, a task in which Hispanists were a fundamental strategic partner. Both aims were complementary because improving the prestige

A. Niño (✉)
Universidad Complutense de Madrid, Madrid, Spain
e-mail: anino@ucm.es

M. Janué i Miret, A. Presas i Puig (eds.), *Science, Culture and National Identity in Francoist Spain, 1939–1959*,
https://doi.org/10.1007/978-3-030-58646-1_14

of the Spanish culture in France would mean an improvement of the position of Spanish as a foreign language in the French education system, and the extent of its teaching would increase the recognition and dissemination of the Spanish cultural heritage.[1]

That basic understanding explains the creation and development, in the first decades of the twentieth century, of several institutions specifically dedicated to fostering academic collaboration between the two countries, promoting the exchange of researchers, professors, and artists. For the French initiative, *Institut Français* had been created in 1912, and *Casa de Velázquez* had started its activities in 1928, both in Madrid.[2] For the Spanish initiative, with the agreement of French authorities, *Institut d'Études Hispaniques* (IEH, for its acronym in French) had been created in 1917, and *Colegio de España* opened its doors in 1934, both in Paris.[3] These institutions enjoyed the official support of the two countries and were specifically dedicated to mutual knowledge and cultural exchange.

The best expression of that implicit pact is found in the creation of the IEH in Paris, in 1917, from a previous *Centre d'Études Franco-Hispaniques* that had been operating in the Sorbonne since 1912. The Spanish authorities put great interest in this institution dedicated to encouraging Spanish culture in a city serving as an intellectual showcase of the European continent, relying on a university as representative as that of Paris. This support also intended to exercise some control over the interpretative discourse circulating in the field of Hispanic studies, which was important for portraying a good image of the country abroad. For the Spanish state, it was the first important initiative in Europe of what we now call cultural diplomacy (on this concept, see the contribution of Janué in this same volume).

The IEH was designed from the beginning as a binational centre, equally dividing the positions of the board of directors between Spanish and French members, with the Spanish ambassador in Paris as the honorary president. The subsidy granted by the Spanish government ensured the participation of Spanish professors and academics to develop the activities, and the Board for Extension of Studies collaborated by selecting the

[1] A detailed analysis of the strategies of French Hispanism and its collaboration with the Spanish authorities can be found in Niño (1988) and Niño (2017).

[2] The history of both institutions can be followed in the works of Delaunay (1994) and Niño (1988).

[3] The history of the IEH can be followed in Niño (2017).

lecturers, professors, and speakers responsible for disseminating an adequate image of Spanish literature, history, and the current situation.

The aims of the centre, stated in its statutes, made up a simple and synthetic programme, representative of the coincidences between the Spanish intellectual elites and the leaders of the expanding French Hispanism:

1st, "to bring together Spanish and French efforts to extend relations and strengthen the bonds of friendship between the intellectuals of the two countries;
2nd, to promote the study of French culture in Spain and the knowledge of Spanish life in France; to assist Spanish academics in fulfilling their mission, providing information, advice, and effective support.
3rd, to extend by all available means Hispanic studies in France".[4]

That programme was implemented to the satisfaction of both parties in the interwar period, with especially brilliant results while the Republican experience lasted.[5] However, the Civil War (1936–39) abruptly interrupted academic exchanges and had devastating effects on some of these institutions. *Casa de Velázquez* was bombed, as it was at the battlefront during the two-and-a-half year-long siege of Madrid. The facilities of *Colegio de España* in Paris were destroyed when they were used as headquarters for German, and later American, troops during World War II. When the cycle of war and destruction ceased and the new political systems stabilized in both countries—Franco dictatorship and the IV French Republic—attempts to re-establish the previous collaboration were strongly influenced by the sequels and ideological clashes inherited from a decade of war and political cataclysms in both countries.

To understand the efforts made by the Francoist authorities to resume collaboration with French Hispanism, the attitude this corporation adopted towards the Spanish regime, and the enormous difficulties encountered, we will consider the two institutions that coincided in Paris and shared a certain binational status: the IEH and *Colegio de España*, both nominally dependent on the Sorbonne but supported by funds provided by the Spanish government through the Cultural Relations Board of the Ministry of Foreign Affairs.

[4] *Inauguración de conferencias del CEF-H,* 1913, p.12.

[5] The list of Spanish professors and speakers who participated in the IEH between 1930 and 1936 can be found in Niño (2017): 82–84 and s.a. (1939).

University Hispanism before the Spanish Civil War

With the outbreak of the Spanish Civil War, the IEH at the Sorbonne suspended the collaboration of Spanish intellectuals and professors, which had been, since its inception, its own reason for being. In the absence of the lectures given by Spanish professors, the institute used, as much as possible, French and Latin American professors, in addition to some diplomatic representatives, who spoke about the evolution of the Castilian language and civilization in New World republics. Any reference to the conflict in the neighbouring country and to the contemporary Spanish culture itself were excluded from their activities. That strict neutrality, consistent with the policy of non-intervention adopted by the French government, was followed not only by the IEH but also by *Colegio de España* in Paris. Its director, Angel Establier, was dismissed by the government of Madrid for not collaborating with the political and propaganda activities of the Republican embassy but remained in office with the support of the Rector of the Sorbonne. The *Colegio* stopped receiving official funds but remained open thanks to a donation of 50,000 francs granted in November 1938 by the Catalan nationalist leader Francesc Cambó, with the approval of the Francoist authorities, and served as a refuge for some intellectuals escaping the war in Spain.[6]

The IEH also maintained equidistant neutrality, this time dictated by the university authorities and consistent with the official policy of its government. When the Republican embassy, directed since November 1936 by Luis Araquistáin, tried to resume collaboration with the IEH, sending academics of proven loyalty to the Republican cause, the Sorbonne authorities strongly rejected the proposal. At that time, it was reasonable to distrust the intentions of the offer by the Spanish ambassador as they were rightly afraid that the IEH would be used for propaganda purposes by the Republic.

The director of the IEH, Ernest Martinenche, justified the scrupulous neutrality of the institution by appealing to a mission above the political struggles of the moment:

> Like the rest of the University Institutes, ours was forbidden to get embroiled in matters of internal policy. In a civil war that was a real nightmare for us, we could do nothing but wish for the end of that tragedy, although nothing

[6] Among them Blas Cabrera, Xavier Zubiri, Puig i Cadafalch, and Pío Baroja.

> could prevent us from reaffirming our faith in eternal Spain, stressing, more than ever, the eminent services the country has offered to humanity in general and France in particular.[7]

However, the IEH was a binational institution that regularly received funds from the Spanish government "as a subsidy for the Spanish cultural propaganda it performs". These funds covered the costs of Spanish university faculty invited every year to the courses and conferences organized there. Even in the fall of 1936, the IEH received a subsidy of 2500 gold pesetas that the Spanish Ministry of State had provided for the third quarter of that year. For this reason, the Republican authorities felt they had the right to request reports from the IEH leadership "about the current disposition of Institute of Hispanic Studies of Paris towards the Government of the Republic and about the activity developed the year just ended" and denounced "the obvious disaffection it shows towards the cause of legal Spain".[8]

Equally representative of that neutrality and equidistance was the attitude followed by Professor Aurelio Viñas, deputy director of the institute, who also served since 1933 as cultural attaché of the Spanish embassy in Paris. Viñas did not take sides in favour of any group, except "that of the Spanish university professors of both sides. He even won the exceptional esteem professed by the enemy brothers of the past".[9]

This abstentionism of the institute not only triggered the protests of the Republican government, but also resulted in the Franco regime not renewing any fund allocation to the IEH at the end of the war. Joan Estelrich (on Estelrich, see Sesma's contribution in this same volume), in a report he wrote to support his candidacy for the position of cultural attaché at the Spanish embassy in Paris, after analysing the behaviour of official Hispanism towards "national Spain", concluded, "The Spanish State should not contribute with financial support to the Institute's expenses, except when ensuring the effective influence in that organization and the control of at least the conferences given by Spanish intellectuals".[10]

[7] Martinenche (1939): 279.

[8] AN, 20010498/183. Letter from the Spanish embassy of January 19, 1939.

[9] AGA, 66/4336, "Nécrologie d'Aurelio Viñas", by Charles V. Aubrun. From 1939 to 1945, he also remained at the IEH, sharing the miseries of his French colleagues.

[10] AMAE, 1.380/24, Joan Estelrich: "Notes on French-Spanish cultural relations", April 11, 1939, p. 4. The author of this report had been a representative for the Catalan Lliga, a

Apart from official attitudes, the Spanish tragedy triggered very different reactions in French Hispanist corporations. Most maintained an attitude of strict neutrality in the conflict, lamenting the bloodbath in a country with which they maintained strong emotional ties. This compassionate attitude was tempered, depending on the case, with more or less sympathy leaning towards one side or the other. Among collaborators of the IEH, there was a clear tendency to show solidarity with the Republican cause, although few university students publicly showed their support for any of the combatants. Many academics and members of the *Institut* and influential personalities from the French intellectual world of conservative orientation or from the extreme right had manifested themselves in favour of "national" Spain, but not a single university Hispanist did so. That was precisely what the person responsible for Francoist propaganda services in France criticized:

> In the two-and-a-half years of war, the hostility against traditional Spain by all, or almost all, French university faculty in collaboration and friendship with the directors of the Institute has been revealed; none of the hundreds of academics, university teachers, and writers who have expressed their sympathy for national Spain belongs to the groups arising from the Institute.[11]

Some prominent Hispanists also showed a more committed attitude towards the Spanish conflict because of their solidarity with exiled Spanish professors. Marcel Bataillon, who was elected professor of the Sorbonne in the fall of 1937, actively took part in supporting Spanish intellectuals exiled in France, although his radical pacifism prevented him from taking public initiatives in favour of the Spanish Republican government. In a letter to his friend Jean Baruzzi, dated December 13, 1938, Bataillon wrote:

> Martinenche is unwavering in his absurd "neutrality", which consists of not listening to any Spaniard in the [Hispanic] Institute and, above all, of not inviting any so long as the war lasts. With that ease of discarding what bothers him, he suggests heading to *Collège de France*, "which is much freer"…I think it would be embarrassing not to have a humanitarian attitude towards

trusted man of Francesc Cambó, founder of the Francoist Propaganda and Press Office in Paris and director of the *Occident* magazine. The report was sent to the Spanish Ministry of Foreign Affairs shortly after the establishment of diplomatic relations between the government of Burgos and France. Vine. Moreno Cantano (2007).

[11] AMAE, 1.380/24, Joan Estelrich: "Notes on relationships…", doc. cit., p. 2.

Dámaso Alonso [a well-known Spanish philologist and poet]; in addition, it would be an act of Franco-Spanish friendship of great value, whatever the result of the events in Spain.[12]

Since 1938, Bataillon was actively involved in helping those professors, his fellow professionals and friends in many cases but did so outside the university structures such as the IEH.[13] The institutions of official Hispanism turned their gaze away from the Spanish tragedy and, what is more significant, the personal drama of many of their former visitors and collaborators.

The Long Post-War Isolation

At the end of the world war, the Franco regime remained the last vestige of fascism in Europe and was subjected to severe international isolation, which also affected the links with international Hispanism (regarding this issue, see the contribution by Janué in this same volume). Bilateral relations with the IV French Republic went through initial moments of great tension—including the closure of the border—while the Spanish authorities feared the possible actions of the Republicans exiled in France, who had great support in the public opinion of that country.

Most in French Hispanism also repudiated the Spanish dictatorial regime, severed ties with the Peninsula, and continued studying only classical Spanish culture, while disdaining the cultural production of Spain under Franco. "Our availabilities in the circle of official university Hispanism are certainly poor, if not miserable",[14] Estelrich recognized. Only a few French Hispanists, but very substantial, continued supporting the Franco regime. Its main exponent was Maurice Legendre, "a strong traditional catholic… a person of all confidence for Spain",[15] but he was

[12] Bataillon, C. (2014): 137.

[13] AGA, 19.924/14. Burgos, June 2, 1938: Note from the Military Police and Information Service to the Minister of National Education reporting on the founding of a Cervantes Institute (or Athenaeum) in Paris to bring together Spanish intellectuals dwelling in that city "whose purpose will be to foster relations between French and Spanish intellectuals. Of course, that foundation is inspired by the reds because reds are all those who sign the call, among which are Cassou, Bataillon, president of the Hispanic Institute, and professors Languevin and Wallon".

[14] AMAE, 1.380/24, Joan Estelrich: "Notes on relationships…", doc. cit. p. 21.

[15] *Ibid.*

more a publicist than a university Hispanist. The favour of the Francoist authorities allowed him to remain in charge of *Casa de Velázquez* in the post-war period, but in that period the institution was limited to surviving in a leased place while the monumental building of the University City remained in ruins, until its reopening in 1959. Robert Ricard, professor of Portuguese studies who had served as general director of education in Morocco during the Vichy regime, also showed his support. These two figures, characterized by their deep conservative Catholicism, would be the most important contacts of the Spanish authorities with French Hispanism, but Maurice Legendre died in 1955, before *Casa de Velázquez* regained its former splendour, and Ricard, nominally the IEH co-director since 1953, delegated at all times the responsibility of the position to Charles V. Aubrun.

During the extensive belic period, the IEH maintained its teaching activity, replacing the subsidy it previously received from the Spanish government with another from *Service des Œuvres Françaises à l'Étranger* under the French Ministry of Foreign Affairs. Bataillon was succeeded as director in 1946 by Gaspar Delpy, "a person of trust", according to Estelrich, with whose rise ""Spanish interest loses nothing in it".[16] However, he did not win either because the collaboration with the Spanish government did not resume during his tenure given the bad relations between the two governments in those years.

In the absence of cooperation with the Spanish organizations, the IEH completed its programme of activities with lecturers who came from Latin American or US universities. The increasingly Latin American orientation, to the detriment of previous Iberism, indicated the disconnection—because of political circumstances—taking place between French Hispanism and the Spanish academic and cultural environment at the time (Janué also addresses aspects related to the Latin-Americanism/Iberism binomial in his contribution in this same volume). An example of this estrangement occurred in 1947 when, in the middle of the diplomatic crisis between the two countries and with the border closed, the IEH organized a course full of symbolism dedicated to analysing the end of Spanish domination in America. The speakers were Latin American ambassadors and personalities such as Ricardo Levene, rector of the University of La Plata, who talked about the great liberators of Spanish America. This

[16] *Ibid*.

Latin American orientation also adopted the perspective most feared by the Francoist authorities described by Establier a few years earlier:

> There are two opposite ways to conceive the intellectual relations between France and the Spanish-American republics (and the same could be said regarding Spain): either in the sense of recognizing Spain for its traditional role, civilizing the New World and source of the best American tradition, or in the sense of the intended "*idées françaises*" of the eighteenth century, as a source of progress and civilization. It is unnecessary to say how convenient it is to restore or reinforce the first concept and combat the second, both in France itself, for which there is no lack of illustrious and influential allies, as in America.[17]

The only sign of continuity with the previous period was the Deputy Director of the IEH, Aurelio Viñas, "the very faithful and effective servant of the University of Paris for thirty-five years".[18] Specialist in the reign of Philip II, he was to help *Hispanisants* who wanted to orient themselves towards historical studies. He was a great traveller, many universities had invited him, and he maintained a network of friendships linking the IEH to world Hispanism. An example of his interpersonal skills was the unsuccessful attempt to mediate the historiographic conflict between Américo Castro and Claudio Sánchez Albornoz. The kind spirit of Viñas allowed him not only to maintain good relations with the Spanish intellectuals from both sides but to serve with loyalty to Bataillon, during his time as director of the IEH, while enjoying the confidence of José Félix de Lequerica, ambassador of the Franco regime in Paris.[19]

The isolation imposed on the Francoist administration is well represented by the attitude of the two most influential French Hispanists then, Bataillon and Jean Sarrailh, who, from their dominant academic positions, maintained ties with the Spanish intellectuals in exile and strongly promoted Latin American studies as an alternative to Hispanism with its former exclusive peninsular orientation.

[17] *Ibid*, p. 22.

[18] AGA, 66/4336, "Nécrologie d'Aurelio Viñas", by Charles V. Aubrun.

[19] AN, 200104987183, letter from Marcel Bataillon to the rector of the University of Paris, dated July 6, 1944, requesting compensation for Viñas as director of a University Institute. See Viñas death memorial on *ABC*, 02/11/1958, where his position in the Sorbonne was described as "an advanced position in the defense and dissemination of all our intellectual values".

Sarrailh held the position of rector of the Academy of Paris between 1947 and 1961, when the power of a rector was enormous, as it covered both the university and middle school. His intervention was decisive to favour the Republican refugees in France because, thanks to his influence, many of them were able to occupy Spanish teaching positions in the lyceums of the capital. Sarrailh had worked closely with the Centre for Historical Studies team of the JAE while working at *Institut Français* in Madrid. Later, he would write a monumental work dedicated to the enlightened Spaniards of the eighteenth century where he would portray the background of that great intellectual renewal movement he himself knew during his stay in Spain during the Silver Age of the Spanish culture. During the German occupation, he took part in the resistance from his position in Toulouse and did not hide his sympathies for the Republican side and his repudiation of the Franco regime.

As rector of Paris, he was determined to maintain frozen relations with the new Spanish regime. On March 16, 1953, and on the initiative of Sarrailh himself, *Conseil de l'Université* amended the statutes of the IEH to abolish the six Spanish members provisioned for its board of directors. As those had to be chosen—consistent with the 1928 regulation—according to the Spanish embassy, it seemed the best thing was to do without them, not to have to seek the opinion of the Spanish authorities. It was a symbolic measure without practical consequences because the board of directors had never met, nor had its members been renewed since its inception.

Another measure showing the principalship's little interest in fostering relations with Spanish institutions was the decision to promote "Latin Americanist" studies, as an alternative to more traditional Iberian Hispanism. In 1952, at the initiative of the rector himself, *Conseil de l'Université* created *Institut des Hautes Études de l'Amérique Latine*, which would absorb some activities and subjects taught until then at the IEH.[20] It had the mission of "putting together the activities relating to Latin American civilisations", acting as a "centre for documentation, teaching, and research".[21] A geographer, Pierre Monbeig, was placed at the head of the new institute, another sign of the marginalization of classical Hispanism. As a multidisciplinary institute, it was a novelty in France to replace the classic notion of specialization in disciplines with specialization according

[20] Sarrailh (1956): 5–8.
[21] "*L'Institut des Hautes Études de l'Amérique Latine…*", s.a. (1964): 197.

to geographical and cultural areas, something then widespread in American universities. It was also an occasion to recognize the autonomous moral personality of Latin America, defined as a geographical unit but also a cultural and spiritual unit.

Bataillon, who had become the great patriarch of French Hispanism for his prestige and position in the great centres of academic decision, also contributed to the new Latin American orientation since his great tour of that continent in 1948. He then began his studies on Las Casas and other Americanist themes. However, discovering New World Hispanism also meant meeting most of the exiled Spanish intellectuals. That was what Bataillon did during his journey, angering the diplomatic representatives of the Franco regime.[22] The controversy between Bataillon and Castiella, then Spanish ambassador to Lima, was significant, especially following an interview the former granted "to an old Spanish red, a deserter of our nationality". The interview was published in the Peruvian newspaper *La Tribuna* and referred to the spiritual wasteland of that "political, religious, and spiritual dictatorship" that ruled in Spain. Castiella felt obliged to come out in defence "of today's Spain and the truth" before the "absurd and far-fetched" statements of "Erasmist Marcel Bataillon".[23] In the exchange of letters that followed, Bataillon told the ambassador:

> as a Hispanized Frenchman and Hispanist who suffers almost as much as the Spanish exiles of physical deprivation from Spain, I aspire to the day when I can return to Spain, invited or not by some university, to enjoy friendship and spiritual exchange with the noblest people in the world without fear of being watched, filed, and used for propaganda purposes.[24]

Castiella responded with a long letter in which he spoke not only as an ambassador but also as a professor at the Spanish university, director of the *Journal of Political Studies*, and even as a former Sorbonne student:

> Don't you think the best that foreigners could do—especially the neighbours—is to leave us alone? All we ask is a little objectivity and impartiality. Unless France wants to continue intervening in the domestic affairs of Spain

[22] See the controversy between Bataillon and the Spanish ambassador in Lima, Fernando María Castiella, in August 1948, reproduced in Bataillon, C. (2014): 270–276.

[23] The reaction and words used by Castiella in the dossier conserved in the AMAE, R.4922/21.

[24] *Ibid.*

(...) I see you—so loving of my homeland, for which you have unforgettable kind words—full of prejudices and misgivings.

This was the tone used by two excellent representatives of the culture of both countries in 1948. Naturally, Bataillon was vetoed by the Spanish authorities for subsequent openings, for example when the rector of Paris proposed for him to be part of the reconstituted board of directors of *Colegio de España*.

Restoring University Cooperation

The re-establishment of relations between French Hispanism and the Spanish authorities was, therefore, slow and difficult, the result of the efforts of the less belligerent Hispanists with the Franco dictatorship and of the latter's interest in having a significant presence again in what was considered "the Mecca of international Hispanism".

Although, as we said, the Sorbonne authorities were not willing to resume university cooperation with the representatives of a despised dictatorship, the IEH co-directors wished to provide their centre with courses given by specialists from outside. With their own resources, they had temporarily hired professors such as Francisco Ayala, Waldo Ross, and Gerardo Diego. However, the creation of the Latin Americanist Institute felt like a dangerous competition, and to address it, the support of the Spanish government was needed. The IEH was experiencing extraordinary growth: when Aubrun occupied the director's post, in 1953, the centre had two professors and four assistants, in addition to Viñas and a Spanish lecturer; when he left the position in 1972, there were 6 professors and 60 *maîtres de conférences* or assistants. Student growth in those years was also spectacular: 100 students enrolled through the 1940s, but enrolment reached 800 in 1952, 1500 in 1955, and 2300 in 1962.[25]

The direction of the IEH was nominally shared between Ricard, the privileged interlocutor of the Spanish embassy, and Aubrun, the true great patron of the IEH, because of the disinterest of the former towards bureaucratic tasks. Thanks to his gift for languages—he was fluent in German and English, in addition to Spanish—Aubrun was the representative of French Hispanism in many countries, particularly in Germany, England, and the United States. He represented the University of Paris at

[25] See Niño (2017): 107. Aubrun (1963a), Aubrun (1963b) and Aubrun (1991).

the first international Hispanic congress held in Oxford and was regularly invited by Columbia University, Tulane in New Orleans, and other American universities, where he spent a quarter every year teaching courses. Aubrun had been personally touched by the time spent as a lecturer at the University of Murcia between 1927 and 1928, where he established a solid friendship with the Murcian poet Jorge Guillén and other professors of that university.

Aubrun was in charge of resuming cooperation with the Spanish authorities. His first aim was to replace Aurelio Viñas after his death in 1958. Aubrun negotiated the way to replace him, getting the Spanish government to renew the bonus granted to a Spanish teacher with a permanent position at the IEH. The negotiation was difficult, and agreeing became complicated. The cultural attaché of the Spanish embassy explained the difficulties because of the ideological positions of the Sorbonne authorities:

> Politically, there are two trends in the Institute. His (Ricard), which we could call entirely orthodox, and that of his immediate collaborator and a great Hispanist, Mr Aubrun, which, without being heretical, we could consider "Erasmist", if you allow me the anachronism. In the cartography of Hispanism, there is, of course, another politically extreme position, represented by rector Sarrailh himself.[26]

Interestingly, although Aubrun's personality was not the most appreciated in the Spanish embassy, he had the advantage of serving as a mediator against an even more hostile rector:

> I want to note, although Ricard is a man with whom we will never find problems, Aubrun is more experienced and reserved, and 4-member meetings will, therefore, be nuanced by a slight "game" on both sides. Personally, I do not see any obstacle or inconvenience in it, and I am almost pleased this is so because the presence of Aubrun is in some way equivalent to a tacit "*nihil obstat*" by rector Sarrailh, which, without this circumstance, would have always been "difficult to deal with". Naturally, as the candidates will be chosen jointly, proposed to you, and generously—and politically—paid by

[26] AGA. 66/4336. Previous cultural advisor, José Luis Messía, letter dated March 6, 1958, to the director general of Cultural Relations.

Spain, it is good to have all the local "*agréments*" set in the assurance no one will sneak in through the hatch.[27]

After arduous negotiations and several resignations—Carlos Clavería, Eugenio Asensio, and Zamora Vicente were proposed as candidates—the professor of the University of Murcia, Manuel Muñoz Cortés, philologist and disciple of Dámaso Alonso, was elected for the position, from October 1960. For Aubrun, it was a way of returning the favour to a university that had received him as a lecturer in his time of formation and with which he maintained close ties. For the Spanish cultural attaché, the hiring was safe because he was a "person of absolute fidelity to the regime". Viñas' replacement would not be officially designated deputy director, a purely honorary position, but would have all its powers. The cultural attaché recounted the negotiation with Aubrun this way:

> Regarding the main goal of correctly guiding this French Hispanism, so abundant, so valuable, but—as you say very well—so prone to the cartoon, to adapt us to its "clichés" and to confuse appearance with essence, the endeavour is not a task to achieve in one day or one year, and I still fear it can never be achieved at all, for I suspect the ultimate causes are to be sought in psychology rather than lack of knowledge. Our duty, however, is to ensure this is as complete and truthful as possible, and I believe, in this respect, the help of Muñoz Cortés can be very useful (...) as you say, this is the main stronghold to be conquered.[28]

The importance of the IEH for the Spanish cultural policy was not in doubt. In the interview with the cultural attaché, Aubrun had described the Institute as "the most important nursery of Hispanists in the world", and the previous cultural attaché explained to the authorities of Madrid that Paris has not only something of the "Holy See of the universal culture" but also of the "Mecca" of Hispanism, at least of European Hispanism.[29]

[27] AGA, 66/4336. Previous cultural advisor, José Luis Messía, letter dated January 30, 1959, to the director general of Cultural Relations.

[28] AGA, 66/4336. Letter from the cultural attaché in Paris, Rafael F. Quintanilla, to the director general of Cultural Relations, June 17, 1960.

[29] AGA, 66/4336. Letter from the cultural advisor, José Luis Messía, to the director general of Cultural Relations of March 21, 1959.

However, the broader operation was the negotiation with the Sorbonne of a new plan to exchange visiting lecturers and professors, funded by the directorate general for Cultural Relations of the Spanish Ministry of Foreign Affairs. The Spanish authorities decided in 1959 to allocate 500,000 francs a year to that plan,[30] equivalent to 50,000 pesetas; this operation sought to regain control of the formation of future Hispanists studying at the IEH. Ricard, Aubrun, the Spanish cultural attaché, Messía, and Joaquín Pérez Villanueva, then director of *Colegio de España*, took part in the negotiation. Manuel Sito Alba, director of the Spanish Library in Paris, not only participated but also took the opportunity to invite these lecturers to give speeches at his institution. The list of the five candidates who had to go to the IEH was negotiated at the beginning of each course by mutual agreement between the two parties. Aubrun insisted that these visiting professors adapt their interventions to the institute's curriculum. As a gesture of goodwill and liberality, the cycle of lectures given by visiting Spanish professors and financed by the Spanish government was inaugurated by Rafael Lapesa, who had been a lecturer during the Second Republic; however, later, the proposal professor was José María Pemán, a known Franco regime advocate.

The assessment by the Spanish embassy of that experience was very positive, and, in 1966, visits to the universities of Toulouse and Bordeaux were considered.

> The point of view of this embassy is that this is one of the most interesting and fruitful activities of our cultural action in France, as it is received in an environment uniquely prepared to assimilate them: 3000 future Spanish language and culture teachers educated by the Sorbonne without costing Spain a penny.[31]

Manuel Muñoz Cortés remained in Paris for two years in the position held by Aurelio Viñas but left to head the Spanish Institute in Munich, a better-paid and more interesting place personally because he was married to a German woman. He was succeeded in October 1962 by Pérez Villanueva, professor of history at the University of Valladolid, who had

[30] It must be understood that these correspond to the values prior to the reform of the French franc of 1960.

[31] AGA 66/4399. The cultural attaché in Paris, Rafael F. Quintanilla, to the head of Cultural Exchange of the Ministry of Foreign Affairs, April 24, 1968. The Spanish ministry allocated 125,000 Fr for each of the four professors invited each year by the IEH.

been general director of universities during the ministry of Ruiz Jiménez and lived in Paris as director of *Colegio de España* at *Cité Internationale*. Pérez Villanueva had begun teaching Spanish history courses at the IEH since 1958, thanks in this case to Ricard's intervention, but he obtained the position of a permanent and tenured professor. The Spanish embassy celebrated the appointment as a victory, aware that the IEH was the world's largest centre of Hispanism and one of the most politicized.

> I need not stress to your excellence not only the cultural but also political importance of this gesture, eloquently showing the gradual change in attitude of the Sorbonne regarding the official representatives of Spain; this will also allow focusing on issues of such historical and political significance, as stated, with the right criteria and scientific rigour characterizing Mr Pérez Villanueva.[32]

After the departure of Pérez Villanueva, because of the assault and closure of *Colegio de España* at *Cité Universitaire* in the summer of 1968, the historian José Antonio Maravall was hired—who had also been the director of *Colegio de España*—but by then, he had changed his personal ideology towards democratic liberalism. Maravall remained in the post until 1972, when the collaboration with the Spanish embassy was definitively interrupted. That year, the IEH underwent a radical transformation because of French university reform: it lost its previous autonomy and was divided between two of the new universities that succeeded the Sorbonne.

Regaining Control of *Colegio de España* in Paris

Maravall, together with Ambassador Aguirre de Cárcer and the cultural attaché Luis Díez del Corral, had been a protagonist of another diplomatic battle, which had to be fought to regain control of *Colegio de España* at the University City of Paris at the end of the 1940s. On April 11, 1941, during the Vichy regime, ambassador Lequerica had taken advantage of political circumstances to impose a modification of the statutes of the *Colegio* in favour of Spanish interests so that the members of the board of directors and the director appointed by the Spanish Republican regime could be revoked. That victory was short-lived: on December 15, 1944, with new authorities in France and by decree, the Rector of the Paris

[32] AGA 66/4336. Dispatch of Ambassador Areilza of November 16, 1961.

Academy and president of the University Council cancelled that modification and reappointed Establier as the centre's director, the same one who had been its director since its inauguration in the Republican era. That prevented, paradoxically, the occupation of *Colegio de España* by Republican exiles, as occurred with many other premises such as consulates, schools, and the Chamber of Commerce in Paris.

Since the reopening of the border in 1948 and the normalization of relations, Ambassador Aguirre de Cárcer undertook the recovery of Spanish organizations that had fallen to the enemies of the regime, which included regaining control of the *Colegio*. After Lequerica's strategy failed to change the legal status of the *Colegio* to ensure the director's appointment, Aguirre de Cárcer was inclined to declare peace with rector Sarrailh and reach a negotiated resolution to change the director of the *Colegio*. The operation required many negotiations and diverse diplomatic pressures, but the new conciliatory attitude of the Spanish authorities achieved its aim, although keeping the formalities imposed by Sarrailh.

> I had to move a lot these days to get to this result because the matter was embroiled. It has asked a lot from these gentlemen to accept a professor from Madrid as Director of a centre as prominent for their leftism as the University City, especially the Rector, reluctant until the end. But thanks to the personal intervention of your French colleague, who has behaved discreetly in the matter, his hostility has been overcome for the time being.[33]

The ambassador had to intervene, summoning the rector of the University City to protest the delay in the appointment of the new director nominated by the Spanish government and threatening reprisals:

> Then he argued against candidate Maravall certain arguments I have openly rejected by refusing to admit them. I have suggested that the procedure followed here and the difficulties posed to the exercise of our cultural action will be rigorously applied in Spain regarding many and important French cultural activities whose free play we will only respect if we get identical treatment here.... It seems to me my very strict and energetic attitude within courtesy has made an impression on their disposition.[34]

[33]AMAE, R. 4921/24. Letter from Díez del Corral to Carlos Cañal, informing of the meeting of the Board of Directors of the University City of Paris on February 14, 1949.

[34]AMAE, R. 4921/24. Telegram from Aguirre de Cárcer to the Minister, January 21, 1949.

The appointment of Maravall as the new director and the agreed resignation of Establier were interpreted by the embassy as a great victory, as it certainly was, given the efforts it took to achieve it. Thus, "Hispanization" of the *Colegio* could begin in 1949, with the introduction of 50 students, an extraordinary loan awarded by the Spanish government to reform the building, and regularly sending speakers to turn the centre into a showcase of cultural activity in the new Spain. Maravall remained in the position until 1954, and prominent personalities of the Spanish cultural world of the time such as Camón Aznar, Manuel Ballesteros, Dr Jiménez Díaz, musician Joaquín Rodrigo, and Pedro Laín Entralgo visited the institution, to teach courses and give seminars.

In his reports, Maravall was optimistic, but he complained about the "suffocating state in which I see myself here as a result of previous management", and the "unpleasant inheritance with which I found myself". He talked about the *Colegio* as

> a great foundation to support good work to expand our culture and our spirit… I believe the *Colegio* is worth considering not only as a residence, so our scholars are more or less well lodged, but also as a basis to develop cultural activities.[35]

The hostile environment permanently surrounding the activities of the *Colegio* hampered those intentions. On several occasions, there were incidents involving Spanish students of "communist" or "separatist" orientation, residents of other *Cité* schools, who put Republican flags on the balcony or distributed political propaganda. The *Colegio* remained as a redoubt of Spanish national Catholicism, oblivious to internationalism and the liberal environment of the rest of the University City, visited only by the Catholic and conservative sector of French Hispanism.

At the IEH, things were not much better. During the entire duration of the exchange sponsored by the Spanish authorities, the management kept secret the origin of the financing, aware of the hostility it could cause among students and among the teaching staff, which very much shows the limitations of a cultural propaganda action that could not be done openly because of political circumstances.

[35] AMAE R. 4921/24. Letter sent to the director general of Cultural Relations to attach to the report of June 3, 1949.

Parallel to the activities of the Spanish professors sent by the Spanish government, cultural life at the IEH was encouraged in the 1950s and 1960s by a group of professors headed by Claude Couffon and Robert Marrast, "well-known for their extreme leftist ideology and their activism against our political regime", according to the Spanish diplomatic authorities.[36] Any personality related to Hispanic culture, currently living in or passing through Paris, was invited by this association to speak with the IEH students. Writers, poets, and professors such as the Ecuadorian Jorge Icaza, Rafael Alberti, and Miguel Delibes were invited. The political orientation of these activities did not go unnoticed in the Spanish embassy. On the occasion of a memorial after the death of Miguel Hernández, the ambassador informed the Ministry of Madrid:

> in recent years, at least, all the Hispanic activities of the abovementioned Group of Studies have limited their topics to only two names, Federico García Lorca and Miguel Hernández, and professors Marrast and Couffon, supporters of the group, are well-known for their political ideology, allowing to interpret, without a doubt, their intentions.[37]

Robert Marrast himself oversaw directing the Spanish theatre group of the Sorbonne. As in American campuses, in the 1960s, a small theatre company was created to represent classical works in the theatrical week celebrated every year. Between February 2 and 15, 1962, for example, they staged two dramatic works by García Lorca and one by Bertolt Brecht, "*Fusils de la Mère Garrare*", a tragedy of Spanish theme, in collaboration with the theatre troupe from the University of Murcia. For the Spanish ambassador, José María de Areilza, the actions of this company "have been characterised to date by their mediocrity and by a clear departure from official Spain".[38]

The political activism of the students and some professors was a constant concern for the Spanish authorities, fearful of any manifestation of opposition to the Spanish regime. The IEH leadership, however, maintained a neutral and equidistant position that had characterized the centre since its origins. That was an attitude the director himself conveyed in the form of methodological advice to his students. Aubrun had a more

[36] AGA, 66/4336. Dispatch of the Spanish ambassador, Paris, November 18, 1961.
[37] AGA, 66/4336. Dispatch of the cultural attaché in Paris, Paris, April 4, 1962.
[38] AGA, 66/4336. Dispatch of the Spanish ambassador, Paris, March 20, 1962.

rational than sentimental conception of Hispanism. The Hispanic cultural tradition was, for him, a repository of ideas and feelings where each generation could find useful materials for their needs, but always from the perspective of humanistic and universal culture. He rejected nationalist narcissism and did not understand how one could speak of a German, Italian, or French Hispanism. Hispanism was an internationalist discipline, not involving any special affection for the cultural tradition of a particular nation. His advice to the young "*Hispanisants*" made this position clear:

> Authentic Hispanists should in no case allow their feelings, Hispanophilia or Hispanophobia, influence their work in relation to the Spanish nation. Their judgements must be formulated within an objective and rational vision, as far as possible. [...] Your own condition as a foreigner allows you to keep your distance with them, judge them from afar and near. Judging them from within would lead to one of the two most common attitudes among certain Spanish fans of Spanish things, either to patriot narcissism or masochism. [...] Let us get rid of dithyrambs and panegyrics of cultural propaganda. We are interested in the things of Spain, not *ad maiorem Hispaniae gloriam* but *usum universale*.[39]

This position suggested a more general evolution caused by political disagreement. The relationship of French Hispanism with its object of attention returned during the Franco regime to a pattern whose origin can be traced back to the end of the nineteenth century, when that corporation consolidated coinciding with the crisis of pessimism affecting the Spanish elites. Then, the relationship was marked by the arrogant and patronizing attitude of those analysing Spanish culture from the outside, on a plane of superiority, matched by the narcissistic reaction of those who feel insecure and dependent on the valuation of others. These behaviour patterns, reinforced by the ideological confrontations of the time, explain the strategies followed by the cultural leaders of the Franco regime in their relations with French Hispanism and their relative failure.

Conclusions

From the two cases studied, some conclusions can be extrapolated to all cultural activities of the Francoist authorities in post-war France. Since the early twentieth century, Spanish diplomacy tried to use the Sorbonne as

[39] Aubrun (1963): 1.

the spearhead of an ambitious plan of "Spanish intellectual influence" across Europe. Until 1936, effectively, Spanish academics and intellectuals invited to Paris by the IEH were the most qualified people to lead the campaign abroad in favour of the prestige of Spanish culture and history. The Spanish Civil War interrupted an exemplary university collaboration that had benefited both the development of French Hispanism and the openness and prestige of the Spanish intellectual class. During the first years of the Franco regime, the rupture and distancing from university Hispanism were maintained. Diplomatic authorities restored, with great difficulty, the old exchange circuits, again sending Spanish professors and students to Paris. Control of *Colegio de España* was regained in 1949, and cooperation with the IEH resumed in 1958. However, the purpose of academic exchanges varied and acquired a more defensive stance. The goal was to present the "true Spanish tradition" in a naturally hostile environment and counteracting the anti-Spanish implications of foreign historiography supposedly influenced by the black legend.

It is hard to believe the subsidy of the Spanish government, re-established after 23 years of suspension, and the presence again, thanks to it, of Spanish professors and academics in the Sorbonne classrooms, really served to guide "that French Hispanism prone to the cartoon", especially when that subsidy and the collaboration with the Spanish government was kept secret by the directors of the IEH so not even the centre's own professors suspected the existence of such patronage. It is significant that part of the teaching staff, including some Spaniards who considered themselves exiled by the regime, actively militated against the Franco dictatorship. The activity of *Colegio de España*, which only reached the Spanish residents themselves and remained an island in an International City comprising over 4000 students from around the world, could not break deep-rooted anti-Spanish stereotypes revived by the repudiation of the Spanish regime and the denunciation activities of the dissident Spanish students.

The Franco regime made real efforts to use French Hispanism as a strategic ally in a policy of cultural diplomacy and invested resources to achieve it. The purpose, as in previous stages, was to use the interest in Spanish culture to improve the image of the country abroad and contribute to the international prestige of the nation. However, they met resistance from the main leaders of university Hispanism, who were opposed to collaboration with a dictatorial regime. When this resistance was overcome and university exchange networks were rebuilt, the result was always influenced by the persistent rejection of a regime and an ideologically biased conception of the Spanish cultural tradition.

Acronyms used

AGA: General Archive of the Administration (Alcalá de Henares, Madrid)
AMAE: Archive of the Ministry of Foreign Affairs (Madrid)
AN: *Archives Nationales* (Pierrefitte-Sur-Seine, Paris)
IEH: *Institut d'Études Hispaniques*

References

Aubrun, Charles V. (1963a): "L'Institut d'Études Hispaniques. Rapport 1961-62", *Annales de l'Université de Paris*, 2, 33e année: 190-196.

Aubrun, Charles V. (1963b): "Carta de marear para hispanistas", *Les Langues Néo-latines*, n° 166: 1.

Aubrun, Charles V. (1991): *Mémoires*, presentadas por Sebastian Neumeister, Marburg, Hitzeroth.

Bataillon, Claude (2014): *Marcel Bataillon: hispanismo y compromiso político*. México, FCE.

Delaunay, Jean-Marc (1994): *Des palais en Espagne. L'Ecole des hautes études hispaniques et la Casa de Velázquez au coeur des relations franco-espagnoles du XXe siècle (1898-1979)*, Madrid, Casa de Velázquez.

Martinenche, Ernest (1939): "L'Institut d'Études Hispaniques de l'Université de Paris de 1935 à 1938", *Annales de l'Université de Paris*, p. 279.

Moreno Cantano, Antonio César (2007): "Delegaciones y oficinas de prensa españolas en el extranjero durante el primer franquismo: el caso francés (1936-1942), *Studia Histórica*, 25, pp. 265-301.

Niño, Antonio (1988): *Cultura y Diplomacia: Los hispanistas franceses y España, 1875-1931*, Madrid, CSIC, Casa de Velázquez y Société des Hispanistes Français.

Niño, Antonio (2017): *Un siglo de Hispanismo en la Sorbona*, Paris, Éditions Hispaniques.

s.a. (1939): *Hommage à Ernest Martinenche: études hispaniques et américaines*. Paris.

s.a. (1964): "L'Institut des Hautes Études de l'Amérique Latine. Extrait du Rapport présenté au Conseil d'Administration pour l'année 1962-1963", *Annales de l'Université de Paris*.

Sarrailh, Jean (1956): "L'Institut des Hautes Études de l'Amérique Latine", *Revue de l'Enseignement Supérieur*, 2.

CHAPTER 15

Pause and Adaptation in the Post-War Period: The Re-establishment of Spanish-German Cultural Diplomacy (1945–1958)

Mariciό Januė i Miret

"*Soft Power*" in Preceding Spanish-German Relations

This article analyses the development of Spanish-German cultural diplomacy from the aftermath of the Second World War until the late 1950s. The term "Germany" refers here to the territory of the Federal Republic of Germany (FRG),[1] as Spain did not establish diplomatic relations with

This text has been developed in the framework of the project, "Ciencia, cultura y nación en España, del desastre de 1898 al fin de la dictadura franquista" / "Science, culture, and nation in Spain from the 1898 'disaster' to the end of the Franco Dictatorship," FFI-HAR 2016-75559 (AEI/FEDER,UE).

[1] For all the acronyms used in the text, see the list at the end.

M. Januė i Miret (✉)
Department of Humanities, Universitat Pompeu Fabra, Barcelona, Spain
e-mail: maria.janue@upf.edu

M. Januė i Miret, A. Presas i Puig (eds.), *Science, Culture and National Identity in Francoist Spain, 1939–1959*,
https://doi.org/10.1007/978-3-030-58646-1_15

the Democratic Republic of Germany (GDR) until 1973. The study focuses on the humanities, the arts, the social sciences and teaching.

Our approach is based on the theoretical premise that science and culture have a relationship with politics characterized by "mutual resources" (*Ressourcen füreinander*), in human, institutional and ideological-discursive relations.[2] In this framework, historical fractures may be able to form processes of adaptation that permit "constructed continuities" (*konstruierte Kontinuitäten*).

In contrast to traditional diplomacy, which relied on the use of coercion and force, cultural diplomacy sought to use the power of attraction and persuasion, becoming a model that inspired admiration for its values and/or its results in various fields of knowledge. This explains why cultural diplomacy is described as a "soft power."[3] It was hoped that investments in the international prestige of the country's image in the fields of culture and science would translate into greater political and economic influence abroad.[4]

In the case of Spanish-German relations, the difficult circumstances faced by the Weimar Republic in the years following the Great War made Spain a priority for German cultural diplomacy because of Spain's wartime neutrality.[5] With National Socialism's rise to power, cultural relations between the two countries were tense due to the mutual distrust with which the Second Spanish Republic and the German Nazi government viewed each other. However, ongoing personal relationships facilitated the warming of cultural contacts again after the outbreak of the Spanish Civil War. With the outbreak of the Second World War, Germany reinforced its cultural diplomacy towards Spain, giving impetus to institutions that would become a model for the establishment of intellectual and academic networks.

Nazi cultural diplomacy promoted German interest in Hispanic studies, which was supported by personal continuity in both countries. Prestigious German scholars of Spanish culture roused the Spanish nation and the Hispanic character in books and articles for academic reasons, public dissemination and political influence.[6] Nazi authorities sought to attract the

[2] We followed Ash (1996: 61–62, 2006: 19–30). See also Antolín Hofrichter (2018): 22–26.
[3] Gienow-Hecht and Donfried (2010): here 21.
[4] Aschmann (1999): 398.
[5] We followed Janué Miret (2015a, b).
[6] Janué Miret (2019).

most important intellectuals and scholars of the day, provided they were in ideological agreement. The Franco regime, for its part, only began to trim its relations with Berlin publicly when it appeared that the Nazis might lose the war.[7]

In the following pages we will examine the consequences of the end of the Second World War on this axis of German-Spanish cultural collaboration.

The Shift in Bilateral Cultural Relations Under Allied Intervention (1945–1948)

The defeat of the Axis powers in the Second World War entailed the dismantling of Spanish-German cultural diplomacy. The provisions of the Allied Control Committee were decisive in this regard. The resolutions with the greatest impact were the obligatory repatriation of German citizens and the freezing of German assets in Spain and, in the cultural sphere, the dissolution of German cultural institutes and German schools in Spain.[8] During this stage, Spain took the approach of trimming back cultural relations with Germany, making it a priority to demonstrate cooperation with the Allies to avoid accusations of fascism that could deepen Spain's isolation.[9]

The Allied powers' embassies in Madrid called for the Spanish government to repatriate German citizens and to detain those identified as "agents of Nazism."[10] The Spanish authorities displayed reluctance to surrender the German nationals as demanded.[11] Efforts were made to recruit German scientists and technicians with the cooperation of the National

[7] García Pérez (1990); Delgado Gómez-Escalonilla (1992): 167; Collado Seidel (1993): 475; Bernecker (2002): 179; Saz Campos (2003): 341–346; Sesma Landrín (2011: 279, 2017).

[8] Delaunay (1989); Collado Seidel (1992, 2001, 2005); García Pérez (1994): 556–569.

[9] Weber (1992a): 17, 37; Bernecker (2002): 179–181; Sanz Díaz (2008, 2010); Ruiz Escudero (2008a): 201–202.

[10] AMAE R.2192-17, Burgos, Resolución del Consejo Aliado, Nota de Prensa, Madrid, 6.5.1946.

[11] Extensive documentation on this issue in AMAE R.2192-17; Weber (1992a: 37–60, 1992b: 217–222); Messenger (2014): 120–131; 162–170.

Institute of Industry (INI) and the Advanced Centre for Scientific Research (CSIC), especially the Juan de la Cierva Trust.[12]

The Allies also called for Spanish authorities to freeze the assets and buildings owned by the German government or the Nazi party, including para-state companies, and put them at the disposition of the Allied embassies as reparations payments.[13]

One of the Allied provisions that most seriously disrupted relations between Spain and Germany in the field of culture was the order to liquidate German cultural institutes. This measure affected both the institutes that Nazi Germany had maintained in Spain and those who worked to promote Spanish culture in Germany. The latter included two cultural centres that stood out for their work at their shared headquarters in Berlin: the Ibero-American Institute (IAI)[14] and, above all, the German-Spanish Society (DSG). During the Nazi period, the DSG had become the main axis of bilateral cultural relations.[15] Following the occupation of Germany in May 1945, Allied troops took over the headquarters of both organizations. Traugott Böhme, a German teacher and archivist who had opposed National Socialism, was appointed commissioned director of the IAI.[16] American occupation forces seized the building that had served as the headquarters of the IAI and the DSG, along with its property and archives.[17]

With regard to German cultural institutes in Spain, the Nazi government had created the Institute of German Culture (DWI) in Madrid in 1941.[18] Together, the DWI in Madrid and the German Cultural Institute (DKI), established in Barcelona in 1942, became the main centres for the dissemination of Nazi culture in Spain. The DWI worked in close cooperation with the CSIC, the German Academic Exchange Service (DAAD), the IAI and the DSG, as well as with German scholars interested in

[12] Presas (2008: 229–235, 2010: 91–98); Weber (1992a): 77–79; Ruiz Escudero (2015): 36, 62. On the CSIC and its discourse in this period, see Antolín Hofrichter (2018): esp. 35–157.

[13] Weber (1992a: 61–69, 1992b: 226).

[14] Liehr et al. (2003).

[15] Janué Miret (2008a, b).

[16] GSTA, I.HA.Rep. 218, 380, IAI activity report from 1.5.1945 until the end of March 1946.

[17] GSTA, I.HA.Rep. 218, 380, Böhme, commissioned director of the IAI, to Major J.W. Taylor. *US Group Control Council Education Branch*, 20.9,1945.

[18] On the DWI, see Hausmann (2001) and Hera Martínez (2014).

Hispanic studies. The DWI had two directors over the course of its history: the first was the specialist in romance languages Theodor Heinermann, who was replaced in 1944 by Karl Vossler, a renowned scholar of Hispanic studies from the University of Munich known for his conservative Catholic beliefs. During the interwar period, Vossler had defended Spain's role as the moral teacher of Europe in his 1929 essay *Die Bedeutung der spanischen Kultur für Europa*. The Nazi party had forced him into early retirement from his university post in 1937 over his opposition to anti-Semitism and fascism. Despite this, he had accepted a role as ambassador for the science of the Reich abroad. Vossler would never assume the post of DWI director; those duties would be carried out until the end of the war by Berthold Beinert, who had been involved in the institute's library and activities since its inception.[19] In the same building as the DWI, a subsidiary of the German Archaeological Institute of Berlin (DAI) was founded in late 1943, under the direction of Helmut Schlunk, a German archaeologist and art historian specialized in the High Middle Ages of the Iberian Peninsula.[20] That building was also the headquarters of the subsidiary of the German Academy of Munich (*Deutsche Akademie*), precursor of the Goethe Institute created in 1951, under the direction of Bernhard Schulz, a staunch member of the Nazi party.[21]

Following the German surrender, the Allied Commission ordered the closure of the DWI and its subsidiary in Barcelona and designated Hans Rothe, a playwright and German translator who had fled the Nazi regime to Spain, to assess the role that both centres had played under the Nazi regime. Rothe called for Beinert to surrender the records and bibliographical collection of the DWI to the Allied Commission, which Beinert sought to avoid.[22] Rothe's interlocutors at the DKI in Barcelona were Erich A. Krotz, head of the centre, and Hans Schlegel, a philologist and prolific translator of classical Spanish dramas into German, who resided in the

[19] Hausmann (2001): 211–215; Rodríguez López (2008): 119–120; Janué Miret (2015a).

[20] PAAA Embassy Madrid 45. [From Hans Rothe] to Mr. Gilbert, 19.9.1945; Hausmann (2001): 236; Maier Allende and Schattner (2010).

[21] PAAA Embassy Madrid 45. [From Hans Rothe], German Institute, 12.9. 1945; PAAA Embassy Madrid 45. [From Hans Rothe] to Mr. Gilbert, Institute 19.9.1945 (2).

[22] PAAA Embassy Madrid 45. [From Hans Rothe], German Institute, 12.9.1945; PAAA Embassy Madrid 45. [From Hans Rothe] to Mr. Gilbert, Institute 19.9.1945 (2).

Catalan capital and was formerly the technical manager of the Unicolor chemical company, associated with I.G. Farben and Bayer.[23]

The Allies also ordered the closure and confiscation of German schools in Spain, a mission that was also supervised by Rothe.[24] The Third Reich had invested large sums of money in the construction and renovation of school buildings in Spain. Of the 12 schools that had opened in Spain, 5 were still operating at the end of the war.

The closure of the Madrid school was particularly controversial.[25] After Hitler's suicide in May 1945, the school's board of directors was replaced by a board consisting entirely of Spaniards. The school's new board of directors instituted Catholic guidelines. At the end of the war, the Allies demanded unconditional surrender of the school, which was opposed by the board of directors. As the board of directors did not back down, the school was unable to open at the beginning of the autumn 1945 term.

The German school of Barcelona acceded more easily to the Allied demands.[26] A few months before the end of the war, Eckhart Stegmann, who had been the school's director since 1939, proposed counteracting the separation from official posts and the Spanish church by abolishing co-education. Following Germany's surrender, Ferdinand (or Fernando) Birk, treasurer of the German School Association (*Schulverein*), head of Unicolor and president of the German Chamber of Commerce for Spain, took a leading role as the representative of the German school in its negotiations with Rothe. The school's board of directors accepted the school's surrender to the Allied Commission in autumn 1945. This made it possible to begin the process of converting the school to an international school.

The proposal for the establishment of an International School provoked a controversy about the role that the French school should play in this project. The Germans rejected this link. In the end, the new International School was installed in the buildings of the former German school in Barcelona, taking over that school's property and infrastructure.

[23] PAAA Embassy Madrid 45. [From Hans Rothe] to Mr. Gilbert, Help for Barcelona Library, 29.10.1945; PAAA Embassy Madrid 45. [From Hans Rothe] to Mr. Gilbert, 5.11.1945.

[24] On the expropriation of German schools in Spain, see documents in AMAE R3361-91; Collado Seidel (2007b): 126–137; Herzner (2019): 159–171.

[25] Documentation about the confiscation and dispute, in PAAA Embassy Madrid 44.

[26] Documentation in PAAA Embassy Madrid 44; PAAA Embassy Madrid 45; PAAA R63877; PAAA R3361-91; AMAE R2192-16. In addition, Chamrad et al. (1994): 117–141.

It was launched in the autumn of 1946 under the name Lycée Français-Collège International.

At the end of that year, the United Nations recommend that all nations withdraw their diplomatic missions from Madrid. In May 1948, the three Western Allies signed an agreement with Spain for a "convention for the elimination of economic potential in Spain that could constitute a threat to the peace and the liquidation of balances and claims for payment between the governments of Spain and Germany."[27] Under this convention, all the rights, titles and current or future interests to property in Spain held by the German government and its agencies were ceded to Spain.

The Cold War as a Framework for the Re-establishment of Relations Marked by Continuity (1949–1953)

In a climate of the highest Cold War tension over Germany and the Berlin Blockade, the division of that country was consolidated with the founding of the FRG, on the one hand, and the GDR on the other. The FRG's first chancellor was the Rhenish Catholic Konrad Adenauer, leader of the conservative Christian Democratic Union (CDU), who would remain in office until late 1963. The Allies permitted the FRG to establish diplomatic missions beginning in the spring of 1951.[28]

Isolation and the mediating role of the United States marked the international position of both Spain and the FRG during this period.[29] The FRG would take a prudent approach to its relations with Spain during the initial phase so as not to arouse concern over its fascist past.[30] In contrast, Spain considered the achievements in the field of cultural relations to be an appropriate alternative to traditional diplomatic relations to improve the country's image.[31]

[27] Weber (1992a: 69–80, 1992b: 228).

[28] Collado Seidel (1993): 477.

[29] Weber (1992a): 209; Collado Seidel (1993): 476.

[30] Collado Seidel (1993): 478; Lehmann (2006): 75–76; Sanz Díaz (2008): 167; Ruiz Escudero (2007: 134, 2008b: 2942, 2015: 26, 29–30, 86).

[31] Delgado Gómez-Escalonilla (1992): 457; Weber (1992a): 189; Moreno Juste (2005): 190; Sanz Díaz (2008): 183; Ruiz Escudero (2008a: 202, 2008b: 2940–2944, 2015: 33, 76).

This period saw changes in the consideration of the Franco regime by the Western Allies, who gradually focused less on the regime's authoritarian features and emphasized its anti-communist nature.[32] Franco's Spain was admitted to the United Nations in 1952 and signed agreements with the United States the following year. Under these circumstances it was Spain that took the initiative. By 1950, exchanges of students and professors were already taking place, and the Ministry of Foreign Relations (MAE) was subsidizing 15 lectureships in Germany, where it would maintain its densest network during the 1950s and 1960s.[33]

In October of that year, the Institute of Hispanic Culture (ICH), associated with the Ministry of Foreign Relations, organized a Congress for Cultural Cooperation in Madrid, featuring some of the "authentic German Hispanists" who had stood out in the previous period for their support of the dictatorships of both countries, including Rudolf Grossmann, Helmut Petriconi and Edmund Schramm.[34] Grossmann was a professor of the University of Hamburg, where he headed that university's Ibero-American Institute (IAI-H), and had co-authored the prestigious Slaby/Grossman Spanish-German dictionary (1932–1937). In a 1941 article published in the journal of the IAI-H, Grossman had compared Spain's historical achievements with those of National Socialist Germany.[35] He had also written about the essence of the Spanish culture and character.[36] Petriconi, professor at the University of Greifswald, had been a member of the DSG's board of directors and on the editorial board of the journal *Ensayos y Estudios*, published by the IAI during the Second World War. In 1941, Petriconi had given a lecture on the image of Spain in the German consciousness to open the German Book Exhibition in Barcelona.[37] He moved to Hamburg after 1945, where he was a colleague of Grossmann. Schramm had conducted a research project under the direction of Petriconi into the Spanish anti-liberal Donoso Cortés and was associated with the German school of Madrid and the German-Spanish Intellectual Exchange Centre, the precursor to the DWI, during the interwar period.[38] He had also been

[32] Sanz Díaz (2008): 167; Ruiz Escudero (2015): 32–33, 84.

[33] Weber (1992a: 192, 1992b: 216); Collado Seidel (1993): 476–477; Sanz Díaz (2008): 159–167; Ruiz Escudero (2007: 135, 2008b: 2944–2947, 2015: 79, 93–97, 110–120).

[34] Bräutigam (1997): 103; Sanz Díaz (2008): 161, 165; Ruiz Escudero (2015): 114–115.

[35] Grossmann (1941).

[36] Grossmann (1942, 1943).

[37] Petriconi (1941).

[38] Schramm (1935).

a member of the Gorres Society (GG), a Catholic-oriented German scientific institute headquartered in Madrid since 1927 that published the prestigious journal about Spanish culture *Spanische Forschungen der Görresgesellschaft*. In late 1936, he had held a conference at the University of Greifswald on the preconditions of the Spanish Civil War in which he defended Franco's military "uprising."[39] During the Second World War, the *Deutsche Informationsstelle*, a department of *the Deutsches Institut für Aussenpolitische Forschung* controlled by the Ministry of Foreign Affairs (AA), commissioned Schramm to write a book demonstrating France's military and propaganda support for the Second Republic during the Spanish Civil War.[40]

In 1951, Antonio María Aguirre González was named Spain's first ambassador to the FRG. A career diplomat, he had served as consul in Hendaya, had studied in Berlin and had been a commercial attaché at the Spanish embassy in that city in 1940.[41] He had been the head of the Spanish government's Mission to the Allied High Commission since July 1950. To head the cultural affairs section, he designated Juan Manuel Castro-Rial, a full professor of international law, Falangist and member of the Blue Division who had studied in Berlin and worked with the IAI. Following the appointment of Aguirre as ambassador, the Spaniards created a cultural office in Bonn to promote the exchange of university students. Similarly, progress was made in the development of German Philology studies in Spain, first provisionally at the University of Salamanca in 1952–1953, then in Madrid, and definitively at both universities in 1954–1955, followed the next year by Barcelona.[42]

Another important step in promoting mutual relations was the establishment of the European Documentation and Information Centre (CEDI) in 1953 with the financial and strategic support of the MAE. Its establishment resulted from the efforts of the Franco regime's Catholic elite to promote relations with Catholic-Integralist and anti-democratic sectors of Western European conservatism.[43] Participants in the congress included many universities and institutes of the CSIC, ministerial

[39] Schramm (1937).

[40] Schramm (1940).

[41] Collado Seidel (1993): 486; Ruiz Escudero (2007: 134, 2015: 29).

[42] Moreno Juste (1993): 461, 465–467; Ruiz Escudero (2007: 134, 2008b: 2955).

[43] Collado Seidel (1993): 480; Moreno Juste (1993: 471–472, 2005); Ruiz Escudero (2015): 75–76.

bureaucrats, diplomats and members of the financial aristocracy.[44] Based on the ideals of the "Christian West" and the under the guise of "organic democracy," their activities were a combination of anti-communist propaganda aimed at the East with anti-democratic agitation aimed at the West.[45] Among the various national centres of the CEDI established abroad, the only group with notable activity was the RFA.[46] This can be explained by the ideological affinities between the pro-Franco Catholic elite and the representatives of the anti-democratic and reactionary conservatism of the German branch of the CEDI, which coincided with the views of the Western Circle (*Abendland-Kreis*), which was close to certain sectors of the CDU and especially the Bavarian Catholic Christian Social Union (CSU).[47] Among the most prominent figures of this department was the jurist Joachim von Merkatz, who had directed the IAI and the DSG during the Nazi dictatorship.[48] In the post-war period, he joined *the conservative German* Party (*Deutsche Partei*) and was a member of the Commission of the Council of Europe; in 1956, Merkatz became the Minister for Justice.[49] Pro-Franco politicians had access to high-level political circles close to the German government through the CEDI.[50]

The FRG government's priorities for mutual relations during this period were the negotiations that began in 1951 to lift the Spanish freeze on German citizens' property not covered by the 1948 seizure agreement. Controversy arose because the Spanish state had liquidated part of the frozen German assets. Claims for the return of the property of the five German schools were pressed especially forcefully.[51] Tension resulting from this issue delayed the appointment of the FRG's first ambassador to Madrid.[52]

[44] Weber (1992a: 203–204, 250, 1994: 1086).

[45] Weber (1994): 1088–1090, 1102; Lehmann (2006): 65–75.

[46] Collado Seidel (1993): 479–481; Moreno Juste (1993): 469–470; Aschmann (2008): 143–146.

[47] Weber (1994): 232–240, 1079, 1095–1096, 1100–1102; Aschmann (2008): 143–146; Ruiz Escudero (2015): 81.

[48] Moreno Juste (1993): 469; Janué Miret (2008a, 2008b).

[49] Weber (1992a): 237, 1994: 1083, 1091, 1094); Lehmann (2006): 73; Aschmann (2008): 144.

[50] Weber (1994): 1103.

[51] Documentation in AMAE R3361-91. See also Weber (1992b): 114–228; Lehmann (2006): 38–40; Ruiz Escudero (2015): 97.

[52] Weber (1992a: 209, 212, 1992b): 212.

However, the FRG's Ministry of Culture provided financial support to the 1951 establishment of the German-Spanish Society of Munich (DSG-München), which would become the largest and most important German-Spanish association.[53] Its board of directors included representatives of the economic sector with close ties to the CSU government of Bavaria.[54] Grossman also sat on the board, as did Hermann Hüffer, historian and former cultural adviser at the German Embassy in Madrid, who would later take a background role due to his Nazi past.[55] In Spain at that time, the Institute of Political Studies (*Instituto de Estudios Políticos*) translated Hüffer's work on mutual cultural relations originally published in German during the Second World War.[56] DSG-München would maintain contact with prominent figures in the Franco regime, including General Moscardó, a "hero" of the Spanish Civil War. During the Second World War, Moscardó had headed the Spanish-German Association (AHG), which enjoyed financial support from the German and Spanish governments.[57]

The FRG's first ambassador to Spain was not appointed until October 1952. The position went to Prince Adalbert of Bavaria, who remained in the office until 1956. Adalbert was the founder and president of DSG-München and fulfilled the Spanish government's criteria as Catholic and conservative, as well as having an ancestral tie to the Spanish royal family.[58] The Embassy's first cultural attaché was Andreas W. Bauer, a disciple of Vossler and German lecturer at the University of Oviedo.[59]

The work of the FRG's first ambassador to Spain provided impetus to the cultural relations between the countries. By 1953, eight of the former

[53] Weber (1992a): 200–204; Lehmann (2006): 76–77; Ruiz Escudero (2015): 190.

[54] Collado Seidel (1993): 481–482; Sanz Díaz (2008): 161; Ruiz Escudero (2015): 81, 164.

[55] Lehmann (2006): 78.

[56] Hüffer (1942, 1951).

[57] Janué Miret (2008a, 2008b).

[58] Weber (1992a: 98–105, 1992b: 212); Collado Seidel (1993): 486–487; Aschmann (2008): 141–142; Ruiz Escudero (2007: 134–135, 2015: 29).

[59] Ruiz Escudero (2008a: 203–204, 2008b: 2944, 2015: 90).

German schools had resumed their educational activities.[60] In 1961, the Madrid school was the largest in Europe.[61]

The GG, which had been dissolved and had its property confiscated by Nazis who considered it hostile to the Third Reich, was also re-established in 1953. The re-establishment initiative was led by the Catholic Hispanist Hans Juretschke, a specialist in Spanish romanticism, with the cooperation of the CSIC.[62] Juretschke had taught German language and literature at the University of Madrid since 1939, and in 1940 had published his doctoral dissertation on Spanish-French relations. In 1942, he had been named a scientific assistant to the Third Reich's embassy in Madrid.[63] He had also worked with the DWI, where he was responsible for the publication of the journal *Investigación y Progreso* and was one of the directors of the *Boletín Bibliográfico*. He had also been a member of the AHG. After the war, he had avoided repatriation, returning to his professorship at the University of Madrid and collaborating with the CSIC.

Also in 1953, the FRG succeeded in having Spain return the property and funds of the DAI library, which would reopen the following year.[64] Schlunk, the institute's founding director, was returned to his post, where he would serve until 1971. Like Juretschke, Schlunk had managed to avoid repatriation and, since 1948, had taught at the University of Valencia and was a researcher at the CSIC.

[60] AMAE R3361-91, Escuelas Alemanas, Expropiaciones, Oficina de Información Diplomática, Registro de corresponsales, Werner Schulz, corresponsal permanente desde Madrid, Frankfurter Allgemeine, 5.12.195[2], 5.1.1953; AMAE R3361-91, Escuelas Alemanas, Expropiaciones, Oficina de Información Diplomática, Registro de corresponsales [7.1954]. On the resumption of the schools, see Chamrad et al. (1994): 141–184; Ruiz Escudero (2008b): 2948; Collado Seidel (2007a): 138–142; Sanz Díaz (2008): 169–171; Herzner (2019): 176–208.

[61] Sanz Díaz (2008): 179.

[62] Sanz Díaz (2008): 171–172; Ruiz Escudero (2015): 148–151.

[63] Vega Cernuda (2003); Jorba (2007); Rodríguez López (2008): 108.

[64] AMAE R3361-91, Escuelas Alemanas, Expropiaciones, Oficina de Información Diplomática, Registro de corresponsales, Enrique Barth, corresponsal permanente desde Madrid, Frankfurter Neue Presse, 28.2.1953, Madrid, 14.3.1953; Sanz Díaz (2008): 172; Maier Allende and Schattner (2010): 345–347; Ruiz Escudero (2015): 121–123.

Taking Back the Initiative: The "German Miracle" vs Spanish Stagnation (1954–1958)

In addition to its "economic miracle" (*Wirtschaftswunder*), the FRG saw significant changes in its international status in the mid-1950s. By 1952, the Convention on General Relations between the Three Powers and the FRG, known as the German Treaty (*Deutschlandsvertrag*) had been signed, restoring almost complete sovereignty to Germany. However, resistance from France delayed the treaty's ratification until the autumn of 1954. The FRG joined NATO a year after the treaty was implemented.[65]

Spain also made progress in its international re-socialization that year, becoming a full member of the United Nations. In the economic sphere, however, Spain's autarchic policy spelled a period of stagnation for that country.[66] In international trade, the FRG was the largest supplier to Spain and the largest purchaser of Spanish goods. In this context, Spain's diplomatic ambition to portray the Franco dictatorship as worthy of integration into the West led it to strive to achieve the political support of the new FRG.[67]

This was the context in which negotiations were conducted for signing the Spanish-German Cultural Agreement.[68] Although the agreement was approved by the Spanish Cabinet in the spring of 1953, Adenauer postponed signing it until after the German parliamentary elections scheduled for September of that year. Germany's claim regarding a prior agreement on the return of cultural property also contributed to the delay in signing the agreement. After finally being signed by both parties in Bonn at the end of 1954, the agreement was ratified by the Spanish parliament late the following year. For the FRG, it was the first Cultural Agreement signed with a non-occupying country, and for Spain it was the first signed with a European power. From that time on, Spain would begin to participate in FRG technical development assistance programmes.[69]

[65] Collado Seidel (1993): 477–478.

[66] We follow Weber (1992a: 134–152, 1992b: 213–215, 226); Sanz Díaz (2005): 252–275, 628, 1077; Ruiz Escudero (2007: 136, 2008b: 2953–2954, 2015: 32).

[67] Weber (1992b): 216–217; Lehmann (2006): 125–174.

[68] Documentation about the negotiations in AMAE R3361-91. See also Weber (1992a): 194–195; Aschmann (1999: 392–398, 2008: 134); Lehmann (2006): 78–83; Sanz Díaz (2008): 174–175; Ruiz Escudero (2008b: 2950–2951, 2015: 124–131).

[69] Sanz Díaz (2008: 177–178, 2010: 375).

The implementation of the Cultural Agreement marked a turning point for the FRG in taking the initiative on cultural relations between the two countries.[70] One example is the 1955 inauguration of the German Library (*Deutsche Bibliothek*) of Barcelona, successor to the DKI, as a result of a provision of the FRG General Consulate.[71] The project received financial support from a group of companies committed to encouraging Spanish-German cultural relations. The German Library would later be incorporated into the network of Goethe Institutes.

Meanwhile, progress had been made in the recovery of the IAI in Berlin. In 1946, the municipality of Berlin had taken over maintenance of what was known at the time as the Latin American Library, although it was renamed the Ibero-American Library (*Ibero-Amerikanische Bibliothek*) in 1954.[72] In 1962, the library would recover its original name of IAI.

Spain also engaged in policies of cultural rapprochement. The most far-reaching would lead to the inauguration of the Spanish Institute (*Instituto de España* or *Spanisches Kulturinstitut*) in Munich in 1956, the precursor to the current Cervantes Institute. The institute sought to disseminate the most agreeable face of the Franco regime by promoting the Spanish language and high culture.[73] The project's new director was philologist Carlos Clavería Lizana, who had been a lecturer in Spanish at the universities of Marburg and Frankfurt in the early 1930s and held an important office in the Interior Ministry's Foreign Press Department in Burgos in 1936.[74]

The German response soon followed, and in the autumn of 1957, during the celebration of the first German Culture Week in post-war Spain, the German Culture Institute was inaugurated in Madrid, sponsored by the AA, which would be incorporated into the network of Goethe Institutes.[75] Its director until 1968 was Werner Brügmann, a specialist in

[70] Weber (1992b): 215–216; Ruiz Escudero (2008a: 204, 2008b: 2948–2949, 2015: 35, 104, 180–188).

[71] Documentation about the Library in PAAA B90-EA 553; PAAA B93-33; PAAA B96-EA 151; PAAA B96-EA 30; PAAA B96-EA 7. See also Ruiz Escudero (2007: 137–138, 2008a: 204, 2008b: 2949, 2015: 35, 143–147).

[72] Weber (1992a): 196–197.

[73] We follow Sanz Díaz (2008): 165; Ruiz Escudero (2008a: 205–209, 2008b: 2951–2952, 2015: 131–138).

[74] Lapesa (1974).

[75] Documentation about the Institute and the German Cultural Week in PAAA B96-150; PAAA B90-EA 553; PAAA B90-EA 647; PAAA B96-EA 29. See also Weber (1992a): 165–166; Aschmann (1999): 398; Sanz Díaz (2008): 173; Ruiz Escudero (2008a: 209–211, 2008b: 2948–2954, 2015: 138–143, 199–201).

Spanish literature and a member of the GG. The leading members of the German community in Madrid were associated with the Institute.

Germany's economic power and international status, together with Spain's aforementioned commercial weakness, enabled Germany to revise the Spanish policy regarding confiscation of German cultural property. In 1957, the Spanish government underwent changes that led to the abandonment of its autarchic economic policy. It was this new government that engaged in negotiations with the FRG for the return of German assets.[76] These negotiations would culminate in the Convention on certain results of the Second World War and the Convention for the rehabilitation of Industrial Property Rights, signed in 1958 by the Spanish and German Foreign Ministers, Fernando Castiella and Heinrich von Brentano (who was close to the Western Circle and the CEDI).[77] These conventions provided final disposition of the German assets in Spain, opening the way for the return of the property belonging to the German schools.[78] Shortly after the signing of these conventions, Brentano would make an official visit to Spain.[79]

The conventions of the spring of 1958 have been interpreted as the end of the post-war period in bilateral German-Spanish relations. The new "unapologetic"[80] phase of German policies towards Spain would also be reflected in cultural diplomacy and, together with a "German-Spanish spring,"[81] in the economic and trade cooperation, would contribute to the stabilization of the Franco dictatorship.[82]

[76] Documentation about the negotiations in PAAA B90-EA 553.

[77] Collado Seidel (1993: 487, 2007b); Sanz Díaz (2008): 168; Ruiz Escudero (2015): 32.

[78] Documentation about these negotiations in PAAA B90-EA 553; PAAA B90-EA 647. Also see references in Chamrad et al. (1994): 120–121.

[79] On the visit, see PAAA B90-EA 647, Aufzeichnung aus Anlass des Besuches des Herrn Bundesministers des Auswärtigen in Spanien vom 7–10 April 1958; PAAA B90-EA 647, Knappstein, Madrid, an das AA, Bonn, Spanienbesuch des Herrn Bundesminister von Brentano, 14.4.1958; Sanz Díaz (2005): 192, 335–338.

[80] Weber (1992a: 162–188, 1992b: 22).

[81] Sanz Díaz (2005): 57, 66, 197, 275, 284–285, 1095.

[82] Niehus (1989): 456; Weber (1992a: 162–188, 1992b: 224); Bernecker (2002): 180; Aschmann (2008): 153–154; Sanz Díaz (2008): 176–183; Ruiz Escudero (2008b: 2955–2956, 2015: 32, 201–202, 205).

Conclusions

The break in bilateral cultural diplomacy following the Second World War was not definitive. However, the post-war period represented a significant rift, at least at the beginning. The assumption of the German government's functions by the Allied Control Council produced a noticeable shift in the cultural relations between Spain and Germany.

In the new context, both countries shared having to face the international community's desire to isolate them as well as being subordinate to American guidelines. This factor had contradictory effects over the course of the period. On the one hand, it would restrain the development of bilateral relations, as both countries sought the support of the Allies, particularly the United States, for their reintegration into the international community. On the other hand, each country represented a resource for the other, albeit ersatz, in terms of the desired international acceptance.

The continued presence of conservative, strongly anti-communist elements in each government, together with changes in the Allies' views over the course of the Cold War, worked in favour of the gradual re-establishment of diplomatic relations, including in the cultural sphere. In the end, the re-establishment of these bilateral relations would prove more significant for Spain than for the FRG. Over the long run, Germany would have greater economic capacity and infrastructure available to develop the relations.

In both countries, the process of re-establishing the mutual cultural diplomacy would be marked by significant continuity of personnel in both the cultural and political spheres. The agents who helped re-establish mutual cultural diplomacy, be they intellectuals, scientists, politicians or a combination of all these professions, came principally from the Catholic-conservative sectors and had very weak or non-existent, commitment to liberal democracy.

However, continuity in the personnel who re-established bilateral cultural diplomacy did not mean they had made no change from earlier times in their discourse. In the post-war period, expressions of fascist ideologies were marginalized. Ideological manifestations were adapted to the new times, highlighting values compatible with those of the western bloc in the context of the Cold War, especially anti-communism. In the German case, we also find examples of what might be called "ideological mutation," that is, a discourse that demonstrates commitment to the liberal-democratic orientation of the new FRG. In many cases, however, the "constructed continuity" would indicate this was more of a "mimesis," an apparent change in discourse stemming from pragmatic reasons rather than from a firm belief in democratic principles.

The prevailing restraint in bilateral cultural diplomacy would start to be reversed after the signing of the German Treaty in 1954. Growing tension in the international order over the course of the Cold War resulted in increasing acceptance and integration of both countries into the western bloc. This trend encouraged intensified cultural diplomacy between the countries.

Beginning in the mid-1950s, the consolidation of the German "economic miracle," in contrast to Spain's economic stagnation, meant that the initiative for bilateral cultural relations was reversed. (Western) Germany would once again take the lead, as it had before the end of the war. The FRG's dynamism during the latter part of the post-war period was reflected in the creation and re-establishment of institutions that, albeit with different discursive content, played an analogous role in cultural diplomacy to the role that had previously been played by the institutes and schools closed in 1945. Similarly, significant progress was made in revitalizing German interest in Hispanic culture, whose discourse was generally in line with the ideas about Hispanism and the Spanish Empire promoted by Franco's regime. During these years, Spain also made some achievements in terms of cultural diplomacy.

The signing of the 1958 conventions for review of the policy on confiscated German assets marks the end of the post-war period in relations between the two countries. The 1960s would see a new "German-Spanish spring," in the economic and trade spheres, accompanied by a new "unapologetic" period of bilateral cultural diplomacy.

Acronyms Used

AA	Ministry of Foreign Affairs (FRG) / *Auswärtiges Amt*
AHG	Spanish-German Association (Madrid) / *Asociación Hispano-Germana (Madrid)*
CDU	Christian Democratic Union / *Christlich Demokratische Union*
CEDI	European Documentation and Information Centre / *Centro Europeo de Documentación e Información*
CSIC	Advanced Centre for Scientific Research / *Centro Superior de Investigaciones Científicas*
CSU	Christian Social Union / *Christlich-Soziale Union*

DAAD	German Academic Exchange Service / *Deutscher Akademischer Austauschdienst*
DAI	German Archeological Institute / *Deutsches Archäologisches Institut*
DKI	German Cultural Institute of Barcelona / *Deutsches Kulturinstitut*
DSG	German-Spanish Society of Berlin / *Deutsch-Spanische Gesellschaft Berlin*
DSG-München	German-Spanish Society of Munich / *Deutsch-Spanische Gesellschaft München*
DWI	German Culture Institute of Madrid / *Deutsches Wissenschaftliches Institut*
FRG	Federal Republic of Germany / *Bundesrepublik Deutschland*
G	Görres Society / *Görresgesellschaft*
GDR	German Democratic Republic / *Deutsche Demokratische Republik*
IAI	Ibero-American Institute of Berlin / *Ibero-amerikanisches Institut*
IAI-H	Ibero-American Institute of Hamburg / *Ibero-amerikanisches Institut Hamburg*
ICH	Institute of Spanish Culture / *Instituto de Cultura Hispánica*
INI	National Institute of Industry / *Instituto Nacional de Industria*
MAE	Ministry of Foreign Affairs (Spain) / *Ministerio de Asuntos Exteriores*
SPD	German Social Democratic Party / *Sozialdemokratische Partei Deutschland*

Archives Consulted

AMAE	Archive of the Ministry of Foreign Affairs, Madrid / Archivo del Ministerio de Asuntos Exteriores, Madrid
GSTA	Geheimes Staatsarchiv Preussischer Kulturbesitz, Berlin
PAAA	Politisches Archiv des Auswärtigen Amts, Berlin

References

Antolín Hofrichter, Andrés (2018), *Fremde Moderne. Wissenschaftspolitik, Geschichtswissenschaft und nationale Narrative unter dem Franco-Regime 1939–1964*, Oldenburg, De Gruyter.

Aschmann, Birgit (1999), *"Treue Freunde"…? Westdeutschland und Spanien 1945–1963*, Stuttgart, Franz Steiner Verlag.

Aschmann, Birgit (2008), "La República Federal de Alemania y la imagen de Alemania en España, 1945–1963", *Ayer*, 69: 129–154.

Ash, Mitchell G. (1996), "Denazifying Scientists—and Science", in Judt, Mattias; Ciesta, Burghard (Eds.), *Technology Transfer Out of Germany After 1945*, Amsterdam, Harwood Academic Publishers.

Ash, Mitchell G. (2006), "Wissenschaftswandlungen und politische Umbrüche im 20. Jahrhundert—was hatten sie miteinander zu tun?", in Bruch, Rüdiger vom; Gerhardt, Uta; Pawliczek, Aleksandra (Eds.), *Kontinuitäten und Diskontinuitäten in der Wissenschaftsgeschichte des 20.* Jahrhunderts, Stuttgart, Franz Steiner Verlag: 19–37.

Bernecker, Walther L. (2002). "Alemania y España en la época del Nacionalsocialismo", in Vega Cernuda, Miguel Ángel; Wegener, Henning, *España y Alemania. Percepciones mutuas de cinco siglos de historia*, Madrid, Editorial Complutense: 155–181.

Bräutigam, Thomas (1997), *Hispanistik im Dritten Reich. Eine wissenschaftliche Studie*, Frankfurt a.M., Vervuert Verlag.

Chamrad, Susanne; Santiago, Donald; Schouwink, Wilfried (1994), *100 Jahre Deutsche Schule Barcelona. Chronik/100 años Colegio Alemán Barcelona. Crónica. Memoria 1993–1994*, Barcelona, Der Vorstand des Schulvereins/Junta Directiva del Colegio.

Collado Seidel, Carlos (1992), "España y los agentes alemanes 1944–1947. Intransigencia y pragmatismo político", *Espacio, Tiempo y Forma*, Serie V, H[a] Contemporánea, t.V: 431–482.

Collado Seidel, Carlos (1993), "En defensa de occidente. Perspectivas en las relaciones del régimen de Franco con los Gobiernos Democristianos de Alemania (1949–1966)", in Tusell, Javier; et al. (Eds.), *El régimen de Franco (1936–1975): política y relaciones exteriores*, volume II, Madrid, UNED: 475–492.

Collado Seidel, Carlos (2001), *Angst vor dem "Vierten Reich". Die Alliierten und die Ausschaltung des deutschen Einflusses in Spanien 1944–1958*, Paderborn/München/Wien/Zürich, Ferdinand Schöning.

Collado Seidel, Carlos (2005), *España, refugio nazi*, Madrid, Temas de Hoy.

Collado Seidel, Carlos (2007a), "Ende und Neuanfang der Deutschen Schulen in Spanien nach 1945", in Altmann, Werner; Vences, Ursula (Eds.), *Por España y el mundo hispánico. Festschrift für Walther L. Bernecker*, Berlin, edition tranvía—Verlag Walter Frey.

Collado Seidel, Carlos (2007b), "La República Federal de Alemania y el restablecimiento de la presencia industrial de anteguerra en España", *Iberoamericana*, VII, 26: 109–122.

Delaunay, Jean-Marc (1989), "La liquidation des avoirs allemands en Espagne (1945–1961)", in Urquijo y Goitia, José Ramón de; Etienvre, Jean Pierre (Eds.), *España, Francia y La comunidad europea*, Madrid, CSIC/Casa de Velázquez: 219–245.

Delgado Gómez-Escalonilla, Lorenzo (1992), *Imperio de papel. Acción cultural y política exterior durante el primer franquismo*, Madrid, Consejo Superior de Investigaciones Científicas.

García Pérez, Rafael (1990), "La idea de la "Nueva Europa" en el pensamiento nacionalista español de la inmediata posguerra, 1939–1944", *Revista del Centro de Estudios Constitucionales*, 5: 203–240.

García Pérez, Rafael (1994), *Franquismo y Tercer Reich. Las relaciones económicas hispano-alemanas durante la segunda Guerra Mundial*, Madrid, Centro de Estudios Constitucionales.

Gienow-Hecht, Jessica; Donfried, Mark C. (2010), "The Model of Cultural Diplomacy. Power, Distance, and the Promise of Civil Society", in Gienow-Hecht, Jessica; Donfried, Mark C. (Eds.), *Searching for a Cultural Diplomacy*, New York/Oxford, Berghahn Books: 13–29, 21.

Grossmann, Rudolf (1941), "Spanien wird wieder Grossmacht", *Ibero-Amerikanische Rundschau*, 6: 126–129.

Grossmann, Rudolf (1942), "Vom Wesen der spanischen Kultur", *Ibero-Amerikanische Rundschau*, 1: 3–6.

Grossmann, Rudolf (1943), *Der spanische Nationalcharakter. Vortrag gehalten in Leipzig am 10. Januar 1943*, Leipzig, Schriften der Deutsch-Spanischen Gesellschaft Zweigstelle Leipzig: 2.

Hausmann, Frank Rutger (2001), "Das Deutsche Wissenschaftliche Institut in Madrid", in *"Auch im Krieg schweigen die Musen nicht": Die Deutschen Wissenschaftlichen Institute im Zweiten Weltkrieg*, Göttingen, Vandenhoeck & Ruprecht, Capítulo: 211–237, 7.

Hera Martínez, Jesús de la (2014), "Ciencia y propaganda: El Instituto Alemán de Cultura de Madrid y sus antecedentes: (1924–1945)", *Iurisprudentia Elegans: Revista de Derecho Político e historia Constitucional*, 1: 57–75, accesible: http://www.iurisprudentiaelegans.es/imagenes/numeros/EL%20INSTITUTO%20ALEMAN%20DE%20CULTURA.pdf (accessed 18.5.2018).

Herzner, Dominik (2019), *Deutsche Auslandschulen in Spanien. Auswärtige Kulturpolitik zwischen Konflikt und Kooperation*, Bielefeld, transcript Verlag.

Hüffer, Hermann (1942), *Deutsch-spanische Kulturbeziehungen in alter und neuer Zeit*, München, edición especial de la revista de la Deutsche Akademie, *Deutsche Kultur im Leben der Völker*, 1, 17.

Hüffer, Hermann (1951), "Las relaciones hispanogermanas durante mildoscientos años (un resumen)", *Revista de Estudios Políticos*, 56: 43–75.

Janué Miret, Mariciò (2008a), "La cultura como instrumento de la influencia alemana en España: la Sociedad Germano-Española de Berlín (1930–1945)", *Ayer*, 69: 21–45.

Janué Miret, Mariciò (2008b), "Imperialismus durch auswärtige Kulturpolitik: die Deutsch-Spanische Gesellschaft als "zwischenstaatlicher Verband" unter dem Nationalsozialismus", *German Studies Review*, 31, 1: 109–132.

Janué Miret, Mariciò (2015a), "Relaciones culturales en el "nuevo orden": La Alemania nazi y la España de Franco", *Hispania*, LXXV, 251: 805–832.

Janué Miret, Mariciò (2015b), "The Role of Culture in German-Spanish Relations During National Socialism", in Fernando Clara, Cláudia Ninhos (Ed.), *Nazi Germany and Southern Europe, 1933–1945*. Science, Culture and Politics, UK, Palgrave Macmillan: 84–104, 201.

Janué Miret, Mariciò (2019), "Contributing to the Cultural "New Order": How German Intellectuals Attributed a Prominent Place for the Spanish Nation", in Maria Björkman, Patrik Lundell y Sven (Ed.), *Intellectual Collaboration with the Third Reich. Treason or Reason?*, Routledge: 214–229.

Jorba, Manuel (2007), "Record de Hans Juretschke (1909–2004) i dels seus estudis Catalans", *Anuari Verdaguer*, 15: 529–537.

Lapesa, Rafael (1974), "Carlos Clavería Lisana (1904–1974)", *Boletín de la Real Academia Española*, Año LXI, Tomo LIV, Cuaderno CCIII, sept-dic: 356–363.

Lehmann, Walter (2006), *Die Bundesrepublik und Franco-Spanien in den 50er Jahren. Ns-Vergangenheit als Bürde?*, München, Oldenburg.

Liehr, Reinhard, Maihold, Günther, Vollmer, Günter (coords) (2003), *Ein Institut und sein General, Wilhelm von Faupel und das Ibero-Amerikanische Institut in der Zeit des Nationalsozialismus*, Frankfurt am Main, Vervuert.

Maier Allende, Jorge, Schattner, Thomas G. (2010), "Das Deutsche Archäologische Institut und die Archäologie der Iberischen Halbinsel/El Instituto Arqueológico Alemán y la arqueología de la península ibérica", in Rebok, Sandra (Ed.), *Traspasar fronteras. Un siglo de intercambio científico entre España y Alemania/Ein Jahrhundert Deutsch—Spanisch Wissenschaftsbeziehungen*, Madrid, Centro de Estudios Constitucionales: 330–357.

Messenger, David A. (2014), *Hunting Nazis in Franco's Spain*, Baton Rouge, Louisiana State University.

Moreno Juste, Antonio (1993), "El Centro Europeo de Documentación e Información. Un intento fallido de aproximación a Europa 1952–1962", in Tusell, Javier, et al. (Eds.), *El régimen de Franco 1936–1975: política y relaciones exteriores*, volume II, Madrid, UNED: 459–474.

Moreno Juste, Antonio (2005), "La política europea de los católicos españoles en los años 40 y 50", in Recio, Sánchez (Ed.), *La Internacional Católica: Pax Romana en la política europea de posguerra*, Madrid, Biblioteca Nueva, 175–211.

Niehus, Gerlinde Freia (1989), *Aussenpolitik im Wandel. Die Aussenpolitik Spaniens von der Diktatur Francos zur parlamentarischen Demokratie*, Frankfurt a.M., Vervuert.

Petriconi, Hellmuth (1941), "Das Spanienbild im deutschen Bewusstsein", *Geist der Zeit. Wesen und Gestalt der Völker. Organ des deutschen Akademischen Austauschdienstes*, 19, 2: 65–75.

Presas, Albert (2008), "Technological Transfer as a Political Weapon: Technological Relations between Germany and Spain from 1918 to the early 1950s," *Journal of Modern European History*, 6, 2: 218–235.

Presas, Albert (2010), "Technoscientific Synergies between Germany and Spain in the Twentieth Century: Continuity Amid Radical Change," *Technology and Culture*, 51, 1: 80–98

Rodríguez López, Carolina (2008), "La Universidad de Madrid como escenario de las relaciones hispano-alemanas en el primer franquismo (1939–1951)", *Ayer*, 69: 101–128.

Ruiz Escudero, Inés (2007), "Las relaciones entre la República Federal Alemana y el régimen franquista tras la Segunda Guerra Mundial a través de la prensa española (1945–1958)", *HAOL*, 12: 133–141, accesible: https://dialnet.unirioja.es/servlet/autor?codigo=2112052 (accessed 5.2.2021).

Ruiz Escudero, Inés (2008a), "La política cultural española en la República Federal de Alemania: el Instituto de España en Munich (1956–1966)", *Investigaciones Históricas*, 28: 201–216, accesible: https://dialnet.unirioja.es/servlet/autor?codigo=2112052 (accessed 5.2.2021).

Ruiz Escudero, Inés (2008b), "Las relaciones hispano-alemanas a través de su política cultural (1951–1958)", in Nicolás Marín, María Encarna; González Martínez, Carmen (Eds.), *Ayeres en discusión: temas clave de Historia Contemporánea hoy*, Murcia, Universidad de Murcia, Servicio de Publicaciones: 2940–2956, accesible: https://dialnet.unirioja.es/servlet/libro?codigo=399225 (accessed 5.2.2021).

Ruiz Escudero, Inés (2015), *Franco y Adenauer: la diplomacia cultural hispano-germana en los años cincuenta*, Valladolid, Universidad de Valladolid.

Sanz Díaz, Carlos (2005), *España y la República Federal de Alemania (1949–1966): política, economía y emigración, entre la Guerra Fría y la distensión*, Madrid, Universidad complutense, Tesis doctoral, accesible https://eprints.ucm.es/7245/1/T28931.pdf (accessed 7.3.2019).

Sanz Díaz, Carlos (2008), "El papel de la política cultural en las relaciones hispano-alemanas, 1949–1966", *Ayer*, 69: 155–185.

Sanz Díaz, Carlos (2010), "Relaciones científico-culturales hispano-alemanas entre 1939 y 1975/Die wissenschaftlich-kulturellen Beziehungen zwischen Spanien und Deutschland von 1939 bis 1974", in Rebok, Sandra (Ed.), *Traspasar Fronteras. Un siglo de intercambio científico/Über Grenzen Hinaus.*

Ein Jahrhundert Deutsch-Spanische Wissenschaftsbeziehungen, Madrid, CSIC/DAAD: 359–381.

Saz Campos, Ismael (2003), *España contra España. Los nacionalismos franquistas*, Madrid, Marcial Pons.

Schramm, Edmund (1935), *Donoso Cortés. Leben und Werk eines spanischen Antiliberalen*, Hamburg, Ibero-Amerikanisches Institut.

Schramm, Edmund (1937), "Die geistigen und politischen Voraussetzungen des spanischen Bürgerkrieges", *Zeitschrift für Politik*, 27: 41–59.

Schramm, Edmund (1940), *Frankreichs Einmischung im spanischen Bürgerkrieg*, Berlin, Deutsche Informationsstelle.

Sesma Landrín, Nicolás (2011), "Importando el Nuevo Orden. El Instituto de Estudios Políticos y la recepción de la cultura fascista y nacionalsocialista en España (1939–1942)", in Gallego, Ferran, Morente Valero, Francisco (coords), *Rebeldes y reaccionarios: intelectuales, fascismo y derecha radical en Europa*, Mataró, El Viejo Topo: 243–280.

Sesma Landrín, Nicolás (2017), "Modernism and Falangism: The Journal *Escorial* and Its Authoritarian Modernizing Project", in Gallego, Margaleff; Morente, Ferran (Eds.), *Welfare as a Totalitarian Tool. Fascist Italy and Francoist Spain in the European Context*, Francisco, Sussex Academic Press: 90–116.

Vega Cernuda, Miguel Ángel (2003), "Introducción", in Vega Cernuda, Miguel Ángel (Ed.), *España y Europa: Estudios de crítica cultural: Obras completas de Hans Juretschke*, Madrid, Editorial Complutense: IX–XXV.

Weber, Petra-María (1992a), *Spanische Deutschlandpolitik 1945–1958. Entsorgung der Vergangenheit*, Saarbrücken-Fort Lauerdale, Verlag breitenbach Publishers.

Weber, Petra-María (1992b), "Política española hacia Alemania, 1945–1958: El impacto político y económico de las relaciones hispano-alemanas", in Bernecker, Walther L. (Ed.), *España y Alemania en la Edad Contemporánea*, Frankfurt am Main: Vervuert Verlag: 209–230.

Weber, Petra-María (1994), "El CEDI: promotor del occidente cristiano y de las relaciones hispano-alemanas de los años cincuenta", *Hispania LIV/3*, 188: 1077–1103.

CHAPTER 16

Un scandale: Franco à l'UNESCO: The Franco Dictatorship and the Struggle for International Representation in the Social Sciences

Nicolás Sesma

Introduction

On 30 November 1952, convened by the *Ligue des droits de l'homme* in Paris' Salle Wagram, a notable cast of intellectuals and political representatives met for an act of protest under the slogan: "Un scandale: Franco à l'UNESCO." Indeed, ten days prior, Francoist Spain had been admitted

This work was funded by the research project HAR2017-85967-P, "The University Labor Service (SUT) in Franco's Spain. A comparative European perspective (1950–1970)," directed by prof. Miguel Ángel Ruiz Carnicer and financed by the Spanish Ministry of Economy and Competitiveness.

N. Sesma (✉)
Université Grenoble Alpes, Saint-Martin-d'Hères, France

Casa de Velázquez, Madrid, Spain
e-mail: nicolas.sesma-landrin@univ-grenoble-alpes.fr

M. Janué i Miret, A. Presas i Puig (eds.), *Science, Culture and National Identity in Francoist Spain, 1939–1959*,
https://doi.org/10.1007/978-3-030-58646-1_16

into the United Nations Educational, Scientific and Cultural Organization. Following speeches by figures of the French resistance and the Spanish exile community, as well as the reading of messages sent for the evening by the Catalan cellist Pau Casals and African American writer Richard Wright, among others, Albert Camus finally took the floor, summarizing the general sentiment of those present with an uncompromising rant:

> ...is not Calderón or Lope de Vega that democracies have just welcomed into their society of educators but, rather, Joseph Goebbels [...] poor Hitler; instead of committing suicide [...] he should have imitated his friend Franco and armed himself with patience. Today, he would be a UNESCO delegate for the education of the Upper Niger.[1]

Both the future Nobel laureate in literature and many other UNESCO collaborators sought to mobilize public opinion in democratic countries and pressure their respective governments to block the entrance of the dictatorship. Not surprisingly, and particularly in France, home to UNESCO headquarters, the organization was considered particularly symbolic of the values that triumphed thanks to the allied victory in the Second World War, something that should necessarily entail the exclusion of the Franco regime due to its "fascist character."[2] However, nothing could prevent the approval of the Spanish candidacy by a wide majority a week later at the Seventh General Conference of the institution, thus ratifying the favourable opinion issued by its Executive Council and by the United Nations (UN) Economic and Social Council.[3]

This progressive inclusion of Francoism in the specialized agencies of the UN, a result of the fact that, in November of 1950, the General Assembly revoked its resolution of condemnation in late 1946,[4] has traditionally been considered the first step towards the full acceptance of dictatorship in the post-war multilateral international system. According to this interpretation, the definitive closure of the so-called "Spanish question" and thus the culmination of this process of rehabilitation would come with the dual signing in 1953 of the Concordant with the Vatican and the

[1] Camus (2017): 190.

[2] United Nations General Assembly. Resolutions (A/RES): Resolution 39 (I). "Relations of Members of the United Nations with Spain," 12 December 1946.

[3] Dulphy (2007): 313–316.

[4] A/RES: Resolution 386 (V). "Relations of States Members and specialized agencies with Spain," 4 November 1950.

accords with the United States as well as through its inclusion as a full member of the UN in December of 1955.[5]

However, in many ways, international acceptance of Franco's Spain by Western powers had already been a reality for quite some time, and not only de facto acceptance, but some cases even officially, as demonstrated by Spain's presence at numerous key technical meetings for the configuration of the post-war world. For example, a delegation from the dictatorship—led by the prestigious engineer Esteve Terradas i Illa—had been invited to the Convention on International Civil Aviation held in Chicago in late 1944, and the resulting agreement, which gave rise to the creation of an International Civil Aviation Organization (ICAO), was signed by the Franco government. In fact, when ICAO sought to join the UN as a specialized agency, the dilemma was put forth of what to do about a founding member of the organization that had been condemned by the UN. This polemic issue was settled by an angry unilateral Spanish withdrawal that was nonetheless a symbolic gesture, for it was preceded by a tacit agreement to continue collaborating through extraofficial channels.[6]

Regarding international cultural circles, both in the implementation of UNESCO and in the first meetings of the European movement, Spanish representation was led by exiled republican academics, but the exclusion of the dictatorship was also not truly complete. In this way, Catholic platforms such as the organization *Pax Romana* had secured contact with conservative European and American intellectuals—its 19th World Congress was held in Salamanca and El Escorial in 1946–. In the same way, none other than William H. Beveridge accepted an invitation to attend the inauguration in March 1946 of a chair on Social Security at the Faculty of Political and Economic Sciences at Madrid Central University. With the creation of the chair and Beveridge's presence, the dean of the faculty and future minister of foreign affairs, Fernando María Castiella, sought to demonstrate that the country was abreast of recent trends in international academic debates, in which the "social question" and disciplines designed to study it, such as sociology, political science, and

[5] See the contribution by Lorenzo DELGADO in this volume.

[6] Sochor (1991): 41–44. "... since Spanish airspace could not be removed from aeronautical maps by the stroke of a pen, some pragmatic way had to be found to accommodate the 'unwelcome guest'. For all practical purposes, Spain was never considered a pariah state within ICAO, and when the time came for it to re-join the Organisation, it was given a seat on the Council [...] In fact, ICAO had been working behind the scenes to maintain relations with that country".

economics, were occupying an ever more prominent role. Meanwhile, it was also about counteracting the work of republican exiles, who through the Union of Spanish University Professors in Exile (UPUEE) and the Declaration of Havana (1943) had also addressed the social question and demonstrated their adherence to the principles that they proposed to solve it in the Atlantic Charter of 1941.[7]

Science and culture thus occupied an ambivalent place within the general political situation and the international strategy of the Franco dictatorship. On one hand, it was not in desperate need of contact with these arenas, as it had been asserted on certain occasions; although they were scarce, it possessed sufficient channels to avoid a complete "intellectual autarky." Likewise, in terms of international technical regulations—maritime law, civil aviation, international payments and so on—circumstances resulted in imposing the pragmatism of having to negotiate with those who effectively controlled the peninsular territory. On the other hand, the Franco regime continued to be considered a political pariah, the last stronghold of defeated fascism, absent from the great international initiatives, which contrasted with the prestige commanded by numerous intellectual figures in exile, considered in their own right to be the authentic incarnation of Spanish culture. In addition, focused on the repression of dissent, internal struggles and high-level diplomacy, neither Franco nor those in his inner circle ever considered academic and cultural aspects one of the priorities for staying in power, and, hence, those in charge of its management, who were fully conscious of the repercussions for international perceptions of the regime, acted with significant autonomy and had to contend with quite limited financial and human resources.

Based on these premises, the objective of this article is to provide an approximation to the implementation and development of the cultural strategies of the Franco dictatorship, as well as their repercussions in domestic and foreign politics, through the specific case of the social sciences. Thus, we will provide an account of the institutionalization of these disciplines in Spain, deeply marked by the clash between the traditional Catholic circles and the world of the *Institución Libre de Enseñanza* (Free Institution of Education, ILE, 1876), each with their own model of national construction. Next, we analyse the situation following the Spanish Civil War, paying particular attention to a Falangist institution, the Institute of Political Studies (IEP). Finally, we examine the performance

[7] Claret (2019): 299–302.

of the Francoist delegation in UNESCO, whose trajectory can help us to understand the functioning of the intellectual apparatus of the dictatorship as a whole and its desire for exclusivity in representing national culture.

What is Spain? Two Competing Projects of National Construction

As is well known, the *Junta para Ampliación de Estudios e Investigaciones Científicas* (Council for the Development of Studies and Scientific Research, JAE, 1907) was the most important effort for renewing the Spanish cultural, social and scientific life during the last century. Inspired by the principles of the ILE and presided over by the Nobel Prize winner in medicine Santiago Ramón y Cajal, the JAE was officially part of the Ministry of Public Education, although it enjoyed a certain autonomy, which allowed it to pursue its own objectives: on one hand, the modernization of scientific and educational structures, through the creation of a series of specialized centres; on the other hand, the formation of a generation of cadres within the great international currents of scientific knowledge, for which a pension system was designed to cover travel and stays abroad.[8]

Although those responsible for the JAE immediately sought to isolate themselves from partisan struggles, they were undoubtedly fully aware of the political implications of their proposals, which necessarily led to democratic rupture with the oligarchic and confessional system of the Restoration (1875–1923). In this sense, despite initially concentrating on a small elite sector from the urban middle class, the desire to generalize its pedagogical model to low-income sectors indicated the JAE's clear vocation to lead a process of nationalization of the population in a new civic culture, a process similar to the conversion of "*Peasants Into Frenchmen*," to use the expression coined by Eugen Weber to refer to what took place under the French Third Republic (1870–1940).[9]

Since its creation, the JAE was accused of being foreign-centred and anti-Spanish by conservative circles and political Catholicism, accusations that were strongly established in the collective imaginary by the campaign to repress and discredit the institution led later on by the Franco

[8] García Velasco (2007): 157–199.

[9] Weber (1976): 303–338.

dictatorship.[10] However, both the JAE and the ILE were, in reality, clearly "patriotic" initiatives in that they upheld nineteenth-century enlightened and revolutionary liberalism and were bearers of a global project for the construction of the "Spanish nation."

Those responsible for the JAE, therefore, were not "heterodox Spaniards,"[11] because to label them thus implies assuming the Francoist vision that the Catholic and traditionalist model theorized by Menéndez Pelayo constituted the only possible Spanish orthodoxy. What was happening was that two projects of national construction were competing to implement their political and cultural ideals in the institutions: the "liberal fatherland" incarnated by the JAE and the "fatherland of the faith" defended by conservative sectors.[12] As Xosé Manoel Nuñez Seixas has stated, "several Spanish nationalist discourses exist."[13] A clear example of this patriotic character of the JAE was the title chosen for the documentary film made between 1929 and 1930 to present to the international public the work by the centres of the JAE: *What is Spain?*

Within this general framework, and although they never had a specific centre, the social sciences occupied a prominent place. Among the pioneers of sociology in Spain were numerous Krausists and institutionalists, such as Gumersindo de Azcárate, Adolfo Posada and particularly Manuel Sales i Ferré, selected in 1899 to hold the first chair in sociology at Madrid Central University. Through this and other intellectual platforms—*Boletín de la Institución Libre de Enseñanza*, *Revista de Occidente*—an accelerated importation of the techniques of empirical sociology occurred, as their capacity to offer more precise diagnoses of the reality of society perfectly suited to aspirations for "scientific interventionism" by the JAE institutions.

Each initiative launched from the ILE environment was responded to by political Catholicism with an equivalent initiative, but one that was structured according to the Catholic and conservative spirit, and the new social sciences were no exception. Thus, Severino Aznar, one of the country's most dynamic Catholic activists, quickly demonstrated great interest in sociological studies, in a reflection of the desire to renew the instruments of Catholic action inspired by the doctrine of Pope Leo XIII, which would lead him to found the "Social Weeks" (1906–1912) and the

[10] Sesma (2013): 386–415.
[11] López Sánchez (2006).
[12] Moreno Luzón (2003): 207–235.
[13] Nuñez Seixas (2018): 267–297.

"Christian Democracy Group" (1918). Similarly, the charismatic Ángel Herrera, president of *Asociación Católica Nacional de Propagandistas* [National Catholic Association of Propagandists] (ACNP, 1908), quickly incorporated economic and sociological topics into the programme of their "Study Circles," the platform to train a new Catholic elite for counteracting the growing presence of ILE figures in positions of responsibility. However, the widespread rejection of empiricism among the Catholic ranks meant that in all these spaces the topics analysed were treated using methodologies that were closer to the history of thought, family law, and particularly social philosophy than to the actual social sciences.

ILE figures and Catholics, like their divergent perspectives in approaching the "social question," clashed on numerous occasions. The most polemic of all of these was the reassignment in 1916 of the aforementioned chair in sociology at Madrid Central University. Among the aspirants were Severino Aznar and JAE Secretary General José Castillejo, before a tribunal which included noted figures in political Catholicism, as well as an individual linked to both the confessional world and the JAE: the priest and expert Arabist Miguel Asín Palacios.

In an undoubtedly controversial decision, the chair was awarded to Severino Aznar with a vote in favour by Asín Palacios, despite the fact that he did not demonstrate being prepared from a methodological point of view. Thus, in a private memorandum—"this note not given to anyone"—Castillejo stated ironically:

> Mr. Aznar modestly avoided speaking of sociological doctrines or books on sociology [...] Mr. Asín has given the vote to his friend: this is for perhaps the most bitter harm for the country [...] There can be no teaching of scientific sociology at Central University until Mr. Aznar dies or retires.[14]

Castillejo's outrage, which ended up leaking to the press and being the subject of a tense correspondence, ultimately led to Asín Palacios's resignation from the JAE. Some years later, this episode would be widely used to demonstrate the supposed sectarian and anti-religious nature of the institution.

Nonetheless, the researchers and grantees of the JAE always included individuals hailing from Catholic university circles, such as Luis Recasens

[14]Archive of the Council for the Development of Studies and Scientific Research (AJAE), File of Julián Ribera (JAE/122-151); Castillejo (1999): 265–285.

Siches, professor at Santiago de Compostela, who would become, throughout the 1920s, the main upholder of interest in the social sciences and their methodological innovations. Indeed, thanks to a grant to expand studies in Berlin and Vienna (1925–1926), where he was able to work with Hermann Heller, Werner Sombart and Hans Kelsen, Recasens would end up becoming the architect of the definitive consolidation in Spain of empirical political science and sociology.[15] For this large task, he had the invaluable help of another two academics linked to the JAE: on one hand, Francisco Ayala, professor at University of La Laguna, who travelled as well to Germany and Austria (1929) to work with Heller and became an expert on sociology and state theory,[16] and, on the other, José Medina Echavarría, a lawyer by training, lecturer at the University of Marburg (1931), professor of philosophy of law at University of Murcia and undoubtedly the most interested in empirical sociology from the beginning.

In July of 1936, José Medina obtained a new grant from the JAE to hone his sociological research techniques at universities of Columbia in New York, and the Social Research Laboratory in Chicago.[17] The trip, however, had to be cancelled in view of the seriousness of the events in Spain: a coup d'état against the legitimate republican government of the Popular Front, which would end up in a long and bloody civil war. The conflict spread to—among many other things—the institutions that had played a central role in the country's progressive scientific normalization during the first third of the century as well as the majority of its protagonists. Those who chose the path of exile also included Recasens, Ayala and Medina, all involved in republican diplomatic services. These social pioneering scientists would be, from now on, "sociologists without their own society."[18]

Exclusively National/Catholic Social Sciences?

The loss of the main proponents of the empirical strand of the social sciences ran parallel to the reinforcement of their normativist practice, as a philosophy and as a reinforcement of the Catholic ministry, also

[15] AJAE, File of Luis Recasens Siches (JAE/121-72).
[16] AJAE, File of Francisco Ayala (JAE/13-611).
[17] AJAE, File of José Medina Echavarría (JAE/98-481).
[18] Gómez Arboleya (1958a): 69; Gómez Arboleya (1958b): 821–839.

institutionalized through the creation in 1943 of the Balmes Institute of Sociology within *Consejo Superior de Investigaciones Científicas* [Higher Council of Scientific Research] (CSIC) the body designed to put the country's scientific production at the service of the Franco regime.[19] Logically, neither the selection of its managerial staff nor the resulting academic outcomes constituted any surprise. Thus, the director of the Institute was none other than Severino Aznar, the incarnation of "sociology at the service of religion and dictatorship,"[20] while its journal, *Revista Internacional de Sociología*, stated in 1948:

> The inability of empirical sociology to establish a social perspective is manifest [...] Contrary to the opinion of Max Weber, classical natural law remains standing.[21]

Undoubtedly, all these circumstances appear to have led us towards a dual conclusion, perfectly grounded in much of historiography. First, the perception of social sciences, and especially empirical sociology, as "intrinsically an 'oppositional science,'" which would have caused its persecution by fascist regimes: "fearful of the power of the empirical sociologist to reveal unpleasant facts about societies."[22] Second, related to the particular case of the Franco dictatorship—at least until the arrival of the technocrats—the interpretation of its first years as a parenthesis from the scientific point of view, given the impossibility of carrying out truly rigorous research and methodologies under doctrinal parameters of "ultramontane and reactionary character."[23] The social sciences were not oblivious to this paradigm because, as Salvador Giner warned, "a reactionary regime that exiled—when it did not physically eliminate—the republican and progressive intellectuals of the country could only be violently hostile to the most delicate of the social sciences: sociology."[24]

Both deductions appear to be confirmed upon analysing the impeccable trajectory followed by exiled social scientists. Recasens Siches, Francisco Ayala and Medina Echavarría never abandoned their commitment to democracy and anti-Francoism—all of them collaborated with republican

[19] Malet (2008): 212.
[20] Álvarez-Uría and Varela (1992): 58.
[21] Larraz (1948): 40.
[22] Turner (1992): 1.
[23] Otero Carvajal (2006): 59.
[24] Giner (1990): 52.

institutions in exile—and brilliantly continued their careers in the Americas, where they made important contributions to the academic consolidation of sociology. In this sense, Medina Echavarría was responsible for the section on sociological works at the Mexican publishing house Fondo de Cultura Económica (FCE),[25] and thus was in charge of coordinating pioneering translations of the works of, among others, Karl Mannheim, Werner Sombart and, particularly, Max Weber, whose 1944 edition of *Economy and Society* gave Spanish-speaking readers access to the German author long before his works were available in other languages.[26]

However, regarding both the classic association between scientific advances and progressive values and the lack of scientific activity during Francoism, the response is necessarily much more complex. Despite the terrible repression it exerted on the cultural world—purge of university, suppression of the JAE—the Franco dictatorship had a scientific policy, as numerous authors have demonstrated.[27] Moreover, it would be erroneous to think that it encouraged and practised a single model of science: the national/Catholic variety. This would entail leaving out of the analysis the notion of "alternative modernization" inherent in fascist projects[28] and represented in the dictatorship by powerful groups. On one hand, military cadres who were aware of the need for technology and applied science for the defence sector and whose economic interests pointed towards the development of a nascent military/industrial complex through the National Institute of Industry.[29] On the other, certain Falangist intellectual sectors that aspired to impose their cultural strategies on the whole nationalist coalition that emerged from the civil war.

Regarding the latter sectors and as we have stated elsewhere,[30] the search for this fascist type of modernization led Falangism to a strategy of reappropriating the legacy of the JAE during the initial years of the dictatorship. The nationalizing discourse of the JAE, conveniently decontextualized and purified of its democratic intent, could be partially assumed by the single party to put it to work in its own nationalization project in a fascist, nationalist and imperialist tone. This process also included the integration of those JAE grantees who had not fallen due to repression or gone into exile. Falangism

[25] Ribes and Morales (2017): 625–635.
[26] Morcillo (2012): 612.
[27] Camprubí (2017): 9–14.
[28] Herf (1985) and Griffin (2007).
[29] Presas (2005): 197–218.
[30] Sesma (2017): 90–116; Saz (2003): 270.

had a serious lack of qualified staff, and, hence, recruiting from this class of high-level cadres was very attractive as a way to counter the merely restorative plans defined by the ACNP and by the monarchists.

The main recipient of this transfer of former grantees was the Institute of Political Studies, an organization initially conceived of as a think tank for the single party and which eventually became one of the "epistemic communities" placed at the service of the dictatorship as a whole, in this case in legal advising, doctrinal creation and improving international perceptions of the regime.[31]

Among the members of the IEP between 1939 and 1956, approximately 30 percent had been beneficiaries of a JAE grant at one time, among them several university students who had been interested in empirical social sciences. To cite just two significant examples: Enrique Gómez Arboleya, who studied at University of Berlin with Nicolai Hartmann and Eduard Spranger in 1934–1935, and Francisco Javier Conde, who worked closely to Carl Schmitt at University of Berlin between 1934 and 1936. These intellectuals came to break with the idea of an ideologically uncontaminated empirical social sciences because "sociologists served the Nazi regimes just as other scholars did,"[32] and they were called on to play important roles as doctrinal agents and external intellectual projection of the dictatorship.[33]

Although the normative sociology of the Catholic variety came to be widely dominant in the post-war period, the IEP provided the empirical perspective, which also included paying attention to the activity of exiles. Thus, the aforementioned translations of the FCE were widely used at the *Revista de Estudios Políticos* (*REP*), IEP's journal, from Gustav Radbruch's *Legal Philosophy* (1944) to Max Weber's *Economy and Society* (1944); however, care was taken to censure the fact that Medina Echavarría was the author of those translations.

With the arrival of Javier Conde at the helm of the IEP in 1948, attention to the social sciences experienced a decided push. In order to renew the legitimizing discourse of the dictatorship and possess effective instruments with which to probe public opinion, it was deemed necessary to assimilate the modern techniques of research and analysis dominant in American production and train the future leaders of the Movement in

[31] See the contribution by Albert PRESAS in this volume.

[32] Turner (1992): 2.

[33] See the contribution by Antonio NIÑO in this volume.

those techniques. To this end, as part of the specialized courses offered at the IEP, some new "sociology courses" (1949–1956) were offered under Enrique Gómez Arboleya; and research grants were also negotiated, which led to the departure of, among others, Juan José Linz (Conde's chair assistant) to Columbia University and Salustiano del Campo to the University of Chicago, both in 1951.[34] Along these same lines, Conde sought to differentiate the Institute from the regime's propaganda and give it a halo of professionalism and scientific credibility. To this end, he brought numerous prestigious authors to the *REP* who were able to project an image of loftiness and intellectual independence.

Of course, Conde was not guided by a desire for liberalization or openness of the political/cultural structures of the state nor a desire for a true rapprochement and dialogue with the outside world but, rather, by the effort to match his doctrinal references and instrumental dialectic to European and global realities, something which ultimately implied an important legitimizing role. Additionally, in the face of Franco's limited attention to academic theorization,[35] these were initiatives that, to a large extent, Conde promoted with almost complete autonomy.

However, Conde's aims initially clashed with intellectual circles that were not willing to accept representatives of the official culture of the dictatorship as valid interlocutors. Thus, in 1950, *Les Temps modernes*—directed by Jean Paul Sartre and undoubtedly the "journal/institution" of the French post-war period—in a monographic issue on the domestic situation in Spain, alluded to Javier Conde as a mere Falangist lackey.[36] UNESCO, for its part, when selecting collaborators for one of their key projects in the early 1950s—a survey of specialists aimed at providing a complete overview by country of the history, trends and methodologies in sociology and political science—once again turned to the community of researchers in exile. The individual in charge of drafting the report "*La science politique en Espagne depuis trente ans*" was once again none other than Recasens Siches, for whom:

> Il y a lieu de signaler que l'Université espagnole ne compte qu'une seule chaire de sociologie [...] et que le professeur qui en est titulaire (M. Severino AZNAR) ne s'occupe pas, à proprement parler, de sociologie, mais de poli-

[34] Pecourt (2014): 130–138; Campo (2001): 161–180.
[35] Reig Tapia (1990): 80.
[36] Ayensa (1950): 2044–2054.

tique sociale […] L'Etat totalitaire de caractère nazi-fasciste créé en Espagne par la révolution militaire de FRANCO a suscité différentes apologies du régime: celles d'Ignacio Maria de LOJENDIO, de Javier CONDE GARCIA et de Laureano (*sic*) SANCHEZ AGESTA.[37]

The Success of Francoist Social Sciences in International Representation

Francoist Spain's admission into UNESCO caused a rapid reversal of this situation. Until that point and although it had never had official recognition, the delegation of the "Spanish republican government in exile" had enjoyed, since the First General Conference held in 1946, significant visibility as a special guest of the institution. However, as we stated previously, as of the Sixth General Conference in 1951, the Francoist representatives were now received as observers, which was a precursor to their eventual admission.[38]

Indeed, just one year later, the new observer Commission named by the dictatorship—headed by the consul in Geneva, José Sebastián de Erice—formalized the membership request, given that the diplomatic legal services confirmed that it could "cast aside any fear of intrusion by the Organization in Spain's internal matters." The cost, they stated, would be limited to "enemies of the Spanish regime and of Spain in general organizing campaigns against Spain on the occasion of its entrance."[39] In this sense, the mobilization of the anti-Fascist press and the doubts expressed by some delegations—spearheaded by the French minister of public information, André Marie—achieved the postponement of the vote regarding new members. However, amid a political climate marked by the Korean War and growing pragmatism in Hispanic-French relations, the vote was finally held, and only four countries were opposed (Mexico, Uruguay, Burma and Yugoslavia), while the British, US and French delegations voted in favour.[40] The Francoist media, which were ordered to "restrain criticisms of UNESCO and its initiatives," to downplay the importance of

[37] Recasens Siches (1950): 274.

[38] Biblioteca de Catalunya. Fons Joan Estelrich (BC.FE). Box 1: "Delasheras. Comment on Unesco conference. Paris, July 14": n.d. (1951).

[39] BC.FE. Box 1: "Order of 6 November 1952 by which the Spanish Observer Commission is named at the General Conference of UNESCO"; "Notes on UNESCO" n.d. (1951).

[40] Martínez Lillo (1993): 313–317.

the Cold War context, and to "silence the admission of Libya and Nepal at the same session,"[41] reported the news with their usual bravado:

> Spain has known of the existence of UNESCO [...] Spain has believed that it cannot deny this initiative its presence, its support, its collaboration [...] Ultimately naïve are men such as Camus [...] who have written to the UNESCO Secretary General to oppose Spain's entrance. Poor individuals, whom God has blinded in the extreme such that they consider themselves to be the spokespersons of universal conscience! [...] nobody was surprised that Spain was admitted to UNESCO.[42]

From then on, the republican exile community lost prominence in international perceptions of Spanish culture, leaving room for the Francoist monopoly over the representation of national sciences. This desire for exclusivity was not simply a product of circumstances or the passage of time at all, but rather, it was a political objective of the regime's authorities from the outset. Thus, as noted in the internal documentation to prepare entrance into UNESCO, one of the issues that was to be granted maximum priority was the purging of the "Spanish emigres" and their substitution with "Spanish nationals":

> ...there is now no genuinely Spanish personnel in the Organization [...] As is known, the Spanish personnel that has worked at the Organization for more or less extensive time periods, were recruited from among exiled intellectual elements and of whom it can be assumed that, at least deep down, are hostile to the current regime [...] for greater security, this personnel can be gradually substituted as the terms of the contracts of each person are met.[43]

Along with control over the officials and their particular "nationalization," the Francoist authorities were well aware that "the efficacy of Spanish participation in UNESCO depended on the correct choice of the permanent delegate." Unsurprisingly, the profile sought was that of a person with "sufficient intellectual rank to be properly received everywhere

[41] BC.FE: Box 1:"Circular of the Subsecretariat of Popular Education. Tétouan Provincial Office," 25 October 1952; "Confidential Telegram by the General Press Office. High Commission of Spain in Morocco," 21 November 1952.

[42] BC.FE. Box 1: File "UNESCO. Spain's admission 1952": Miguel Zelayeta,"Spain has been at UNESCO", n.d. (1952);"Spain at UNESCO," *España*, 20 November 1952.

[43] BC.FE. Box 1: "Spain's participation in UNESCO. Organizational outline," December 1951.

and to know languages, write well, know how to speak in public, know the world of addressing international affairs, and in short master the internal structure of UNESCO in all of its services."[44] In this sense and although the individual officially designated to occupy the post was then Spanish Ambassador in Paris José de Rojas y Moreno, the Count of Casa Rojas (1893–1973), the true promoter of the activities of the delegation was Joan Estelrich i Artigues (1896–1958). Named deputy permanent delegate in April of 1953, Estelrich had for a long time been organizing the participation of the Francoist committee at UNESCO and preparing the Parisian cultural environment for its arrival. Thus, in previous years and in close contact with the cultural attaché at the Embassy, Luis Díez del Corral (another former grantee of the JAE and noted member of the IEP) had handled the visits to Spain by numerous conservative French intellectuals, such as Henri Massis and Thierry Maulnier.[45]

Undoubtedly a determining figure of the political and cultural Catalanism of the beginning of the century, "*our* former separatist Joan Estelrich," as he was referred to with bitter irony by the Catalan writer *Gaziel* upon learning of his designation,[46] was the ideal person to lead the Spanish delegation. Not surprisingly, to his command of French was added a profound knowledge of the functioning of transnational organizations, for he had moved in the circles of the League of Nations during the 1920s and was a recognized cultural promoter, collaborator and founder of numerous periodicals including *La Veu de Mallorca* and *La Veu de Catalunya*.[47] Regarding his political loyalties, his connection with the dictatorship was the product of the events of the Spanish Civil War, when, as he would disclose in his personal diary: "victory of the military is the lesser of two evils" in view of the outbreak of the social revolution in Catalonia following the failure of the coup d'état in July of 1936.[48] Like many other former members of the Lliga Regionalista headed by Francesc Cambó, Estelrich had decided then to collaborate with the propaganda bodies of the Francoist cause, being one of the creators of the Press Office of the

[44] Ibid.

[45] BC.FE. Box 1: Ministry of Foreign Affairs. General Office of Foreign Policy, "Naming of deputy delegate to UNESCO," 28 April 1953. Box 2: Correspondence between Luis Díez del Corral and Joan Estelrich, 26 January, 8 and 11 March, 5 and 13 April and 23 May 1949.

[46] Gaziel (2010): 291.

[47] Pla (2015): 9–20.

[48] Estelrich (2012): 248.

Spanish State in Paris and the director of the journal *Occident* (also in Paris) as well as the newspaper *España* (in Tangier). The Francoist authorities never fully trusted him, as with the majority of those conservative Catalanists who until the war had followed what Ferran Valls i Taberner called "the false route"; however, they were well aware that Estelrich also had no other option and that his concurrence notably improved the image of the regime abroad.[49]

At the end of the day, as with the case of Javier Conde, Estelrich made virtue of this apparently heterodox profile. Despite the limited means available to him, he was soon elected to UNESCO's Executive Council (1954), and, although always in coordination with the ministerial services, the truth is that he was able to take decisions with remarkable autonomy when "Preparing the tasks and advising the different persons who would occasionally and temporarily intervene, designated by Spain, in any of the bodies or initiatives and enterprises of UNESCO."[50]

An example of this was the treatment received by the social sciences, whose research and teaching methodologies were the subject of periodic surveys by the corresponding department of UNESCO, presided over by the Nobel Peace Prize laureate Alva Myrdal. Thus, the Ministry of Foreign Affairs considered "clearly indispensable that Spanish professionals and specialists intervene in all these enterprises", but remarked in its directives that "intervention by Spain will increase the importance of the Catholic criterion within UNESCO."[51] Something which clearly pointed to a promotion of the Balmes Institute of the CSIC and the methodologies and authors that cultivated the social sciences in a national/Catholic sense. However, Estelrich and the permanent delegation always chose to resort to the Institute of Political Studies, knowing that its academic production would be better received by UNESCO's directors. In this sense, in a preparatory memorandum for the activities of 1954, Estelrich stated with respect to the social sciences section that "the training of a working group presided over by Don Javier Conde would enable Spain's effective collaboration in the services of this department."[52]

Indeed, articulated around the aforementioned Javier Conde and Enrique Gómez Arboleya, the Spanish Political Science Association and its

[49] Marín (2018): 600–603 and Massot (1998): 65–172.

[50] BC.FE. Box 2: "Note for the Secretary," 13 November 1957.

[51] BC.FE. Box 1: "Organization of Spain's intervention at UNESCO. Complementary notes" (1952).

[52] BC.FE. Box 2: Joan Estelrich, "Study of the program and budget of UNESCO" (August 1954).

sociological counterpart were created, and both were recognized by UNESCO in 1955 as associations of scientific interest. As a result, they were formally incorporated into the structures of the organization, which was a great success for the official culture of Francoism and its ambition to be the exclusive representative of national scientific activity.[53] That same year, the permanent delegation would once again turn to the IEP regarding a possible Spanish collaboration for a manuscript on the "Study of the central planning functions in modern states," a task that was ultimately taken on by the noted member of the Institute Manuel Pérez Olea.[54] Many other commissions for publications and projects would arrive in subsequent years, culminating in the development by the Falangist centre of the Spanish version of UNESCO's *Dictionary of the Social Sciences*.[55] Evidence of the importance of the IEP's work in the process of substituting exiled intellectuals for Francoist authors as the primary representatives of Spanish culture was reflected in an internal report by the Institute comparing the situation of 1950—the date of the appearance of the aforementioned report by Luis Recasens Siches on the state of Spanish political science—and 1960—when the IEP's participation in a study of similar characteristics was requested through the figure of Carlos Ollero.[56] The report concluded:

> Of the credit that our fatherland can enjoy in the educated scientific opinion of the world, a large part is due to the actions of the Institute [...] As an example, the Institute of Political Science of Berlin (Europe's most important scientific/political centre) is going to publish a volume on the state of political science in the Western world. The Institute has been asked to draft the chapter regarding Spain and has fulfilled the task, logically emphasizing the work carried out by Spain and above all by the Institute. UNESCO published a similar volume some years ago. The chapter on Spain was written by one of our most sectarian exiles. The book that will soon appear invalidates and overpowers the previous one.[57]

[53] *REP* (1961): 147–151. The directors of the Association were as follows: president: Francisco Javier Conde; secretary: Carlos Ollero; spokespersons: Luis Diez del Corral, Eduardo García de Enterría, Torcuato Fernández Miranda and José Luis Villar Palasí.

[54] *Archivo General de la Administración* [General Archive of the Administration] (AGA), IEP Collection, Box 26/4034, Letter from Francisco Javier Conde to Eduardo García de Enterría, 5 January 1956.

[55] *REP* (1958–1959): 3–245. Campo (1975).

[56] Ollero (1960): 81–102.

[57] AGA, Section on Presidency, General Secretariat of the Movement, Box 51/18540, "Note," 24 June 1960.

Meanwhile, and although we cannot go into the necessary level of detail, this promotion of the empirical social sciences and its methodologies for analysing the situation of society ended up becoming a double-edged sword for the Francoist regime. It was based on its praxis that many of the young collaborators of the IEP and the other institutions of the Movement, such as Manuel Jiménez de Parga, Javier Pradera, or Pablo Lucas Verdú, began their progressive evolution towards democratic or social democratic positions, disenchanted by the contrast between the country's true social reality—known through their investigations and initiatives such as the *Servicio Universitario de Trabajo [The University Labor Service]* of Father José María Llanos—and the triumphalist conformity of the official discourse of the dictatorship and its single party.

Conclusions

Scientific activity was not completely deactivated during the Franco dictatorship, as has sometimes been suggested. Nor, in our opinion, was there a single model of science that was practised—a reactionary, national/Catholic model—but rather, empirical research, at least in the field of the social sciences, was also present and played a notable role in the normalization of the dictatorship's participation in certain international circles, such as UNESCO and its milieu. This circumstance does not suggest a less negative view of the regime from a political point of view, but rather, it is simply a matter of whose analysis is essential for understanding the regime's strategies—internal and external—and the composition of its political and intellectual personnel.

The Francoist notion of this scientific activity was always very instrumental. Thus, at least with regard to the social sciences, their promotion was largely conceived of as a way to arrogate the exclusivity of the representation of national culture, in clear competition with the world of the republican exiles, who represented a democratic/liberal-style nationalization. Along these same lines, a correct calculation also lay behind the support for empirical research by the Spanish delegation to UNESCO, which was that this type of academic production would be better received by the institution's scientific leadership.

Last but not least, and in contrast with the widespread view that attributes decision-making to the dictator in nearly all political and diplomatic spheres, as well as the perception of the Francoist political class as mediocre and incompetent, the study of the social sciences during this period once again demonstrates the existence of a series of personalities, such as Francisco Javier Conde and Joan Estelrich, who possessed high intellectual status, an enormous capacity to interpret contexts, international connections and the ability for *working towards Franco* adopting initiatives autonomously.

References

Álvarez-Uría, Fernando y Varela, Julia (1992): "Colegios invisibles y relaciones de poder en el proceso de institucionalización de la sociología española", en Reyes, Román (Ed.): *Las Ciencias Sociales en España. Historia inmediata, crítica y perspectivas. Tomo 1. Sociología*, Madrid, UCM: 57–82.

Ayensa, Alfonso (1950): "La culture espagnole sous le régime de Franco", *Les Temps modernes*, 55: 2044–2054.

Campo, Salustiano del (Coord.) (2001): *Historia de la sociología española*, Barcelona, Ariel: 161–180.

Campo, Salustiano del (Coord.) (1975): *Diccionario de ciencias sociales*, Madrid, IEP.

Camprubí, Lino (2017): *Los ingenieros de Franco. Ciencia, catolicismo y guerra fría en el Estado franquista*, Barcelona, Crítica.

Camus, Albert (2017): "L'Espagne et la culture. 1952", en *Conférences et discours (1936–1958)*, Paris, Folio.

Castillejo, David (Comp.) (1999): *Epistolario de José Castillejo. III. Fatalidad y porvenir, 1913–1937*, Madrid, Editorial Castalia.

Claret, Jaume (2019): "La Unión de Profesores Universitarios Españoles en el Exilio", *Historia y Memoria de la Educación*, 9: 295–317.

Dulphy, Anne (2007): "L'adhesion à l'UNESCO en 1952, une étape controversée dans la légitimation internationale de l'Espagne franquiste", en *60 ans d'histoire de l'UNESCO*, Paris, Maison de l'UNESCO: 313–316.

Estelrich, Joan (2012): *Dietaris*, Barcelona, Quaderns Crema.

García Velasco, José (2007): "Un proyecto para la modernización de España", en *El laboratorio de España. La Junta para Ampliación de Estudios e Investigaciones Científicas, 1907–1939*, Madrid, Sociedad Española de Conmemoraciones Culturales: 157–199.

Gaziel, Agustí Calvet (2010): *Meditacions en el desert (1946–1953)*, Barcelona, La Magrana.

Giner, Salvador (1990): "La sociología española durante la dictadura franquista", en Moreno Fernández, Luis y Giner, Salvador (Coords.), *Sociología en España*, Madrid, CSIC: 51–69.

Gómez Arboleya, Enrique (1958a): "Sociología en España", *Revista de Estudios Políticos*, 98: 47–83.

Gómez Arboleya, Enrique (1958b): "Spain" en Roucek, Joseph S. (Ed.), *Contemporary Sociology*, New York, Philosophical Library: 821–839.

Griffin (2007): *Modernism and Fascism. The Sense of a Beginning under Mussolini and Hitler*, Basingstoke, Palgrave Macmillan.

Herf (1985): *Reactionary Modernism: Technology, Culture and Politics in Weimar and the Third Reich*, Cambridge, Cambridge University Press.

Larraz, José (1948): "Un esquema crítico de la sociología", *Revista Internacional de Sociología*, 24: 31–42.

López Sánchez, José María (2006): *Heterodoxos españoles. El Centro de Estudios Históricos, 1910–1936*, Madrid, Marcial Pons.

Malet, Antoni (2008): "Las primeras décadas del CSIC: Investigación y ciencia para el franquismo", en Romero de Pablos, Ana y Santesmases, María Jesús (Eds.), *Un siglo de política científica en España*, Madrid, Fundación BBVA: 211–256.

Marín, Martí (2018): "Existí un catalanisme franquiste? Vint anys després", en VVAA.: *El catalanisme davant del feixisme (1919–2018)*, Maçanet de la Selva, Gregal: 593–614.

Martínez Lillo, Pedro Antonio (1993): "La normalización de las relaciones diplomáticas hispano-francesas después de la Segunda Guerra Mundial (septiembre de 1950—enero de 1951)", *Mélanges de la Casa de Velázquez*, 29: 307–326.

Massot, Josep (1998): "Joan Estelrich, propagandista de Franco a París", en *Tres escriptors davant la guerra civil: George Bernanos, Joan Estelrich, Llorenç Villalonga*, Barcelona, Abadia de Montserrat: 65–172.

Morcillo, Álvaro (2012): "Aviso a los navegantes. La traducción al español de Economía y sociedad de Max Weber", *Estudios Sociológicos*, 90: 609–640.

Moreno Luzón, Javier (2003): "Memoria de la nación liberal", *Ayer*, 52: 207–235.

Nuñez Seixas, Xosé M. (2018): "Nación y nacionalismos en España: siglos XIX y XX", en Álvarez Junco, José y Schubert, Adrian (Coords): *Nueva historia de la España contemporánea (1808–2018)*, Barcelona, Galaxia Gutenberg: 267–297.

Ollero, Carlos (1960): "Die politische Wissenschaft in Spanien", en *Politische Forschung*, Koln y Opladen, Westdeutscher Verlag: pp. 81–102.

Otero Carvajal, Luis Enrique (2006): "La destrucción de la ciencia en España", en VV.AA., *La destrucción de la ciencia en España. Depuración universitaria en el franquismo*, Madrid, Universidad Complutense de Madrid: 15–72.

Pecourt, Juan (2014): "La formación del espacio sociológico durante el segundo franquismo", *Historia Social*, 79: 129–145.

Pla, Xavier (Ed.) (2015): *El món d'ahir de Joan Estelrich. Dietaris, cultura i acció política*, València, PUV.

Presas, Albert (2005): "Science on the Periphery. The Spanish Reception of Nuclear Energy: An Attempt at Modernity?", *Minerva*, 43: 197–218.

Recasens Siches, Luis (1950): "La science politique en Espagne depuis trente ans", en *La Science politique contemporaine. Contribution à la recherche, la méthode et l'enseignement*, Liège, UNESCO: 270–287.

Reig Tapia, Alberto (1990): "Aproximación a la teoría del caudillaje en Francisco Javier Conde", *Revista de Estudios Políticos*, 69: 61–82.

REP (1961): "Asociación Española de Ciencia Política", *Revista de Estudios Políticos*, 119: 147–151.

REP (1958–1959): "Terminología de las Ciencias Sociales", *Revista de Estudios Políticos*, 102–103: 3–245.

Ribes, Alberto J. y Morales, Juan Jesús (2017): "Los sociólogos sin sociedad. La sociología del exilio español", en Balibrea, Mª Paz (Coord.), *Líneas de fuga: hacia otra historiografía del exilio republicano*, Madrid, Siglo XXI: 625–635.

Saz, Ismael (2003): *España contra España. Los nacionalismos franquistas*, Madrid, Marcial Pons.

Sesma, Nicolás (2013): "¡Muera la intelectualidad traidora! La crítica franquista al universo de la Edad de Plata", en Moreno Luzón, Javier y Martínez López, Fernando (Eds.): *La Institución Libre de Enseñanza y Francisco Giner de los Ríos: nuevas perspectivas. Tomo I. Reformismo liberal. La Institución Libre de Enseñanza y la política española*, Madrid, Fundación Giner de los Ríos/Acción Cultural Española: 386–415.

Sesma, Nicolás (2017): "Modernism and Falangism. The Journal *Escorial* and Its Authoritarian Modernizing Project", en Gallego, Ferran y Morente, Francisco (Eds.), *The Last Survivor. Cultural and Social Projects Underlying Spanish Fascism, 1931–1975*, Brighton, Sussex University Press: 90–116.

Sochor, Eugene (1991): *The Politics of International Aviation*, Hampshire, Macmillan Press.

Turner, Stephen P. (1992): "Sociology and Fascism in the Interwar Period. The Myth and its Frame", en Turner, Stephen P. y Käsler, Dirk (Eds.), *Sociology Responds to Fascism*, London and New York, Routledge: 1–13.

Weber, Eugen (1976): *Peasants into Frenchmen: The Modernization of Rural France, 1870–1914*, Stanford, Stanford University Press.

CHAPTER 17

Welcome to the Future! Science as a Tool for American Geopolitics in 1950s' Spain

Lorenzo Delgado Gómez-Escalonilla

At its dawn, the Franco regime tried to carry out an autarkic economic policy and industrialization associated with military objectives,[1] which in turn fed the development of a home-grown scientific and technological policy to give them sustenance. Such aspirations, similar to those of other countries at that time, required strong investments, an industry with the capacity to face such challenges, and a firm commitment to training

This work has been carried out within the framework of two research projects, funded by the Junta de Castilla y León [*External assistance and industrial and scientific and technical modernization in Castilla-León: Analysis of local potential and global options, c.1950-present* (SA241P18)] and the Ministry of Science, Innovation and Universities [*Agents of the Spanish Innovation System: Learning strategies and dissemination of knowledge during the economic transition, 1959–1986* (PGC2018-098057-A-I00)].

[1] San Román (1999).

L. Delgado Gómez-Escalonilla (✉)
Institute of History, CCHS-CSIC, Madrid, Spain
e-mail: lorenzo.delgado@cchs.csic.es

M. Janué i Miret, A. Presas i Puig (eds.), *Science, Culture and National Identity in Francoist Spain, 1939–1959*,
https://doi.org/10.1007/978-3-030-58646-1_17

competent scientists and technicians. None of these goals were within the reach of the Spain that came out of a bloody civil war, with its aftermath of infrastructure destruction. In addition, the scientific community that had been developing during the first third of the twentieth century and that still lacked a long-established tradition in terms of knowledge generation and technological development was subjected to the repression and purification of its members carried out by the dictatorship that then governed the destiny of the country.[2]

The formula to overcome such shortcomings was to resort to an external partner that acted as a catalytic agent, a role assigned at that time to Germany.[3] The commitment to having rapport with the Axis countries and with Germany as a necessary collaborator mediated the initiatives undertaken in the Second World War, with the Juan de la Cierva Board of Trustees of the CSIC at the forefront of that orientation.[4] Since 1944, German support could not be counted upon because Germany had to concentrate all its energies and resources in a war conflagration whose evolution was increasingly adverse.

The post-war period and the international isolation of the Spanish dictatorship predicted a bleak horizon for such projects. They subsisted in an attenuated form by utilizing a group of top German scientists, with whom it was expected to obtain advances in areas such as radar, aircraft, submarines, light weapons, optics, and vehicles. However, the *know-how* of German engineers and technology did not flourish in an environment with a shortage of resources and financial means, poor human capital formation, and an industry weighed down by its backwardness regarding equipment.[5] In the mid-1950s, achieving modernization of the industrial and scientific fabric became dependent on cooperation with the United States. The Cold War dialectic and the confrontation with the Soviet Union made real what a few years before would have been inconceivable.

Although the key to the Spanish–American relationship was the geographic potential of the Iberian country for the US military deployment in Europe, that link caused a chain reaction that affected other scenarios. The purchase of American technology and the facilities to attract companies and investments of identical origin soon became key factors to revive the

[2] Claret Miranda (2006); Otero Carvajal et alii (2006).
[3] García Pérez (1994).
[4] López (1999); Íd. (2008):95–98.
[5] Presas i Puig (2008).

Spanish industrial fabric.[6] The asymmetric nature of the relationship established with the powerful Americans also meant that this process was more dependent on them than the regime was willing to recognize. In addition to the changes that were introduced in the business world, "Americanization" had its correlate in the scientific field.

Strategic Interests and Spanish Access to US knowledge Circuits

The United States made an outstanding contribution in the generation of human capital formation circuits into the post-war period. The *Fulbright Program*, established in 1946, promoted educational exchange abroad using foreign currency that the American government owned due to its loans to Allied countries during the world conflict.[7] Shortly after, the *Foreign Leader Program* was launched, which allowed groups with influence (politicians, officials, journalists, social leaders, etc.) to make "immersion" stays in the United States.[8] The *United States Information and Educational Exchange Program* extended, since 1948, the radius of action of those channels. The *European Recovery Program* promoted the technical and managerial training of the leaders of the business world—public and private—through the so-called productivity missions.[9] The creation of the North Atlantic Treaty Organization (NATO) led to a progressive convergence between the armed forces of the countries that formed it, encouraged by the *Mutual Defense Assistance Program*.[10] All those routes of interrelation forged epistemic communities and future international experts who played key roles in the processes of production and transnational circulation of the "semantics of modernization" with an American stamp. Such experiences, in short, set up a breeding ground of elites favourable towards the United States and its leadership.[11]

Spain began to have access to those training, information, and knowledge transfer circuits, with some delay, following the bilateral agreements of 1953. Such agreements allowed the United States to establish a set of

[6] Calvo (2001); Tascón Fernández (2003); Álvaro (2012); Puig and Álvaro (2004).
[7] Lebovic (2013).
[8] Scott-Smith (2008).
[9] Gourvish and Tiratsoo (1998); Barjot (2002).
[10] Mott (2002); Weber (2016).
[11] Haas (1992); Adler (1992); Scott-Smith (2006).

military bases in Spain in exchange for economic and military aid. From the American perspective, rapport with a dictatorship set up with the complicity of fascism was only due to security reasons. The importance of having those bases to strengthen the American defensive presence in Western Europe justified that relationship *contra natura*. After all, Francoism was a lesser evil with respect to the communist threat. The US foreign policy planners did not ignore that the association with that political regime eroded its democratic image. However, strategic realism trumped ideological coherence at that time. Francoism, on the other hand, received American support in exchange for subordinating itself to the United States' military interests.[12] Previously, the images that were shown to the Spanish public of the United States were contaminated by the anti-Americanism of the Francoist leaders and of the government-controlled media.[13] Beginning with those bilateral treaties, in an exercise of political trickery favoured by the Cold War dialectic, those reviled Americans suddenly became "allies". Geopolitics forged strange travelling companions.

In the following years, the Americans developed a persuasive strategy aimed at convincing the Spaniards of the benefits of bilateral collaboration. The main objective, in its initial phase, was to win wills among the decision-making sectors within the country's elite. Their willingness was crucial to the construction of the military bases and their subsequent operation. Priority attention to the elites was not different from the policy deployed in other Western European countries, with the caveat that Spain was governed by a dictatorship—a characteristic that it only shared with Portugal in that regional framework. On the other hand, given the history of anti-Americanism in the pro-Franco media, more intensive treatment was required. Additionally, it was sought that the majority of the population to welcome foreign forces, which could be achieved by associating the friendship with the Americans with an improvement in Spanish living conditions. The need to maintain a close relationship with the Spanish government to preserve the advantages of its military operation and, at the same time, prevent the United States from being perceived as a supporter of Franco's regime made the objectives of that propaganda machine not easily compatible.[14] The final prospect of that persuasive strategy was to promote a progressive opening of the country that entailed alignment

[12] Viñas (2003).

[13] Fernández de Miguel (2012); Niño Rodríguez (2012).

[14] Delgado Gómez-Escalonilla (2009).

with other Western nations, although without rushing to avoid arousing the reluctance of the Spanish authorities or of the most recalcitrant nationalist sectors.

One of the initiatives undertaken to achieve favourable reception in the Iberian country was the resumption of contacts in Research and Development (R&D). The formation of cadres, the transfer of knowledge and methods, and the provision of instruments and updated publications, together with help accessing international organizations, composed a battery of stimuli for scientific and technical development. The improvement of business and professional training favoured the renewal of the economic fabric, and the resources dedicated to military training brought the instruction and operating systems of a sector of the armed forces nearer to NATO standards. Through those channels, the United States promoted the immersion of some of the country's leaders in its "informal empire"[15]: military, businessmen, scientists and technicians, political and social leaders, and those in charge of the media, mainly. This was a process similar to that previously undertaken in other Western European countries.

From the perspective of the Spanish government, aid for the formation of human capital was conceived as one of the compensations for military collaboration with the United States. In that sense, they showed their preference for directing that action towards the cadres of the scientific and technical system, rather than to orient it towards candidates from the humanities and social sciences over which their North American counterparts expressed a clear interest.[16] Although the Spanish authorities wished to improve the training of personnel that should contribute to economic and industrial development, the North American authorities emphasized the promotion of knowledge of the society and ideals of the United States and its government.

At the beginning of the 1950s, action guides (*Country Papers*) began to be prepared that outlined the behaviour to be followed by the United States Information Service (USIS), an organization responsible for projecting a positive image of the United States and winning the support of the Spanish elite. By that time, there was still no *Exchange Program*, that is, Spain was not included among the European recipients of human

[15] Scott-Smith (2008): 423.

[16] "Minutes of the Board of Trustees of the JRC", June 26 and October 23, 1956; "Informative note", September 24, 1956. AGA-MAEC, R-11501/6.

capital training programmes in American university and research centres. In the absence of this, newsletters and specialized publications, radio broadcasts, documentaries and activities organized by the libraries of its centres were used to attract these elites.[17]

The American interest in integrating Spain into its defensive apparatus made the "leaders of public life" the main *target groups* of American politics and demanded more direct action.[18] On the one hand, it was necessary to reduce Spanish isolation and favour the exchange of ideas, information, and people as well as reach out to the main organizations of the Western bloc. On the other hand, those groups had to be convinced that no one was in a better position than the United States to promote the renewal of methods and knowledge that the country required for development.[19] After the agreements of 1953, the collaboration of those elite was the top priority; therefore, training experiences in the United States were focused on them.[20]

Military, Scientific, and Technical Training

The first programmes that were applied in Spain were aimed at attracting "opinion makers" (*Foreign Leader Program*) or had the generic mission of disseminating knowledge of the United States among Spanish society (*Educational Exchange Program*). From the point of view of scientific and technical training, the most relevant were the *Military Assistance Training Program* and the *Technical Exchange Program*, initiated in 1954 and included in the 1953 conventions. Both programmes had their

[17] In mid-1953, the American deployment in this area comprised five delegations (in Madrid, Barcelona, Seville, Bilbao, and Valencia), which employed a total of 18 Americans and 69 Spaniards. Most of the members were in the Spanish capital, with Barcelona as the second focus of action. In addition, in 1951, the Institute of North American Studies in Barcelona was founded by a private initiative, but with the support of the Embassy, and somewhat later created the Center for North American Studies in Valencia (1958) and the Hispano-North American Institute of Culture in Madrid (1961).

[18] Those leaders of public life included "government, military and religious circles, professionals, world of business and finance, and landowners". "Country Paper for Spain", February 1951. NACP, RG 59, Lot Files, Office of Western European Affairs (LF-OWEA), 1942–1958, Spain, box 10.

[19] "US Government's Overseas Information Program in Spain", August 30, 1952. NACP, RG 59, Decimal Files (DF), 1950–1954, Spain, box 2399, 511.52/8-3052.

[20] "US Policy toward Spain", June 1953; "1954–1955 IIA Prospectus for Spain", April 23, 1953. NACP, RG 59, LF-OWEA, 1942–1958, Spain, box 3 and DF 511.52/4-2353.

predecessors in initiatives developed within the framework of NATO and the Marshall Plan.

In keeping with the centrality of the strategic modernizing link in the bilateral relationship, Francoist leaders sought to use American aid to their armed forces across the board, an aspiration that did not correspond with the purposes of the US government and the resources they were willing to use. The contribution of the United States was much more modest than that expected by the Spanish authorities, but it was vital to take a part of that group out from the obsolescence in which it found itself (a diagnosis with which their North American, French, or British counterparts agreed). The technological backwardness and instructional methods of the Spanish Armed Forces were a problem for guaranteeing the optimal incorporation of the material and armament ceded by the United States. Therefore, one of the priorities was the technical and tactical training of the personnel that would be responsible for their use and maintenance. Simultaneously, these military cadres were familiarized with the operational procedures of the Western bloc, preparing them to guarantee the security of national borders and to cooperate in the defence of the bases established on the Iberian Peninsula.[21]

In the following years, the articulation of a teaching circuit that had as its main target military and technical personnel of the Air Force and Navy, under the supervision of the Military Assistance Advisory Group (MAAG) Spain, was favoured.[22] That formative experience was positively valued by both parties, in contrast to the discrepancies that existed regarding the supplied military equipment. The Spanish military considered that part of this equipment was outdated or received late, and had to assume restrictions on its employment as it was confirmed in the Ifni-Sahara War. Thereafter, the US Army became the doctrinal, organic, and operational reference of the Spanish Armed Forces.[23]

American aid was essential for the take-off of jet aviation. The Talavera reactor school was nourished by the equipment supplied by the United States, and its professors received their first flight and aeronautical technique courses at the American air base Fürstenfeldbruck in the German Federal Republic. Shortly afterwards, other United States Air Force

[21] León-Aguinaga and Delgado Gómez-Escalonilla (2021): 55–70.

[22] A broader vision of the development of these training programmes in León Aguinaga (2019).

[23] Cardona (2008): 189.

(USAF) installations in the United States and Europe became habitual destinations for Spanish military personnel and technicians. Most of the courses they attended taught specialties such as piloting, maintenance, weaponry, construction, electronics, communications, and meteorology. USAF instructors also trained a crew of jet fighters who would become the wings of Spanish interception.

Similar actions were carried out with the future crews and maintenance rungs of the ships that were to be modernized with American aid (about 30 Navy ships) or to be incorporated into its fleet (several minesweepers and a pair of destroyers). The officers and specialists involved regularly visited specialty schools for sonar, submarines, and minesweepers, as well as arsenals and the Marine Corps school, with the idea of reinforcing the Spanish contribution to the defence of Rota. Another important facet of that process was training Spanish crews in their own territorial waters since the summer of 1955, with the support of the Sixth Fleet, which included them in their tactical exercises in the area.

Actions with the Army proceeded with less haste because their contribution to American strategic designs was not critical. Preference was given to the tactical use of the most technologically advanced incorporated material (battle tanks, towed and self-propelled howitzers, anti-tank guns, etc.) as well as logistics management, maintenance, and repairs. The most common destinations were American Army facilities in the German Federal Republic, with courses focused on cavalry, artillery, and engineering, with the American centres taking over in 1955 and expanding the specialties to infantry, armoured cavalry, transmissions, and logistics.

There were also soldiers, although in a much smaller volume, who participated in the civil technical assistance programme. Its original purpose was to ensure efficient use of the US military apparatus in Spain. To achieve this, the country had to become stable by improving its economic resources and by increasing productivity and operating infrastructure. That would also show to the elites of the regime and, by extension, to society as a whole the positive effects of American aid. The National Industrial Productivity Commission was the agency responsible for selecting the beneficiaries of that programme in the industry and services sectors, which received the bulk of the aid. The Army High Staff was integrated in its Advisory Council, and MAAG was one of its interlocutors in the North American representation.[24]

It was therefore about an open channel for military and civilian personnel coming from strategic industrial sectors for Mutual Defence, especially from the Junta de Energía Nuclear (JEN) and Instituto Nacional de Técnica Aeronáutica (INTA), closely linked to the armed forces.[25] Both in JEN and in INTA, the military and economic sides mixed with the scientific and technical levels, which favoured the preference given to training candidates in nuclear physics and aeronautical engineering. The American connection was particularly relevant. The Unites States was at the forefront of such knowledge, and its defence industry became a relevant sector of the domestic economy and a pioneer in innovation,[26] a model of conduct for the Francoist leaders.

The Spanish interest in nuclear energy dates back to the first years post–SecondWorld War, when surveys were carried out to study the value of existing uranium deposits in the country and groups of specialists conducting stays abroad began to be favoured. The University of Chicago was a reception centre. JEN stimulated that process, which was reinforced by the signing of the 1953 treaties. After the first world conference on this subject (Geneva, 1955), with a large Spanish contingent, a bilateral agreement was signed with the United States on civil uses of atomic energy and nuclear research reactors under the framework of the "Atoms for Peace" programme. Two years later, a new agreement included the possibility of acquiring experimental reactors—whether by transfer or by collaborating in their construction—together with the fuel necessary for their operation, which allowed the training of personnel in nuclear applications for medicine, agriculture, and industry.[27] After that, there was a growing transfer of research to private companies, also facilitating the integration of Spanish

[24] "Terms of Reference of Joint US Military Group (Spain) and Military Assistance Advisory Group (Spain)", January 14, 1954. NACP, RG 218, Joint Secretariat, Central Files, 1959–1976, box 78. The military publications helped to promote management and organizational techniques advertised by the aforementioned commission and by the courses of the National Institute for Rationalization of Labor.

[25] Vid. "Technical Assistance Program Annex" in "End of Tour Report of Chief, MAAG Spain for period 1 July 1956–1930 June 1958 (U)", August 29, 1958. NACP, RG 334, SCC, box 6.

[26] Leslie (1993); Reynolds (2010): 380.

[27] A succinct description of the bilateral agreements reached in this area in "Note on collaboration in the nuclear field between Spain and the United States", June 18, 1970. AGA-MAEC, R-12190/4.

scientists and experts in international forums. Somewhat later, nuclear energy was transferred from the laboratory to the market, entering the commercial phase of construction of nuclear power plants in Spain and with direct training of engineers and technicians of Spanish companies in North American centres and industries.[28]

Also in aeronautical research, as in nuclear energy, North American interests in investments in its own R&D system converged with parallel demand from the Spanish government. Along with this, there was a strategic element derived from the bilateral link: the Iberian Peninsula was conceived by the United States as an air deployment base and, later, as a platform for tracking its space vehicles. Prior to the signing of the bilateral agreement of 1953, the Air Force tried unsuccessfully to acquire American technology and equipment to renew its obsolete material, while INTA probed the possibility of sending some of its technicians there.[29] However, it took until the mid-1950s for groups of aeronautical engineers to travel to the United States with some regularity to expand their training, with some linked to INTA and others to companies such as Construcciones Aeronáuticas S.A. Those contacts were branching out, from the receipt of funds from the Air Research and Development Command for the development of projects to the sending of more Spanish engineers to North American centres through grants from the National Aeronautics and Space Administration (NASA)-European Space Research Organisation (ESRO) agreement.[30] That mechanism of knowledge transfer had effects on the modernization of Spanish aviation, both military and commercial, and a few years later, it extended to the field of space research. In March 1960, an agreement was signed for the establishment of a station on the island of Gran Canaria for monitoring and communicating space vehicles; notably, this station played a role in the Mercury and Gemini programmes. Both governments delegated the application of this agreement to NASA and INTA. This agreement was followed by others throughout that decade, building new facilities and extending the radius of scientific and technical training received by Spanish staff.[31]

[28] Romero de Pablos and Sánchez Ron (2001); Presas i Puig (2005, 2017); Herrán and Roqué (2012); Soler Ferrán (2017); Rubio-Varas and De La Torre (2018).

[29] Delgado Gómez-Escalonilla and León Aguinaga (2018): 91–92.

[30] Roca Rosell and Sánchez Ron (1993); Sánchez Ron (1997).

[31] Information on the evolution of these agreements in AGA-MAEC, R-12690/5.

With regard to the primary sector, the management of the technical assistance programme belonged to the Ministry of Agriculture, having smaller funds than in the other two productive sectors. US aid opened new possibilities in the field of agronomic research, which until then had remained in scientific destitution as a result of international isolation. Although the number of scholarships granted was still limited, the stays of engineers and technicians made it possible to resume training and exchange abroad, which, before the end of the decade of the 1950s, was transferred to the organization of the Spanish Agricultural Extension Service. Consequences of that influence were the implantation of the American model and a rupture with the national tradition in the field of research and applied technologies, reaffirmed by the maintenance of that link through the National Institute of Agronomic Research, whose implications became more evident in the decade of the 1960s.[32]

In the previously mentioned areas, the relationships of Spanish organizations with their related parties on the other side of the Atlantic were destined to last, marking their subsequent scientific, technological, and, in many cases, economic trajectory. The movement of Spanish professionals to the United States within the framework of that programme resulted in a broader training dynamic, which also included a large presence of technicians and officials from other official institutions related to the transport network, public works, engineering, and cadres of armament, electrical, and steel factories, as well as business leaders of other branches. Another one of the fundamental dimensions of that cooperation was the attention given to the renewal of teaching in business schools, where the leaders of Spanish capitalism began to be trained.[33]

In the selection of candidates, preference was given to their contribution to the modernization of the large transport infrastructure—rail, ports, roads, and airports—the development of the energy sector—hydroelectric, thermal, and nuclear energy stations—the increase in industrial productivity—in the mining, steel, or arms sectors—and in agricultural sectors—linked to the importation of fertilizers and machinery. One of the main instruments of that training circuit was the "productivity missions", among which those dedicated to organization and business management, construction and urban planning, market research and advertising, and

[32] Fernández Prieto (2007): 323–326 and 338–344; Díaz-Geada and Lanero Táboas (2015).

[33] Puig and Álvaro (2004); Puig (2005).

food and textiles stood out. Technical assistance was, in short, an instrument for the training of cadres that, in addition, was fundamental in the professionalization of business management.[34]

The figures on the volume of personnel who participated in these circuits during the 1950s are still incomplete. Military instruction was by far the most numerous: the Air Force alone accounted for more than 1500 beneficiaries until 1963,[35] and the Navy was possibly on par; the numbers for the Army were somewhat lower. The technical assistance programme mobilized almost another 1000 people until 1962. More imprecise is the calculation of other circuits that contributed to that training, bilateral programmes (educational exchange), and Spanish (CSIC, JRC, Institute of Hispanic Culture, etc.) as well as North American channels (American Field Service, the Amo and Doctor Castroviejo Foundations, along with several universities and centres in that country). The number of those who expanded their knowledge in the United States by such means could be several hundred more, on whose specialties we have insufficient data,[36] and all this without counting the courses taught in Spain by Americans or by Spaniards who had received training in the United States. The *Fulbright Program* joined that catalogue of training opportunities at the end of the decade; however, its actual operation began in the 1960s.

Scientific Leadership as a Propaganda Instrument

The cooperation programmes were a privileged means to transfer to the Spanish people, especially their leading cadres, the North American commitment to the modernization of the country and the improvement in their standard of living. Throughout a series of monographs that appeared from 1954 until the early 1960s in *Noticias de Actualidad (NA)*—the reference magazine of the USIS in Spain and the main information channel of the Embassy—the adoption of American scientific and production methods was extolled as the formula for the development of the country.

[34] Álvaro Moya (2011): 19; Íd. (2012): 67–69.

[35] "List of courses and orientation visits made abroad from 1954–1963". AHEA, A10869.

[36] An incomplete list is available, covering from 1946 to the end of 1956, which computes a total of 523 scholarships for Spaniards in the United States (not including the military training programme or the technical assistance programme). Among the main receiving centres were the University of California, Columbia University, Harvard University, New York University, University of Chicago, and Stanford University, and the Massachusetts Institute of Technology. Ruiz Fornells (1956).

The advances made in the numerous sectors that benefited from American aid were repeatedly alluded to while mentioning the Spanish specialists who went to the United States "to see how factories are run there, how mines are exploited, how they cultivate the fields, and how the railroads work".[37]

In that campaign to show the Spaniards, especially to those with greater decision-making power, the multiple advantages of bilateral collaboration and of the American presence in Spain, the dissemination of the scientific leadership of the United States occupied a relevant role. Its prestige had increased considerably following the Allied victory in the world conflict. The images of American scientific and technical advances, often applied to the military sector, created a remarkable fascination in a society struck by poverty and national-Catholic suffocation. Knowledge of such advances was, in any case, very shallow. Since the 1950s, when the possible installation of American bases in Spain began to take shape more clearly, the propaganda machine of that country began to devote a growing space to the dissemination of the various manifestations of such leadership. This was reflected in the information that appeared in *NA* or in the USIS documentaries—disseminated in ministries, training centres, schools, barracks, and so on.[38]

USIS documentaries in Spain, 1954 and 1957

	1954	*1957*
The American Leadership	257	350
- Scientific	109	171
- Productive	109	110
- International	21	48
- Anticommunism	5	15
- Military	3	6

Source: León Aguinaga (2012: 215)

Among the contents that those media collected, news about industrial applications or improvements in the quality of life derived from scientific advances were frequent. There were even fixed sections in *NA* that lasted

[37] "The human factor, the main binder", *NA* (*Spanish-American Economic Cooperation, 1953–1956*), October 1, 1956: 6.

[38] More information on these propaganda channels in León Aguinaga (2009).

for some time on "America invents for the world", "Science and Industry", and "Science and Space", together with the spread of biographical notes of US Nobel Prize winners as examples of the country's scientific leadership. Those reports and images of North American origin constituted, since the mid-1950s, the almost exclusive source of scientific and technological information of the NO-DO, the graphic news organization whose screening was mandatory in all Spanish cinema screens.[39] In addition to the multiple allusions contained in the usual *News* of short duration, in the series *Imágenes* (Images), which was screened as a single documentary of greater duration, the programmes dedicated to American science and technology had a prominent presence. In all those propaganda vehicles, there were two major themes: atomic energy and, somewhat later, advances in the space race.

NO-DO Images

Premiere	*Image*	*News*
1946	85	The Atobomba
1949	214	Traversing space
	249	Space lords
1-III-1953	426	Prodigious technique
19-IV-1954	484	Atomic energy. Development of nuclear physics
17-II-1958	684	Projectiles and rockets. The mastery of space
20-X-1958	719	Nuclear energy in Spain
24-XI-1958	724	Towards the Moon. Space exploration
9-XI-1959	774	Nuclear energy. Disclosure of atomic physics
28-XII-1959	781	Friendly cooperation (Spain and the United States).
15-V-1961	853	Man to space
25-IX-1961	872	Space explorers
11-I-1965	1044	The first atomic merchant ship (Savannah)
24-VII-1967	1176	Wings and flight (from the factory to the air)
31-VII-1967	1177	Time and face
8-I-1968	1200	The "Daedalus". A helicopter carrier for our Navy
29-I-1968	1203	The mastery of the air

Source: Prepared by the author

Since the beginning of the 1950s, successive articles that provided accounts of the opportunities offered by nuclear energy to industry, medicine, or agriculture appeared in *NA*, but it was after the launch of the "Atoms for Peace" campaign in 1953 when the frequency of the news on

[39] Ordóñez and Ramírez (2008): 272.

the subject increased. This showed the centrality acquired by the nuclear issue in relations between states during the Cold War.[40] The objective was twofold: to underscore American leadership and its image as a "champion of peace", expressed in its commitment to the investigation of the civil applications of that energy, in contrast to the Soviet "militarization", as shown by its secret atomic trials. Simultaneously, the mobile teams of the USIS screened documentaries such as *The atom and agriculture*, *The atom and biological science*, or *The atom and medicine*. Such deployment was crowned with the presentation of *Atoms for Peace*, which could be seen in Spain for the first time at massive exhibition fairs in Barcelona and Valencia in 1955. The following year, a special monograph appeared in *NA* on the peaceful use of the atom,[41] and until the mid-1960s, the successive American advances in nuclear matters—almost all related to peaceful purposes—continued to be told, sprinkled with the narration of some "exploits", such as those carried out by the nuclear submarines that crossed the Arctic Circle submerged and circumnavigated the globe.[42]

In 1964 in Madrid, the Atomic Energy Commission (AEC), in coordination with JEN, presented *Atoms in action*, which had circulated in several countries in Europe, Asia, Latin America, and Africa. The exhibition combined informative with demonstrative facets and included a training component, with Spaniards who had completed their education in North American centres acting as technical instructors.[43] It was, somehow, the swan song of the immaculate image of nuclear energy in Spain. Two years later, the accident in Palomares occurred, with the discovery by the Spanish public that American military aircraft flew over the Spanish sky loaded with atomic bombs. From then on, there was growing concern about the lack of information on such a safety-sensitive matter.[44] After that event, nuclear issues almost disappeared from American propaganda in Spain, which was not at odds with continuing to support the interests of its companies in this sector that almost took over the Spanish market.[45]

Along with the atomic challenge, the space race was the other great field of scientific and technological competition between the two superpowers in their rivalry for international hegemony. In 1958, the first cover of *NA*

[40] Hewlett and Holl (1989); Krige (2006).
[41] *NA*, June 6, 1955.
[42] León Aguinaga (2012): 219–221.
[43] "Spanish-American Atomic Exhibition", *Atlántico*, March 1964: 31–35.
[44] Moreno Izquierdo (2016).
[45] Rubio-Varas and De La Torre (2017).

dedicated to the subject was published,[46] and news about propellant rockets, the design of manned space capsules, and even about a possible trip to the Moon began to appear regularly. The idea was to counteract the "Sputnik effect", noting that the launch of the Soviet satellite into orbit was one of many notable achievements that were taking place in the scientific realm, in which the United States continued to be the world leader. In that line were measures such as the approval of the *National Defense Education Act* and the creation of NASA.[47] The topic gained presence in the following years in USIS information channels, through documentaries and brochures, and especially in the *NA* pages: space installations, scientists, and a series of satellites flew over the Earth with tasks that included studying atmospheric composition, solar radiation, and meteorology, providing telecommunications and military and civil observation, locating natural resources, aiding sea and air navigation, and so on (Pioneer, Vanguard, Discoverer, Tiros, Explorer, etc.). As with atomic energy, the space race mobilized the two superpowers that competed before world opinion to demonstrate their scientific and technological capabilities, in a struggle with both ideological and economic ramifications.[48]

At the beginning of the 1960s, there was a more direct implication with that window to the future, when stations monitoring those space inventions were installed in Spain. In April 1961, Soviet astronaut Yuri Gagarin, aboard the Vostok I ship, became the first human being to travel to space and complete an orbit around the Earth. The American propaganda machine reacted by increasing information about its space achievements.[49] A special place was devoted to the Mercury project that allowed astronaut John Glenn to repeat Gagarin's feat, and that was accompanied a few months after with a NASA-sponsored exhibition showing a life-size-scale model of his space capsule.[50] In the following years, the abundance of

[46] "Science explores the sky", *NA*, April 28, 1958.

[47] Barksdale Clowes (1981); Divine (1993); Launius (1994).

[48] About US government propaganda campaigns dedicated to spreading its scientific and technical advances ("Atoms for Peace" and "Open Skies"), see Osgood (2006): 153–213 y 323–353.

[49] Only a few weeks after that event, full-page illustrations of the satellites and rockets sent to space by the United States with their respective characteristics appeared. *NA*, May 1, 1961: 20–23.

[50] "The Mercury project. First North American in orbit"; "The spacecraft that traverses Spain", *NA*, March 1 and June 1, 1962: 12–13 and 7. A neat pamphlet that described the American progress in this field was also published. *US. Space exploration. 1962.*

news on the subject continued to arouse a mixture of fascination and curiosity in Spanish society, which was marked by the media impact of events such as the Gemini V astronauts' visit to Spain in 1965. The symbology of North American scientific and technological pre-eminence reached a fundamental milestone with the arrival of men on the Moon through the Apollo 11 mission, whose crew also made stops in Spain during their subsequent world tour.

Conclusions

Since the mid-1950s, new opportunities opened for Spanish professional cadres to expand their training in the United States. Although their proportions were still modest, those channels allowed articulating an incipient network of knowledge diffusion and methods that came from the other side of the Atlantic and that, gradually, would be assimilated into and adapted to the Spanish context. Those transfers were accompanied by the establishment of professional contacts, the dissemination of publications and specialized journals, and the translation of reference works, together with a growing bilateral collaboration in various sectors. The aspiration to convert the beneficiaries of those programmes into "transmitters" of the American model was always present as one of the main motivations of the US government. That was not a uniform or linear process: there were sectors where American influence was more intense, others where it found itself with better conditions to take root, and others, in short, where it did not succeed. The impact of all that activity is partially known; however, there is still a way to go to obtain an accurate idea of its overall dimensions and effects.

In parallel to the push to guarantee its leadership, the United States assumed that scientific internationalization was a way to win allies and capture the sympathies of relevant groups from other countries. In addition to its direct action, the United States supported the proliferation of international organizations in a wide range of sectors, in line with the influence that those issues acquired on the global agenda. American conceptions and methods were transferred to the Organization for Economic Cooperation and Development (OECD), United Nations Educational, Scientific and Cultural Organization (UNESCO), the World Bank (WB), philanthropic foundations, and so on, which formed an "International

Development Community".[51] The effects of this transnational process also came to Spain, although sometimes with rhythms and intensities different from those of other Western European countries.

At the same time, this action was combined with information towards wider audiences to convince them of the ability of American science to model the future. The propaganda story of those feats, together with the experiences of those who went to United States research centres and universities to complete their training, contributed to enhancing the role of American science in 1950s Spain. The leadership of the United States in the Western bloc, reviled a few years before, was assumed as something natural, making that country the unequivocal reference for any attempt to boost scientific and technical research. In the end, Spain was incorporated like other European countries into that "informal empire", based not only on dependence of US power but also on the desire for emulation.

In the following decade, the technocratic project of joining the developmental economic policy with an educational and scientific policy that provided qualified personnel to grease the locomotive of Spanish growth gained strength. That orientation, based on the Copernican turn of economic policy encouraged by the Stabilisation Plan of 1959, favoured a greater rapport with the ideas on modernization and the formation of human capital coming from the United States. The reforms undertaken by technocratic leaders once again had a privileged interlocutor in the United States, either directly or through organizations such as the OECD and the WB.[52] The roots of that process, in any case, were created in the path developed in the preceding years.

Acronyms Used

AEC	Atomic Energy Commission
AGA-MAEC	Archivo General de la Administración-Ministerio de Asuntos Exteriores y Cooperación/General Archive of the Administration/Ministry of Foreign Affairs and Cooperation
AHEA	Archivo Histórico del Ejército del Aire/Historical Archive of the Air Force

[51] Miller (2006); Krige (2008); Frey et al. (2014).

[52] Delgado Gómez-Escalonilla (2015); Martín García and Delgado Gómez-Escalonilla (2020).

CSIC	Consejo Superior de Investigaciones Científicas/Higher Council for Scientific Research
INTA	Instituto Nacional de Técnica Aeronáutica/National Institute of Aeronautical Technology
JEN	Junta de Energía Nuclear/Nuclear Energy Board
JRC	Junta de Relaciones Culturales/Board of Cultural Relations
MAAG	Military Assistance Advisory Group
NA	Noticias de Actualidad
NACP	National Archives of the United States-College Park
NASA	National Aeronautics and Space Administration
NATO	North Atlantic Treaty Organization
OECD	Organization for Economic Cooperation and Development
R&D	Research and Development
UNESCO	United Nations Educational, Scientific and Cultural Organization
USAF	US Air Force
USIS	United States Information Service
WB	World Bank

References

Adler, Emanuel (1992): "Conclusion: Epistemic communities, world order, and the creation of a reflective research program", *International Organization*, 46/1: 367–390.

Álvaro Moya, Adoración (2011): "Guerra Fría y formación de capital humano durante el franquismo. Un balance sobre el programa estadounidense de ayuda técnica (1953–1963)", *Historia del Presente*, 17: 13–25.

Álvaro Moya, Adoración (2012): *La inversión directa estadounidense en España. Un estudio desde la perspectiva empresarial (c. 1900–1975)*, Madrid, Banco de España.

Barjot, Dominique (ed.) (2002): *Catching up with America. Productivity Missions and the Diffusion of American Economic and Technological Influence after the Second World War*, Paris, Presses de l'Université de Paris-Sorbonne.

Barksdale Clowes, Barbara (1981): *Brainpower for the Cold War: The Sputnik Crisis and National Defense Education Act of 1958*, Westport-CT, Greenwood Press.

Calvo, Óscar (2001): "¡Bienvenido Míster Marshall! La ayuda económica americana y la economía española en la década de 1950", *Revista de Historia Económica*, n° extraordinario: 253–275.

Cardona, Gabriel (2008): *El poder militar en el Franquismo*, Barcelona, Flor del Viento.

Claret Miranda, Jaume (2006): *El atroz desmoche. La destrucción de la Universidad española por el franquismo, 1936–1945*, Barcelona, Crítica.

Delgado Gómez-Escalonilla, Lorenzo (2009): "La maquinaria de la persuasión. Política informativa y cultural de Estados Unidos hacia España", *Ayer*, 75: 97–132.

Delgado Gómez-Escalonilla, Lorenzo (2015): "Modernizadores y tecnócratas. Estados Unidos ante la política educativa y científica de la España del desarrollo", *Historia y Política*, 34: 113–146.

Delgado Gómez-Escalonilla, Lorenzo and León Aguinaga, Pablo (2018): "Fly Me to the Cold War: the United States and the Modernization of the Spanish Air Force in the 1950s", in Stoker, Donald and Westermann, Edward B. (eds.): *Air Force Advising and Assistance: Developing Airpower in Client States*, Warwick, Helion & Company: 89–117.

Díaz-Geada, Alba y Lanero Táboas, Daniel (2015): "Modelos de modernización para el desarrollismo: el influjo de las propuestas estadounidenses en el Servicio de Extensión Agraria (1955–1975)", *Revista Complutense de Historia de América*, 41: 71–94.

Divine, Robert A. (1993): *The Sputnik Challenge: Eisenhower's Response to the Soviet Satellite*, New York, Oxford University Press.

Fernández de Miguel, Daniel (2012): *El enemigo yanqui. Las raíces conservadoras del antiamericanismo español*, Madrid, Genueve Ediciones.

Fernández Prieto, Lourenzo (2007): *El apagón tecnológico del franquismo. Estado e innovación en la agricultura española del siglo xx*, Valencia, Tirant Lo Blanch.

Frey, Marc; Kunkel, Sönke, and Unger, Corinna (eds.) (2014): *International Organizations and Development*, Basingstoke, Palgrave Macmillan.

García Pérez, Rafael (1994): *Franquismo y Tercer Reich. Las relaciones económicas hispano-alemanas durante la Segunda Guerra Mundial*, Madrid, Centro de Estudios Constitucionales.

Gourvish, Terry and Tiratsoo, Nick (eds.) (1998): *Missionaries and Managers: American influences on European management education, 1945–69*, Manchester-New York, Manchester University Press.

Haas, Peter M. (1992): "Introduction: epistemic communities and international policy coordination." *International Organization*, 46/1: 1–35.

Herrán, Néstor y Roqué, Xavier (eds.) (2012): *La física en la dictadura. Físicos, cultura y poder en España 1939–1975*, Barcelona, Universitat Autònoma de Barcelona.

Hewlett, Richard G. and Holl, Jack M. (1989): *Atoms for Peace and War, 1953–1961. Eisenhower and the Atomic Energy Commission*, Berkeley, University of California Press.

Krige, John (2006): "Atoms for Peace, Scientific Internationalism, and Scientific Intelligence", en Krige, John and Barth, Kai-Henrik (eds.): *Global Power Knowledge. Science and Technology in International Affairs*, Osiris 21, Chicago, University of Chicago Press: 161–181.

Krige, John (2008): *American Hegemony and the Postwar Reconstruction of Science in Europe*, Cambridge, MA, The MIT Press.

Launius, Roger (1994): *NASA, A History of the US Civil Space Program*, Malabar-Fla., Krieger.

Lebovic, Sam (2013): "From War Junk to Educational Exchange: The World War II Origins of the Fulbright Program and the Foundations of American Cultural Globalism, 1945–1950", *Diplomatic History*, 37/2: 280–312.

León Aguinaga, Pablo (2009): "Los canales de la propaganda norteamericana en España, 1945–1960", *Ayer*, 75: 133–158.

León Aguinaga, Pablo (2012): "*Faith in the USA*. El mensaje de la diplomacia pública americana en España (1948–1960)", en Niño, Antonio y Montero, José Antonio (eds.): *Guerra Fría y Propaganda. Estados Unidos y su cruzada cultural en Europa y América Latina*, Madrid, Biblioteca Nueva: 197–234.

León Aguinaga, Pablo (2019): "Los programas de formación para la «Mutua Defensa» entre España y Estados Unidos en los años cincuenta", *Ayer*, 116: 49–76.

León Aguinaga, Pablo and Delgado Gómez-Escalonilla, Lorenzo (2021): "The Deployment of U.S. Military Assistance to Franco's Spain: Limited Modernization and Strategic Dependence", *Cold War History*, 21/1: 55–70.

Leslie, Stuart W. (1993): *The Cold War and American Science: The Military-Industrial-Academic Complex at MIT and Stanford*, New York, Columbia University Press.

López, Santiago (1999): "El Patronato Juan de la Cierva (1939–1960), 3ª Parte: La investigación científica y tecnológica", *Arbor*, 637, CLXII: 1–32.

López, Santiago (2008): "Las ciencias aplicadas y las técnicas: la Fundación Nacional de Investigaciones Científicas y Ensayos de Reformas y el Patronato Juan de la Cierva del CSIC (1931–1961)", en Romero de Pablos, Ana y Santesmases, María Jesús (eds.): *Cien años de política científica en España*, Bilbao, Fundación BBVA: 79–106.

Martín García, Óscar J. and Delgado Gómez-Escalonilla, Lorenzo (eds.) (2020): *Teaching Modernization. Spanish and Latin American Educational Reform in the Cold War*, New York, Berghahn Books.

Miller, Clark A. (2006): "'An Effective Instrument of Peace'. Scientific Cooperation as an Instrument of U.S. Foreign Policy, 1938–1950", en Krige, John and Barth, Kai-Henrik (eds.): *Global Power Knowledge. Science and Technology in International Affairs*, Osiris 21, Chicago, University of Chicago Press: 133–160.

Moreno Izquierdo, Rafael (2016): *La historia secreta de las bombas de Palomares*, Barcelona, Crítica.

Mott, William H., IV (2002): *United States Military Assistance. An Empirical Perspective*, Westport and London, Greenwood Press.
Niño Rodríguez, Antonio (2012): "Los dilemas de la propaganda americana en la España franquista", en Niño, Antonio y Montero, José Antonio (eds.): *Guerra Fría y Propaganda. Estados Unidos y su cruzada cultural en Europa y América Latina*, Madrid, Biblioteca Nueva: 155–196.
Ordóñez, Javier y Ramírez, Felipe E. (2008): "Los públicos de la ciencia española: un estudio del NO-DO", en Romero de Pablos, Ana y Santesmases, María Jesús (eds.): *Cien años de política científica en España*, Bilbao, Fundación BBVA: 257–292.
Osgood, Kenneth (2006): *Total Cold War: Eisenhower's Secret Propaganda Battle at Home and Abroad*, Lawrence, University of Kansas Press.
Otero Carvajal, Luis Enrique et alii (2006): *La destrucción de la ciencia en España: depuración universitaria en el franquismo*, Madrid, Editorial Complutense.
Presas i Puig, Albert (2005): "Science on the periphery. The Spanish reception of nuclear energy: An attempt at modernity?", *Minerva*, 43/2: 197–218.
Presas i Puig, Albert (2008): "La inmediata posguerra y la relación científica y técnica con Alemania", en Romero de Pablos, Ana y Santesmases, María Jesús (eds.): *Cien años de política científica en España*, Bilbao, Fundación BBVA: 173–209.
Presas i Puig, Albert (2017): "Human Capital and Physics Research for the Spanish Nuclear Program", en Rubio-Varas, Mª. del Mar and De La Torre, Joseba (eds.): *The History of Nuclear Energy in Spain. Governance, Business and Finance*, London, Palgrave Macmillan: 97–117.
Puig, Nuria (2005): "La ayuda económica de Estados Unidos y la *americanización* de los empresarios españoles", en Delgado, Lorenzo y Elizalde, Mª. Dolores (eds.): *España y Estados Unidos en el siglo XX*, Madrid, CSIC: 181–205.
Puig, Nuria y Álvaro, Adoración (2004): "La guerra fría y los empresarios españoles. La articulación de los intereses económicos de Estados Unidos en España, 1950–1975", *Revista de Historia Económica*, 22/2: 387–424.
Reynolds, David (2010): "Science, technology, and the Cold War", en Leffler, Melvyn P. and Westad, Odd Arne (eds.): *The Cambridge History of the Cold War*, Vol. III *Endings*, Cambridge, Cambridge University Press: 378–399.
Roca Rosell, Antoni y Sánchez Ron, José Manuel (1993), *Ciencia y Aeronáutica*, Madrid, Algaida/INTA.
Romero de Pablos, Ana y Sánchez Ron, José Manuel (2001): *Energía nuclear en España: de la JEN al CIEMAT*, Madrid, CIEMAT.
Rubio-Varas, Mª. del Mar and De La Torre, Joseba (eds.) (2017): *The History of Nuclear Energy in Spain. Governance, Business and Finance*, London, Palgrave Macmillan.
Rubio-Varas, Mª. del Mar y De la Torre, Joseba (2018): "*American Nuclear Training*: científicos, ingenieros y empresarios españoles en los Estados Unidos

del desarrollo atómico", en CAMPRUBÍ, Lino, ROQUÉ, Xavier y SÁEZ DE ADANA, Francisco (eds.): *De la Guerra Fría al calentamiento global. Estadops Unidos, España y el nuevo orden científico mundial*, Madrid, Catarata: 85–109.

Ruiz Fornells, Enrique (1956): *Estudiantes españoles en los Estados Unidos. Diez años de intercambio*, Madrid, Asociación Cultural Hispano-Norteamericana.

Sánchez Ron, José Manuel (1997): *INTA. 50 años de Ciencia y Técnica Aeroespacial*, Madrid, Ministerio de Defensa/Doce Calles/INTA.

San Román, Elena (1999): *Ejército e industria. El nacimiento del INI*, Barcelona, Crítica.

Scott-Smith, Giles (2006): "Building a community around the Paix Americana. The US government and exchange programmes during the 1950s", in Laville, Helen and Wilford, Hugh (eds.): *The US Government, Citizen Groups and the Cold War. The state-private network*, London & New York, Routledge: 83–99.

Scott-Smith, Giles (2008): *Networks of Empire: The US State Department's Foreign Leader Program in the Netherlands, France, and Britain 1950–70*, Brussels, P.I.E. Peter Lang.

Soler Ferrán, Pablo (2017): *El inicio de la ciencia nuclear en España*, Madrid, Sociedad Nuclear Española.

Tascón Fernández, Julio (2003): "Capital internacional antes de la «internacionalización del capital» en España, 1936–1959", en Sánchez Recio, Glicerio y Tascón Fernández, Julio (coords.): *Los empresarios de Franco: Política y economía en España, 1936–1957*, Barcelona, Crítica: 281–306.

Viñas, Ángel (2003): *En las garras del águila. Los pactos con Estados Unidos, de Francisco Franco a Felipe González (1945–1995)*, Barcelona, Crítica.

Weber, Nathaniel R. (2016): *United States Military Assistance Groups during the Cold War, 1945–1965*, Texas A&M University.

Index[1]

[1] Note: Page numbers followed by 'n' refer to notes.

M. Janué i Miret, A. Presas i Puig (eds.), *Science, Culture and National Identity in Francoist Spain, 1939–1959*,
https://doi.org/10.1007/978-3-030-58646-1

C

J

N